Air Power in the Maritime Environment

This book explores the mingling of two rather different perspectives, those of the naval and aeronautical schools of thought, and the impact that they had upon one another in natural, professional and geopolitical settings. To explain the manner in which air power was incorporated into warfare between 1914 and 1945 it studies the deeds of practitioners, the limitations of technology, the realities of combat and the varying institutional dynamics and strategic priorities of the major maritime powers. It is underpinned by an appreciation of the geostrategic setting of the key maritime states, while addressing the challenges of operating in this multifaceted environment and the major technological developments which enabled air power to play an ever greater role in the maritime sphere. The potential for air power to influence warfare in the maritime environment was fully realised during the Second World War and its impact is demonstrated through an analysis of a wide range of the fleet operations and how it was utilised in the defence of trade and sea lanes. As such this book will be of interest to both naval and air power historians and those wanting a fuller perspective on maritime strategy in this period.

Dr David Gates is a Fellow of the Royal Historical Society and the author of numerous publications on aspects of the history of war and on contemporary military matters. Among other distinctions, he has been awarded the Royal Air Force's Salmond Prize in Air Power Studies.

Dr Ben Jones is Senior Lecturer at Portsmouth Business School at the Royal Air Force College, Cranwell, UK.

Air Power in the Maritime Environment

Environment

The World Wars

David Gates and Ben Jones

Routledge
Taylor & Francis Group

LONDON AND NEW YORK

First published 2016 by Routledge

2 Park Square, Milton Park, Abingdon, Oxfordshire OX14 4RN
711 Third Avenue, New York, NY 10017

Routledge is an imprint of the Taylor & Francis Group, an informa business

First issued in paperback 2017

British Library Cataloguing in Publication Data
A catalogue record for this book is available from the British Library

Library of Congress Cataloging-in-Publication Data
Names: Gates, David, author. | Jones, Ben, 1974 – author.
Title: Air power in the maritime environment : the World Wars / by David Gates and Ben Jones.
Description: Burlington, VT : Ashgate Publishing Company, [2016] | Includes bibliographical references and index.
Identifiers: LCCN 2015038120
Subjects: LCSH: Air power – History – 20th century. | Naval aviation – History – 20th century. | World War, 1914–1918 – Aerial operations. | World War, 1914–1918 – Naval operations. | World War, 1939–1945 – Aerial operations. | World War, 1939–1945 – Naval operations.
Classification: LCC VG90. G38 2016 | DDC 358.4/0309041 – dc23
LC record available at http://lccn.loc.gov/2015038120

ISBN: 978-1-4094-2907-4 (hbk)
ISBN: 978-0-8153-6676-8 (pbk)

Typeset in Bembo
by Apex CoVantage, LLC

Contents

Glossary of abbreviations

AASF	Advanced Air Striking Force
ASV	Air-to-Surface-Vessel (radar)
ASW	Anti-Submarine Warfare
BOA	(US) Bureau of Aeronautics
BEF	British Expeditionary Force
CAM	Catapult Aircraft Merchantman
CDC	(US) Caribbean Defense Command
CEP	Circular Error Probable
CSBS	Course Setting Bomb Sight
FAA	Fleet Air Arm
FCS	Fighter Catapult Ships
HF/DF	High Frequency/Direction Finding
HMAS	His Majesty's Australian Ship
HMS	His Majesty's Ship
IJN	Imperial Japanese Navy
MAC	Merchant Aircraft Carrier
RAF	Royal Air Force (British)
RAI	*Regia Aeronautica Italiana* [Italian Royal Air Force]
RFC	Royal Flying Corps
RMI	*Regia Marina Italiana* [Italian Royal Navy]
RNAS	Royal Naval Air Service
UK	United Kingdom of Great Britain and Northern Ireland
UP	Unrotated Projectile System
USAAF	United States Army Air Forces
WAC	(UK) Western Approaches Command

Preface

The maturation of aviation is a good illustration of how so much that occurs in battle can be moulded by developments and decisions that occur across broad swathes of time and space. As so many Europeans discovered or were reminded in the summer of 1914, military capabilities can take years and decades to evolve, yet political intentions can be transformed in a matter of hours and days. The specific manner in which air power was incorporated into warfare owed rather less to visionaries and theorists than it did to the deeds of practitioners, to the limitations of technology, to the realities of combat and to the varying institutional dynamics and often mutable strategic priorities of individual states, most notably Austria-Hungary, Britain, France, Germany, Italy, Japan, Russia, Turkey and the United States of America. As the title of this book's opening section – 'Where the Sea Meets the Sky' – suggests, this study explores the mingling of two rather different perspectives, those of the naval and aeronautical schools of thought, and the impact that they had upon one another in natural, professional and geopolitical settings that were often complex in the extreme.

In compiling this study we have drawn on a wide range of archival and secondary sources, many of which are novel items. We are particularly indebted to the staff of the Library of the Royal Air Force College, Cranwell, for their practical assistance in this respect.

David Gates
Ben Jones

1 The First World War

Where the sea meets the sky

One of the attributes accredited to air power per se is the notion of ubiquity. Whereas the Earth's surface comprises areas of water and dry land, the whole of the planet is covered by the atmosphere; in theory, there is no point that cannot be reached through the medium of the air. However, the ability of an aerial platform to attain a specific location in practice is a function of, above all, technology, as is its capacity to perform an allotted task once there. Moreover, another basic attribute of air power is the constraint known as impermanence, which largely stems from an ineluctable reliance on bases. Even space-going platforms cannot entirely slip the bonds that link them to activities on Earth, if only because it is from them that they derive their very reason for being. Aerospace power as such is ultimately and unavoidably reliant on surface installations and their concomitant personnel; without these, it could not exist, let alone operate.

The very fact that 70.7 per cent of the Earth's surface consists of sea raises questions as to whether these supporting assets might best be located ashore or afloat. In any case, just as air power per se is reliant upon facilities and activities on the Earth's surface, so too is naval power as such ultimately dependent upon installations and enterprises that are rooted in terra firma. As the British soldier and military theorist Charles Callwell observed in 1905:

> The great aim in naval warfare is to secure . . . command of the sea. That object is either attained by destroying the fleets of the enemy . . . or else by driving them into port and then mounting guard over them so as to be in a position to fall upon them in superior force should they dare to emerge . . . . But to carry any such scheme out, the stronger navy is to a certain extent tied to the land. It must have its depots and repairing-stations; and if the ports where these have been established be . . . [inadequately] defended, a raid by a . . . [seemingly] insignificant . . . force may destroy the stores, may wreck the docks and may place the fleets depending on them in a position of difficulty, if . . . not . . . absolute impotence.[1]

---

1 Charles Edward Callwell, *Military Operations and Maritime Preponderance: Their Relations and Interdependence* (London and Edinburgh, 1905), pp. 163–4.

Indeed Callwell pointed to several instances of sea power being compromised by land forces. These ranged from the seizure of the Dutch fleet in 1795 by French cavalrymen (who approached the ships over the frozen River Texel) to the siege and capture of the tsar's Asiatic naval base, Port Arthur, during the Russo-Japanese War of 1904–05.[2] Especially from around 1900, as rigid and semi-rigid airships, notably Zeppelins, began to appear, armed forces were increasingly obliged to consider what part, if any, air power might also play in military matters and whether it might best be assimilated into existing units or configured as an independent entity. The response varied from one country to another, depending on the dynamics within and between a state's political and military institutions, and on its contrasting strategic predilections and material interests. For the great maritime nations especially, the advent of controlled, powered flight presented both opportunities and problems.

So convoluted and hazardous did maritime environments prove that even those who were nominally naval aviators were pardonably daunted by them and pondered whether their assets and endeavours might not be better put to use elsewhere. Then as now, operating air power in a nautical context had its peculiarities, essentially because of two sets of factors. Firstly, besides running those risks inherent in flying, aviators also faced many of the ambient perils of seafaring, including motion sickness. Secondly, the cockpit of any crewed aircraft forms both a human and a technological interface with its surroundings. The more complex those surroundings, the greater the demands placed upon personnel and technology alike.

The fundamental but challenging process of navigation, for one, was frequently that much harder in maritime settings. Often discernible even in darkness, shorelines were and remain useful guides for aircraft. As late as 1942, when limited navigational capabilities were still hampering the conduct of aerial operations, German population centres such as Lübeck and Rostock were the first to be selected as targets for nocturnal raids by the RAF's Bomber Command largely because, located on Germany's littoral, they were that much easier to find. But whereas aircraft would – assuming the Earth's surface could be seen clearly enough – normally use topographical distinctions to help them get their bearings, stretches of water that lay out of sight of any landmass were, to all intents and purposes, featureless, even in conditions well-suited to the human eye. In such circumstances, navigators had to rely purely upon mental computations that were informed by rudimentary instruments, such as compasses, charts, chronometers and sextants. In any event adopting and maintaining a given flight path involves reconciling a shifting number of environmental variables, whereby miscalculations amounting to no more than a fraction of a degree can culminate in wide margins of error. Even a slight but constant cross-wind, for instance, might blow an aircraft significantly off course, with the deviation

2 Ibid., pp. 126–45.

being more or less proportional to the flight's length. As transmarine voyages in general and transoceanic ones in particular extend perforce over hundreds if not thousands of miles, this could lead to aviators being all at sea in every sense of the term. Roland Garros's non-stop flight across the Mediterranean in 1913; the transatlantic crossing by Alcock and Brown in 1919; Charles Lindbergh's journey from New York to Paris in 1927; the flights by Italian pilots, notably Italo Balbo, that criss-crossed the Mediterranean and the Atlantic between 1928 and 1933: all of these were demonstrations of navigational prowess as much as they were feats of fortitude.

Air power's introduction and maturation helped make warfare more a matter of zones than fronts. This effect was particularly pronounced in the maritime environment, where, by means of not only aircraft but also submarines, which were refined alongside them, the third dimension could be exploited to a degree that was simply not possible in operations confined to land masses. Whilst the Earth's entire surface is enveloped by the atmosphere, most of it is covered in water. Although the division of the latter into oceans and seas is essentially as arbitrary as it is nominal, it does yield a rough guide to the proportions of potential theatres of war in maritime settings. They are enormous. Whereas the Atlantic is adjudged to encompass 35,657,000 square miles (92,373,000 square kilometres) and covers a quarter of the globe, the Pacific, with all of 69,356,000 square miles (179,679,000 square kilometres), is almost twice as large again. Extending across a fifth of the world's surface, the Indian Ocean by contrast is deemed to comprise 28,532,000 square miles (73,917,000 square kilometres), while the Arctic, though covering somewhat less than four per cent of the globe, amounts to no fewer than 5,439,000 square miles (14,090,000 square kilometres) of water, making it over five times the size of the Caribbean and over twenty-four times that of the North Sea (223,000 square miles or 575,000 square kilometres). Sprawling over 1,068,000 square miles (2,766,000 square kilometres), the Caribbean is nonetheless comparable in size to the Mediterranean, which amounts to 971,000 square miles (2,516,000 square kilometres), while the Sea of Japan, with its 389,000 square miles (1,008,000 square kilometres), has well over twice the area of the Baltic, which extends over 153,000 square miles (397,000 square kilometres).[3] Moreover, stark as it is, the contrast between the ice floes of the Arctic and the warmth that prevails in the Caribbean is but one illustration of the variety of climatic conditions that can be found across these immense stretches of water.

Extremes of heat and cold and other meteorological and environmental phenomena can profoundly affect the performance of both people and technology. For example, there is a propensity for temperature inversion to occur in some areas where there is an abundance of surface water or a paucity of it. Layers of air with significantly differing temperatures cause the refraction of light, effectively

---

3 *Philip's World Atlas* (London, 1993), p. xiv.

making the atmosphere into a lens for any human eye peering through it. Above all, aviators, mariners and those traversing deserts might thereby experience a 'false' horizon – whereby it is very hard to gauge precisely where, amidst the flickering glare or shimmering haze, the Earth's surface ends and the sky begins – or might even see images of objects that actually lie beyond the horizon. In fact extraordinary refraction can delude observers with a variety of 'inferior' and 'superior' mirages: the shape, orientation, proximity, proportions and dimensions of things can be more apparent than real, with the result that non-existent phenomena might be perceived and real ones overlooked.

The use of camouflage that enables military personnel and equipment to blend into the background and thereby evade detection has saved countless lives. By the same token, however, 'soft' horizons and other tricks of the light have proved fatal, not least for many of those aboard the RMS *Titanic*, which foundered in 1912 after striking an iceberg on a clear, starlit night in waters that were as calm as a millpond. This freakish occurrence surely owed much to the deceptive seascape formed where the shifting, frigid Labrador Current meanders near to the warm Gulf Stream.[4]

At other points in time and space, atmospheric conditions have proved both beneficial and troublesome. In the first decades of flight, maritime reconnaissance from the air was often conducted at fairly low altitudes, between 1,000 and 1,500 feet. From such heights and over suitably clear water, aviators were not only more likely to discern minefields and submarines but could also identify vessels on the surface more readily since, meteorological circumstances permitting, they would appear silhouetted against the horizon. Scene of most clashes between ships of the *Kaiserliche Marine*[5] and the Royal Navy during the Great War, the North Sea – which covers an area almost twice the size of the entire British Isles – is notorious for its fog banks and leaden skies. These can reduce visibility dramatically. Reconnoitring for Admiral David Beatty's battlecruisers as the Battle of Jutland unfolded in 1916, Flight Lieutenant Frederick Rutland was to find he could take his spotter plane no higher than 900 feet, so dense and low was the cloud base. This, in turn, slashed the line of sight from his cockpit to the horizon to some 3,000 yards and, with it, Rutland's view of the approaching German warships. Although these blazed away at the fugacious interloper, Rutland escaped unscathed.[6]

The aircraft most commonly used by navies during the war were, like Rutland's, floatplanes. These struggled to contend with even a moderate swell. Yet even when relatively calm the waters of the North Sea are apt to be choppy, certainly when compared to areas of the Mediterranean, for one. Cold, dank conditions can also adversely affect machinery, particularly electrical circuitry.

4   See: Tim Maltin, *Titanic: A Very Deceiving Night* (Wilsford, 2012).

5   The Imperial German Navy.

6   Beatty's dispatch to Jellicoe, 12 June 1916, Document 164, *The Beatty Papers: Volume I, 1902–1908*, Bryan Ranft (ed.), (Aldershot, 1989), pp. 325–6.

Of the nine seaplanes that were lowered from tenders into the North Sea to raid the Zeppelin base at Cuxhaven on Christmas Day 1914, two were unable to get their engines to start. The rest experienced difficulty in executing the mission because of fog and low cloud, variables that exacerbated the human eye's inherent constraints.[7] Whereas this particular attack did inflict a little damage, however, similar operations against Zeppelin sheds over the next two years proved more embarrassing than effective, essentially because of engine malfunctions and the challenge of taking off from the blustery, heaving surface of the North Sea. Of the eleven Sopwith *Babies* deployed for one such raid on 4 July 1916, for instance, three suffered engine failure, four sustained broken propellers and one capsized.[8]

Part of the problem was the sheer fragility of many early aircraft. Just as mechanical propulsion mechanisms had unshackled water-going vessels from Nature's winds and currents, so too did suitable aero-engines spawn the possibility of powered, controlled flight in machines that were heavier than air. However, this still required the reconciliation of factors that normally varied inversely with one another. Where military platforms were concerned, foremost among these were the demands for firepower, protection and agility.

The first hurdle to be surmounted was that of developing engines that were sufficiently powerful and dependable to lift and keep a winged aircraft and any cargo aloft. Much of the overall take-off weight of early planes consisted of their propulsion systems and fuel stocks. This left little scope for accommodating people, weaponry, desirable aids – notably radio sets, navigational devices, life preservers and parachutes – or armour. Indeed whereas the shift from timber to metal in ship-building had allowed for larger, sturdier vessels, early aircraft were perforce relatively small, with frames constructed from light materials, notably wooden struts that were screwed and wired together and covered in fabric. Besides the number of propellers and the composition, shape, pitch and quantity of their blades, the degree to which wing, tail, fuselage and undercarriage assemblies were streamlined and motors were tuned numbered among the factors that could affect the aerodynamic performance of planes, notably the manoeuvrability, speed, range and operating ceiling. Cockpits – which were essentially open to the elements with, at most, a tiny windscreen – were cramped and could become bitterly cold. Other ambient hazards included noise, vibration, bird strikes and the inverse variation of, on the one hand, atmospheric pressure and oxygen levels with, on the other, altitude.

Buffeted by the winds and precipitation and stressed and strained by manoeuvres, not least the impact of landing, such primitive machines often posed less of a threat to any foe than to their occupants. The latter's corporal and psychological characteristics were, furthermore, key determinants of aviation technology's

---

7   Walter Raleigh, *The War in the Air: Being the Story of the Part Played in the Great War by the Royal Air Force: Volume One* (Oxford, 1922), p. 403.

8   R.D. Layman, *Naval Aviation in the First World War: Its Impact and Influence* (London, 2002), p. 161.

ultimate utility, for the limits of human endurance remain fairly static and are soon reached. Containing little, if anything, in the way of automated systems, early aircraft needed people to not only fly them but also operate equipment, such as weapons, cameras or communication devices, and to carry out tasks such as navigation and observation. Besides the ground crews who refuelled, re-armed and maintained them, most military aircraft needed at least two occupants for their efficient operation. Designed accordingly, many machines were difficult to adapt to changing conditions or to the taking on of additional functions. As a spotter aircraft, the Royal Navy's Short 184 seaplane, for instance, normally carried a pilot and an observer. When called on to act as a torpedo-bomber, however, the aircraft's engine was too puny to cope with its enhanced load: either the observer had to be discarded, leaving the pilot to do the work of two people singlehandedly, or the amount of fuel aboard had to be cut dramatically.[9]

Together with the details of its operating environment, a given platform's innate characteristics help determine its survivability. Combatants have always derived a degree of protection from the limitations of both their opponents' weapons and their capacity to use them to best effect. Factors such as visibility, speed, manoeuvrability and resilience thereby make up an aircraft's passive defences. The three types of aerial platform used in the Great War – tethered observation balloons, fixed-wing aircraft and dirigibles – had corresponding strengths and weaknesses in this respect.

Airships, for instance, only floated because of their intrinsic buoyancy, which was preserved through the displacement of the air through a suitably large bladder full of hydrogen, an extremely flammable gas. Ostensibly, such machines were very vulnerable, not least to incendiary bullets. On the other hand, in the war's early stages they had the skies largely to themselves. Integrated air-defences were still embryonic and posed a negligible, countervailing threat. Even if a hostile dirigible was spotted, focussing firepower on it was problematic; there was a lack of customized anti-aircraft weaponry, while the few fighters that were available had scant chance of intercepting, let alone engaging, such intruders, if only because their quarry had higher operating ceilings and superior endurance. Although Flight Sub-Lieutenant Rex Warneford shot down *L37* over Ghent as it returned from an aborted raid on England in June 1915,[10] it was not until the end of 1916 that Britain's air-defences proved a real hazard for Zeppelins: shortly after a fighter shot one down – the first to be destroyed by an aeroplane in British airspace – on 3 September, two others fell victim to anti-aircraft fire.

In any event Zeppelins had substantial opportunity costs attached to them. Constructing such rigid airships devoured time and material resources. Although they had a relatively lengthy reach and could carry a moderately

---

9 For details of the Short 184, see: Owen Thetford, *British Naval Aircraft since 1912* (London, 1982), pp. 278–83.

10 H.A. Jones, *The War in the Air: Being the Story of the Part Played in the Great War by the Royal Air Force: Volume Two* (Oxford, 1928), pp. 352–3.

large burden, besides being increasingly vulnerable to enemy action, they were barely airworthy in unfavourable weather. Their non-rigid brethren, by contrast, were comparatively easy and inexpensive to build, as demonstrated by the Royal Navy, which, as early as 1915, turned out and fielded several 'blimps' of the Submarine Scout type in the space of a few months. Equipped with wireless for instantaneous communications and endowed with appreciable endurance, these craft were useful for monitoring coastlines especially but, on the other hand, were too unstable to be effective weapon platforms, particularly against relatively small targets like submarines, and could be very ponderous in adverse weather.

As far as winged machines are concerned, there were three categories: seaplanes, flying boats and aeroplanes. Seaplanes had proportionately large, heavy floats in place of wheeled undercarriages, whereas the flying boat took off and landed with its fuselage in the water. Whilst, riding high on their waterskis, the former could have a nasal engine, propulsion plants had to be mounted uppermost on flying boats to keep the propellers clear of waves. Besides adding to its drag coefficient more than wheels did, however, pontoons eroded an aircraft's overall load-bearing capacity. Consequently, flying boats were typically much larger than floatplanes, accommodating heavier armaments and more fuel, which gave them superior firepower and endurance, respectively. But building and then maintaining such sophisticated craft was commensurately more costly and complex, too. Keeping the fuselage's lower reaches absolutely watertight could, together with other aspects of routine servicing, only be carried out ashore. This called for customized facilities that blended some of the technicians and other assets of boatyards – notably concrete slipways and submersible trailers – with those of aerodromes, such as hangars.

The operability of water-going aircraft, particularly floatplanes, was in any case largely dependent upon not just the atmospheric conditions but also the state of the sea. Flying boats were normally based in sheltered harbours, where they could at worst be moored alongside jetties or lighters, whereas seaplanes were usually housed aboard vessels and hoisted on and off with cranes as required. After that of the British, the tsar's navy had the biggest force of such aircraft, not least because they could be employed with comparative ease in the Black Sea. Tideless, landlocked and sheltered at many points by mountain ranges, this large expanse of water has a rather singular microclimate and hydrochemistry. For much of the year many stretches of it are uncommonly serene. Here, Russian warships might often safely tow floatplanes in their wakes, such was the sea state. The conditions that prevailed in the North Sea were very different, however. Certainly, launching aircraft from the Grand Fleet's main seaplane-tender, HMS *Campania*, was impracticable whenever the waves rose above five feet.[11] The

11 Report from Commanding Officer, HMS *Campania*, to Commander-in-Chief, Grand Fleet, 14 June 1915, The National Archives (TNA), AIR 1/436/15/279/1.

calmer waters of the Mediterranean offered a more promising environment and thought was duly given to raiding the Austrian fleet's anchorage at Pola (Pula) in the Adriatic with floatplanes.[12]

If the launch of aircraft from the sea was potentially troublesome, their recovery was more problematic still. If only because aeroplanes could be manufactured faster and more cheaply than flying boats or seaplanes, the temptation to resort to using them was appreciable. In 1910 Eugene Ely managed – by the skin of his teeth – to fly a befittingly light biplane off a stage erected over the bows of the cruiser USS *Birmingham*.[13] Similar structures subsequently appeared on some other capital ships, usually atop of gun-turrets. Although this adaptation ostensibly enhanced such vessels' capabilities, in most circumstances the ramps were impediments, not least to the turrets being used for their primary purpose. Furthermore, the amount of aviation these platforms accommodated was perforce trifling and comprised small, flimsy machines that had little firepower or endurance. Many of them were configured as floatplanes and were launched astride wheeled trolleys. They could then land alongside and be winched back onto the ship. Otherwise, once aloft, they were irretrievable and their only salvation lay in reaching airstrips that were commensurably close by.

It plainly made more sense to support shipping with aviation units that were based at these selfsame airfields, where the arming, fuelling and servicing of their machines could be undertaken far more easily. Operating from runways, aircraft could, moreover, be of the larger, heavier types with superior reach and payloads. Indeed the availability of comparatively unlimited space made it possible to have numerous machines that, customized for different roles, could proffer a spectrum of capabilities. These might range from reconnaissance sorties to combat operations as diverse as torpedo attacks and the defence, with fighter-bombers, of the airfield itself.

As we shall see, the quest for strategic sites for naval bases was one of the cardinal dynamics in the overseas aggrandizement of the established and emerging powers. Most of these outposts came to accommodate aircraft, too. Unsinkable aerodromes clearly had much to recommend them, particularly those located where the Earth's surface comprises – as in the case of Pearl Harbor, Midway, Madeira and Iceland – far more water than land, or gives rise to nautical pinch points, notably Singapore, Panama, Gibraltar, Malta, the Dardanelles, Crete, Suez, Cuba and the English Channel. However, adversaries could verify the position of airfields with relative ease, whereas ships, be they friends or foes, might remain elusive. If tethered to installations ashore, what was nominally nautical air power might too easily prove unable to provide timely or indeed any support to a fleet at sea, if only because of insufficient reach.

---

12 Minute by Wing Commander, Royal Navy, 26 January 1917, TNA, AIR 1/145/15/66.
13 Norman Polmar, *Aircraft Carriers: A History of Carrier Aviation and its Influence on World Events, Volume One, 1909–1945* (Washington, DC, 2006), pp. 3–5.

Choreographing discrete forces could in any circumstances prove difficult, not least because the very flexibility of land-based aviation could spawn
hydra-headed command structures that reflected the competing demands for
its services. What were the paramount commitments of air power in maritime
settings to be and who should ultimately control it? For navies, the alternative
or complementary approach to looking to shore-based aircraft for aid was that
of housing aviation aboard fleets. Encumbering cruisers and battleships with
ramps for aeroplanes was, at best, an inflexible solution and, at worst, a counterproductive one. Designing vessels around aircraft – rather than the reverse – was
far more promising. This spawned a new school of nautical architecture that
yielded, initially, depot ships for seaplanes and, subsequently, flat-topped aircraft
carriers, the first being HMS *Argus*, a converted merchantman.[14]

The first seaplane tenders were almost all forged from existing vessels, too. The
Royal Navy, for example, retrofitted a mercantilist hull and also acquired and
adapted a Cunard liner and a packet steamer. These vessels became HMS *Ark
Royal*, *Campania* and *Ben-My-Chree*, respectively. Likewise, the Germans converted
a commercial ship, the *Santa Elena*, into a seaplane tender. She was prominent in
the Baltic campaign, not least Operation 'Albion', the amphibious attack on the
islands guarding the Gulf of Riga during 1917. Between April and September
of that year the *Kaiserliche Marine*, its ships especially constrained by the barrage,
also relied on seaplanes to harass shipping along the English Channel and Belgian
coastline. Armed with torpedoes, they enjoyed some success, sinking three vessels
on 9 September alone. Damage was only inflicted in 11 of the 43 raids they carried out, however.[15] The aircraft concerned used Zeebrugge harbour rather than a
depot ship as a base. Indeed with most of their surface vessels bottled up in ports,
the Germans generally found it cheaper and simpler to rely on shore-based aviation
rather than on seaplanes. Particularly for a country that had in any case to emphasize
land operations and the army's concomitant needs, developing and manufacturing
such machines and providing their crews with specialist training were dauntingly
time-consuming and expensive. Arguably the most cost-effective seaplane used by
the *Kaiserliche Marine* during the Great War was the so-called '*Wölfchen*', a single
FF.33 that acted as a spotting plane for the auxiliary cruiser *Wolf* during its prolonged
foray into the Indian and Pacific Oceans between 1916 and 1918.[16]

By the time of the war's outbreak, both Britain's army and navy had aviation
components, the Royal Flying Corps (RFC) and the Royal Naval Air Service
(RNAS), respectively.[17] But any desire to bolster air power per se ineluctably

14 For the design of *Argus*, see: Norman Friedman, *British Carrier Aviation: The Evolution of the Ships and
their Aircraft* (London, 1988), pp. 65–7.

15 Layman, *Naval Aviation*, p. 63.

16 Ibid., pp. 92–3.

17 The RFC had originally comprised Military and Naval Wings. Indeed the RNAS was to be habitually described as 'The Naval Wing of the RFC' in the Navy List until the formation of the Royal Air
Force on 1 April 1918.

raised thorny questions about resource-allocations within, firstly, the national economy – the fountainhead of defence – and, thereafter, between the armed services and their various branches. Theoretical solutions to security conundrums comprise deterrence and, should that fail, a triad of defensive measures: counterforce, passive defence (which includes arms–control) and active defence. However, the need to create a balanced structure of forces affects each component of this defensive triad, complicating research and procurement decisions as well as operational matters.

The geostrategic setting

The peculiar geostrategic position of Britain – the world's pre-eminent maritime power throughout the nineteenth and for much of the twentieth centuries – made the opportunities and problems presented by the advent of military aviation more convoluted and pressing than they were for some of her rivals. Britain's insular nature rendered her that much more difficult to invade but, conversely, also bedevilled any bids she made to project power abroad. If only because of the limitations inherent in forces that are confined to suitably deep and expansive stretches of water, maritime mastery normally proved a necessary condition for victory but an insufficient one by itself. Usually, the reach of British military might had to be extended inland through the landing of soldiers or marines on distant shores.

This process, too, however, was problematic. 'If 100,000 or any large proportion of that number of regular troops are necessary to guard [Britain] against invasion, no force is available for garrisons of places on which the safety of . . . [Britain's global] communications depend', noted Captain John Colomb in 1880, as the Europeans' 'Scramble for Africa' and other far-flung possessions gathered pace.[18] In the symbiotic relationship between Britain's armed forces, the Royal Navy, he observed:

> furnishes the patrolling . . . force, while the army secures to it its bases. . . . To leave the naval force responsible for the protection of its bases would be to tie its hands. It would be "using the fleet to maintain its arsenals instead of the arsenals to maintain the fleet". . . . If naval protection without [an army's] . . . protection be productive of danger to the Empire, great disaster may also be expected to result from attempting to hold distant possessions . . . [with an army], . . . that might be completely isolated . . . from its sources of supply and reserves for want of naval protection of its communications with the Imperial base.[19]

18   John Charles Colomb, *The Defence of Great and Greater Britain: Sketches of the Naval, Military and Political Aspects* (London, 1880), pp. 73–4.

19   Ibid., pp. 57–63.

Sir Edward Grey, later Viscount Fallodon, Britain's foreign secretary from 1905 to 1916, once likened her army to 'a projectile fired by the Royal Navy'.[20] This was neither disparaging nor inappropriate; British troops were unavoidably reliant upon shipping for any deployment overseas and, in many instances, depended for their survival upon a maritime umbilical cord.[21] Writing in 1897, Callwell expressed the view that no recent conflict better exhibited 'the relations between sea-power and successful military operations on shore than the prolonged struggle of Great Britain, aided by Portugal and Spain, against the giant forces of Napoleon south of the Pyrenees'.[22] Certainly, transoceanic operations had ineluctably formed a large part of the British army's undertakings in the Napoleonic and Crimean Wars, during which several of the grander offensives were mounted specifically so that Britain's maritime supremacy might be preserved or consolidated. The last of these was a huge amphibious undertaking, the siege of Sevastopol, 1854–55.[23]

Throughout the 1800s many of Britain's disposable forces had had to be dedicated to guarding her overseas outposts (not least pivotal naval bases), the sea lanes that connected them with the homeland and the mercantilist circle upon which her wealth and concomitant creditworthiness depended. By the 1900s industrialization and colonization were dramatically increasing the extent, volume and value of water-borne trade, adding to the Royal Navy's actual and potential commitments, while Britain's army, obliged to find adequate garrisons to police and protect an empire that ultimately stretched across a third of the globe, could spare relatively few soldiers for offensive expeditions. Although scope for amphibious operations clearly enhanced the punch of Britain's maritime power and bedevilled the strategic calculations of putative foes, much of her land forces' potency stemmed from the flexibility afforded by her dominance of the waves.

The Royal Navy's cornerstone stratagem in the 1800s was that of close blockade, whereby adversaries were denied control of transmarine lines of communication. While commercial shipping was either regulated or eradicated, enemy battle-fleets were neutralized. If not destroyed, the latter were kept dispersed and contained, often bottled up in several ports by forces that had secured at least local superiority. Few opponents were able and willing to endure the consequences for their seaborne trade and coastal settlements. Certainly, in the Napoleonic Wars and the related Anglo-American conflict of 1812–15

20 John Arbuthnot Fisher, *Memories of Admiral of the Fleet Lord Fisher* (London, 1919), p. 18.

21 See, for instance, Christopher D. Hall, *Wellington's Navy: Sea Power and the Peninsular War, 1807–1814* (London, 2004).

22 Charles Edward Callwell, *The Effect of Maritime Command on Land Campaigns Since Waterloo* (London, 1897), p. 6. For a summary of naval operations during the Peninsular campaign and the wider Napoleonic Wars, see: David Gates, *The Napoleonic Wars, 1803–1815* (London, 2003), pp. 38–48.

23 See: Alexander William Kinglake, *The Invasion of the Crimea* (London, 1863–69), vol. 2, pp. 221–41; and Andrew Lambert, *The Crimean War: British Grand Strategy Against Russia, 1853–56* (London, 2011).

especially – during which the Royal Navy sealed off much of the coastlines of two continents – London's policy of maritime blockade was to have far-reaching ramifications.[24] Imperial Germany rightly anticipated and feared that, in the event of war, Britain would again resort to such measures.

To further insure against the massing of hostile – or even potentially hostile – warships in inconvenient places, the British had also been prepared to execute pre-emptive naval or amphibious attacks on anchorages and shipwrights, the foremost cases being the annihilation of the Danish Navy in its berth at Copenhagen in both 1801 and 1807, and the Walcheren expedition of 1809, which deprived Napoleon use of the shipyards in the Scheldt estuary.[25] Although the mounting of a similar blow against the growing *Kaiserliche Marine* was advocated in 1908 by Admiral 'Jacky' Fisher for one,[26] it was not until the Franco-German armistice of July 1940 that such desperate measures were actually resorted to again. Unwilling to run the risk that France's fleet – then the fourth largest in the world – would fall under Nazi control, the British prime minister, Winston Churchill, ordered the destruction of the powerful French warships moored at Mers-El-Kébir, Algeria, and the impounding of others.[27]

Some thirty years before this terrible incident, the geographer Halford Mackinder cautioned that: 'In the presence of vast Powers, broad based on the resources of half continents, Britain could not again become mistress of the seas. Much depends upon the continuance of a lead won under earlier conditions. Should the sources of wealth . . . upon which the navy is founded run dry, the imperial security of Britain will be lost'.[28] In any event even the world's foremost navy occasionally feared that it simply was not large and sophisticated enough. Colomb once recalled 'a feeling of national insecurity, created by the fall of . . . confidence . . . in "the wooden walls" . . . to protect [the British Isles] from invasion'. The substitution, he continued, 'of no amount of thickness of armour for wood can win back that . . . bygone trust so rudely shaken by a practical application of the change produced by steam [propulsion]'.[29]

In the age of sail, winds and currents had been a crucial consideration for shipping. Whereas an adverse, little or no wind could delay movements, a tempest might scatter or even wreck a fleet. The principal sea lanes were inevitably aligned with the Earth's least volatile air-streams, seeking to avoid zones such as

24 See: Andrew Lambert, *The Challenge: Britain Against America in the Naval War of 1812* (London, 2012); Gates, *Napoleonic Wars*, pp. 153–63.

25 For details of Copenhagen, see: Christopher D. Hall, *British Strategy in the Napoleonic War, 1803–15* (Manchester, 1992), pp. 157–63. For details of the Walcheren operation, see: Gordon C. Bond, *The Grand Expedition: The British Invasion of Holland in 1809* (Athens, GA, 1979).

26 See: Fisher, *Memories*, pp. 3–5, 18–19; Hew Strachan, *The First World War: Volume One: Call To Arms* (Oxford, 2001), p. 394.

27 S.W. Roskill, *Churchill and the Admirals* (Barnsley, 2004), pp. 151–8.

28 Halford John Mackinder, *Britain and the British Seas* (Oxford, 1907), p. 358.

29 Colomb, *Defence*, p. 16.

the doldrums in the equatorial Atlantic, where unpredictable breezes and sudden storms and calms prevail, and the roaring forties, turbulent tracts south of the equator. As steam replaced sail, the movements of vessels across the oceans became less predictable.

Yet steamers, unlike sailing vessels, needed to replenish their fuel stocks. Insofar that it dictated that they stay within reach of fresh supplies, this limited their operating range. Similarly, as, in ship-construction, steel supplanted wood – a comparatively cheap, plentiful and workable material – repair and maintenance facilities had to become more versatile. This in turn increased their demand for energy, further swelling the strategic importance of fossil fuels. Indeed a major interdiction operation of the Great War was that mounted by the Russians against Turkey's coal supplies, most of which originated from around Zonguldak and were brought by ship to industrial and population centres further west. The seaplane tenders *Imperator Nikolai* and *Imperator Alexander I* formed the core of a flotilla that, in February 1916, attacked the port of Zonguldak and blockaded the Bosporus, sinking the collier *Jamingard*, the largest freighter lost in the entire war.

For some states during this era the availability (or otherwise) of coaling-stations indubitably determined the direction and extent of their expansion overseas. If only because of the sheer size of her domestic mining industry and collier fleet, this was seldom to prove a restraint on Britain. Welsh coal, among others, was particularly sought after for steam generation especially and Cardiff and other ports flourished on the resulting trade. However, the incipient progression from coal-fired to oil-fired boilers in shipping – not least in warships – was to have enormous strategic ramifications. Increasingly, British sea power relied on petroleum, virtually all of which was imported from distant shores, notably those of the Gulf and the Caribbean. Similarly, oil fuelled and lubricated the internal-combustion engine, without which powered, controlled flight was impossible.

Thus far, Britain had been largely indifferent to events in the Middle East in general and Arabia in particular, her principal concerns here being access to the region's nautical arteries, notably the Suez Canal. However, as industrialization, mechanization and colonization proceeded apace, the demand for raw materials alone redoubled the importance of transoceanic commerce, adding to the Royal Navy's burdens. Alongside the major powers' deliberate quest to corner overseas markets and resources, there was another that was more precautionary than avaricious: states would intervene in regions of negligible intrinsic value simply to keep potential rivals at bay, Britain's intermittent dabbling in Afghanistan and remote corners of Africa being good cases in point. This situation also spawned some outlandish strategic schemes, notably Germany's Operational Plan Three of 1903–07, which envisaged amphibious assaults along the USA's eastern seaboard involving 60 ships, among them scores of replenishment colliers, and 120,000 troops. Seizure of the Panama Canal once it opened and the taking of Puerto Rico were also considered by Germany as the Great War loomed, while, above all, her attempts in 1917 to induce Mexico – and Japan, if she could be enticed away from the Allies – to attack America precipitated the USA's embroilment in the conflict.[30]

So immense did the extent of Britain's interests and commitments overseas become in the late 1800s that it threatened to bring her into conflict with several, if not all, of the other major powers simultaneously. London simply could not afford to amass sufficient martial strength everywhere and, from this juncture especially, pursued various diplomatic efforts to keep manageable any threat that did emerge. Some success had been achieved in recent decades through arms-limitation agreements, notably clauses within the Treaty of Paris that had ended the Crimean War in 1856. After the French and British fleets had suppressed the tsar's navy in both the Baltic and Black Seas during that conflict, Russia had been required to demilitarize the coastline and waters of the latter. However, whereas military capabilities can take decades to develop or, if lost, retrieve, political intentions can alter literally overnight. New technology or shifts in the international balance of power could render any accord obsolete or unenforceable. Although rights of passage for warships through the Bosporus and the Dardanelles – the so-called Straits Question – were to remain a *cause célèbre* for decades thereafter, resurfacing in the Treaty of Lausanne (1923) and the Montreux Convention of 1936, Russia repudiated the arms-limitation clauses of the Treaty of Paris in the wake of the Franco-Prussian War of 1870–71.

Whereas they were comprehensively defeated by the Germans on land, the French easily retained their maritime supremacy throughout this debacle. Within a few years, however, Gabriel Charmes in particular and the *Jeune Ecole* in general were pondering how they might successfully confront Britain's naval strength in a conflict that, though intermittently anticipated, never actually materialized. Besides trying to contain Russian sea power, the Paris Congress of 1856 had endeavoured to codify just-war concepts with regard to the conduct of naval blockades and *guerre de course*. But the *Jeune Ecole* concluded that the only way France and other states might hope to counter Britain's mastery of the waves was through asymmetrical warfare, namely such stratagems as unheralded attacks on commercial shipping, the use of 'auxiliary' warships – that, as adapted merchantmen, often appeared innocuous – and the nocturnal bombardment of ports.

Britain's Naval Defence Act of 1889 sought to formalize the maritime superiority that she had long enjoyed by setting a crude ratio of military might, the two-power standard: the Royal Navy's battleships were to be at least equal in numbers to the combined strength of the next two largest navies in the world, namely those of France and Russia.[31] Whilst Whitehall anticipated that this policy would help deter other states from attempting to make significant alterations to the balance of sea power, detractors and dissenters dismissed it as

---

30 Copies of most of Germany's contingency plans for war with the USA are preserved in the collections of the Bundesarchiv, notably in the Admiralty files RM 5/879, RM 5/879k (Karten) and RM 5/885.

31 Paul Kennedy, *The Rise and Fall of British Naval Mastery* (London, 1976), pp. 178–9.

counterproductive, arguing that it would inaugurate an arms-race. However, the act was more of a symptom of competition than a cause. Through his publication of *The Influence of Sea Power Upon History, 1660–1783* (1890) and the *Influence of Sea Power Upon the French Revolution, 1793–1812* (1892) the American naval officer and academic Alfred Thayer Mahan established himself as the leading maritime philosopher in the era of so-called 'New Imperialism'. Although more historical than technical, Mahan's analysis caught policy-makers' imaginations in an epoch of assertive nationalism and, by linking sea power to precepts of national policy and strategy, accelerated an existing trend towards maritime rivalry.

By establishing a rough index of military might, the Naval Defence Act sought to give British planners a means of quantifying that competition. Yet the strategic environment was a bewilderingly dynamic one, not least because nautical technology was evolving and the fear of obsolescence was as significant as the actuality. Designers such as John Holland, Maxime Labeuf and Simon Lake had been working on the manufacture of primitive submarines for some time and, by 1900, such craft were gradually being integrated into the leading navies. In 1913 Krupps, the German industrial giant, incorporated some of Lake's ideas into a variant that, fitted with an engine developed by Rudolph Diesel, had far greater endurance than its forebears, extending the reach of patrols.

Typically, submersibles from then on drew motive power from electric batteries when under water and diesel engines when on the surface. The latter mode of operation was plainly the more energy-efficient of the two, permitting the submarine to travel further and faster when on the waves than when beneath them. Although guns were added to the decks of many, the cardinal armament of most submarines was the locomotive torpedo, since it could be fired both on and under the surface. The first such weapon, the so-called Luppis-Whitehead 'Fish', had appeared in the late 1860s and, by the 1880s, designs with the 'Whitehead Secret' – a depth-controlling mechanism that ingeniously combined a pendulum with a hydrostatic valve – were being adopted by navies the world over.[32] The dimensions, range, payload and speed of torpedoes all steadily increased in the decades leading up to the Great War, as did the sophistication of moored mines, the use of which had already become a widespread tactic by the early 1900s. However, the very size and weight of most torpedoes made them too burdensome to be mounted on aircraft when such platforms first appeared. Torpedoes were, moreover, inflexible weapons and employing them called for bespoke planes with specially trained aviators. Among other inhibitions, a large torpedo dropped from the sky might plunge to 70 feet before levelling off, making them unsuitable for use in relatively shallow waters.

---

32 David Gates, *Warfare in the Nineteenth Century* (London and New York, 2001), p. 182. To the older generations of seamen, the term 'torpedo' denoted various forms of what were later commonly known as mines. The popular metaphor for locomotive torpedoes among British sailors and airmen was 'Fish', whereas German submariners, for instance, often referred to them as '*Aale*', eels.

Above the waves, capital ships in the early days of steel and steam were essentially gun-platforms, as in bygone days. If they were to remain capable of absorbing any punishment while trading salvoes with similar adversaries, the density and distribution of their armour had to undergo refinements, just as their firepower had to be increased if it was to continue to penetrate the protective cladding on opposing vessels. Large-calibre ordnance and denser and more abundant armour added to the size and weight of vessels, however, reducing their speed and manoeuvrability, characteristics that could prove crucial in determining a platform's survivability in an engagement. This in turn raised the question of propulsion mechanisms. On the eve of the Great War better turbine machinery – first developed in the mid-1880s – was proliferating and oil-fired boilers were beginning to supplant coal-fired ones. Although the calorific value of fuel varies from one type to another, as a rule of thumb a capital ship would, steaming hard, devour around a ton of coal for every mile it travelled. (The German auxiliary cruiser *Wolf*, a converted freighter, was sluggish but, with a bunker capable of holding 8,000 tons of coal, had an immense range.) On average, a thoroughbred warship needed replenishing every eight or nine days and might require 2,000 tons of coal to fill its bunkers. Bringing this aboard from supporting colliers was often done by hand and was a filthy, back-breaking, time-consuming job. Oil, a liquid that could be pumped into storage tanks and combustion mechanisms alike, was a superior fuel in every respect and allowed for the development of better engines that endowed even battleships with speeds of 24 knots, despite them being weighed down with ever bigger guns and more armour plate.

Battlecruisers – a design that maximized punch and speed at the expense of armoured protection – were faster still and, more to the point, were capable of steaming very hard for prolonged periods. Indeed until the maturation of naval aviation in the 1930s such platforms were the best means available of projecting power relatively rapidly within expansive, maritime environments and were therefore of particular use in countering a *guerre de course*. This eventuality was one that the British Empire above all anticipated and, consequently, the Royal Navy was pre-eminent in developing the concept and actuality of the battlecruiser. When used intelligently, the combination of pace and formidable firepower that this class of ship embodied could prove lethal to even sizeable surface raiders, as the Germans were to discover when, in December 1914, they tried to attack the British wireless and coaling station on the Falklands, only to encounter HMS *Inflexible* and *Invincible*. Of the five German vessels involved in the action, only the *Dresden*, which had turbine engines, managed to outrun the pursuing battlecruisers.[33] Whereas they were fitted with 12" guns, later British battlecruisers, such as HMS *Lion*, had 13.5" weapons, the selfsame calibre of

33　Jon T. Sumida, 'Sturdee: Falkland Islands, 1914', in Eric Grove (ed.), *Great Battles of the Royal Navy* (London, 1994), pp. 161–8.

ordnance initially mounted on the Royal Navy's dreadnoughts in 1909. The armament on battleships launched just before the First World War was, however, larger still. As 13.5" guns gave way to 15" ones, the reach and penetrating power of projectiles increased commensurately. Whereas the 13.5" shell weighed 1,400 lbs, the 15" was 520 lbs heavier.[34] Nevertheless, the guns were capable of hurling them over distances of eleven miles or more.

Focussing firepower on objects so far away that they could scarcely be seen by unaided human eyes called for new targeting mechanisms. There was obviously a role for spotter planes here. Yet similar problems were experienced with regard to the gathering and dissemination of information that was essential if vessels – and, increasingly, aircraft – were to fight in concert. Although an engagement involving several warships might easily meander over an area amounting to scores of square miles of sea, there was neither any radar surveillance nor much in the way of dependable, instantaneous communication systems, while manoeuvres continued to be cued by signals made with light semaphores and flags. The capacity to exchange fire at ever greater ranges further complicated the design and configuration of armour plating as well, insofar that vertical, lateral sheathing, while affording some protection against projectiles approaching more or less horizontally, could not safeguard ships from shells that had more pronounced parabolic trajectories. Likewise, as the threat from mines and torpedoes increased, so, too, did the need to protect warships below the waterline.

Informal and formal naval arms-control and arms-limitation processes were also made more intricate by the emergence of new actors on the global stage, notably the Japanese, Americans, Germans and Italians, and the denouement of others, namely the once mighty Spanish, Austro-Hungarian, Turkish and Chinese Empires. Indeed the Great War was essentially precipitated by shifts in the international balance of power that had become manifest around the turn of the century. In 1880 the US Navy, for instance, was the twelfth largest in the world; by the early 1900s, it was the third. This expansion called for a commensurate increase in the number of coaling-stations and bases. American interest in acquiring enclaves in the Pacific initially focussed on the Samoan and Hawaiian Islands. London, Washington and Berlin all vied for control of Samoa, which, in 1899, was officially divided between America and Germany, Britain accepting compensation elsewhere.[35] Whereas Samoa was a gateway to the southern Pacific, Hawaii, 2,300 miles off California, formed an intersection on trade routes to and from China and the Philippines. In 1887 the USA renewed a treaty with Hawaii's native monarchy that had long since granted America exclusive commercial access to the kingdom in return for pledges of protection against third parties. Entitlement to a naval base (Pearl Harbor) on Oahu formed

---

34 David K. Brown, *The Grand Fleet: Warship Design and Development, 1906–1922* (London, 1999), p. 37.

35 Samuel E. Morison, Henry Steele Commager, William E. Leuchtenburg, *A Concise History of the American Republic* (New York, 1977), p. 479.

part of this accord. In 1893 Hawaii's queen was overthrown and a provisional government sought the islands' outright annexation by the USA. This quest was eventually fulfilled, not least because of geostrategic considerations: Washington was increasingly fearful of Japanese interlopers in the Hawaiian archipelago at a juncture when it formed America's principal stepping-stone to the newest and grandest of her outposts in the Pacific, the Philippines.[36]

She acquired this constellation of 7,000 islands as a result of the Spanish–American conflict of 1898. This was waged exclusively in maritime settings, namely the Philippines and Cuba, and largely decided by the Americans' emphatic naval victories at Manila Bay and Santiago de Cuba.[37] The triumph of a republic of the New World over a monarchy of the Old, the war ended Spain's empire in the Pacific and Caribbean, transforming the USA into a colonial power with global influence. The Spaniards, who had played such an enormous part in the colonization of the Western Hemisphere, were ousted, losing Cuba (initially to revolutionaries) and ceding Guam – the largest and most southerly of the Marianas – and Puerto Rico to the USA.

'Uncle Sam' also purchased the Philippines. As in Cuba, rebellion had engulfed these prior to the war and, notwithstanding their ruthless attempts, the Spanish had failed to quell it. The USA eventually did so with an operation that was as urgent as it was brutal.[38] For Washington's overriding fear here was that Tokyo and Berlin in particular might seek to partition the vast archipelago, thereby acquiring more potential bases in the Pacific and a larger share of the trade that was flourishing across its basin and rim. Indeed since Japan's victory in the Sino-Japanese War of 1894–95, the European powers, sensing that China's ramshackle empire was on the verge of disintegration, had been striving to secure their own spheres of influence in the Orient. The 'Scramble for Concessions', as it is known, contradicted Washington's 'open-door' policy towards China, imperilling American commercial interests and diminishing the value of the Philippines. It also threatened to provoke violent confrontations between the various parties involved, not least disgruntled elements of the indigenous population. After the Europeans united with one another and the USA in the face of the Boxer Rebellion of 1900, tensions over China abated somewhat; for the time being, America's stance seemed to have gained the tacit support at least of the other major powers. However, although the 'open-door' approach was to remain the basis of the USA's dealings with her for the next thirty years, China escaped partition essentially because those who favoured her dismemberment could not agree on how to divide the spoils. In fact Russia and Japan were to go to war over this very issue as early as 1904.

The arbitrator of the treaty that ended that conflict was, tellingly, the American president, Theodore Roosevelt, who was awarded the Nobel Peace Prize

36 Ibid.
37 See: Robert Conroy, *The Battle of Manila Bay: The Spanish–American War in the Philippines* (New York, 1968); and David F. Trask, *The War With Spain in 1898* (New York, 1981), pp. 257–66.
38 See: Morison et al., *American Republic*, pp. 488–92.

for his efforts. However, by helping to forestall China's demise at the hands of outsiders, the USA's 'open-door' policy boxed in the Japanese in particular. This augured ill: the Mukden Incident of 1931, which gave Tokyo a pretext upon which to seize the whole of Manchuria, and the all-out (but undeclared) Sino-Japanese War that was to erupt in 1937 formed milestones on the road that, in 1941, was to lead to the Japanese attack on Pearl Harbor.

The minutiae of that audacious raid highlighted just how far air power in general and military aviation in particular had evolved since its advent only a few decades before. The attack was also a very stark illustration of a fundamental attribute of air power that is so obvious as to be frequently overlooked: the constraint known as impermanence. This largely stems from an inescapable reliance on facilities and personnel that are located on the Earth's surface and without which aviation could not exist, let alone function. The proximity, quality and security of aerodromes are critical factors in determining the sustainability of aerial operations. Although by the First World War's conclusion the USA, Japan and other major states had acquired a number of airfields within the Pacific basin, a combination of air power's innate limitations and the sheer immensity of that ocean dictated that, however extensive, a network of fixed bases was unlikely to prove adequate in the event of another armed conflict between leading maritime nations. Unsinkable aircraft carriers had to be supplemented with floating ones if air power's reach was to be extended over the high seas. On the other hand, replenishing and protecting mobile facilities was a peculiarly complex task. Among other measures, reducing sea-based air power's impermanence necessitated the development of bespoke logistical support for carriers in the form of fleet trains. In the event it was to be the Americans who proved more adept at this than the Japanese.

Theodore Roosevelt was one of many influential figures who fell under Mahan's spell. He served as assistant secretary of the navy in 1897 and, during his two terms as president (1901–09), the US fleet continued to expand markedly, acquiring ten new battleships and four cruisers. Actively seeking American domination of the Caribbean and Pacific, Roosevelt also developed the navy's essential network of coaling-stations and bases. He advocated both the annexation of Hawaii and the taking of the Philippines long before he entered the White House and, once president, championed a tremendous feat of engineering, the building of an interoceanic channel through the isthmus of Central America.

The conflict with Spain in 1898 had underscored the desirability of a less circuitous route between the USA's eastern and western seaboards. For ships small enough to squeeze through its locks, the Panama Canal, just 40 miles in length, sliced 6,000 miles off the voyage around South America.[39] However, this

---

39 Built between 1887 and 1895, Germany's Kiel Canal, which linked the Baltic and North Seas, had to be widened between 1907 and 1914 to accommodate dreadnoughts. See: Strachan, *Call To Arms*, pp. 52, 406. Similarly, the locks of the Panama Canal effectively imposed limits on the dimensions of American warship designs. The spaciousness of dry docks in which vessels might be repaired and serviced also became an important consideration for all major naval powers, their availability – or otherwise – shaping patterns of operations by big capital ships especially.

man-made waterway was not just an instrument but also a symbol of America's swelling maritime might. The canal's construction was authorized by the Hay-Pauncefote Treaty of 1901, whereby Britain discarded her long-standing interest in such a project, granting the USA exclusive rights to build and manage a channel which, like that at Suez, would in principle be open to all nations on equal terms. If only implicitly, Washington's entitlement to fortify the waterway was also acknowledged and, in 1903, 'Uncle Sam' duly secured a lease in perpetuity over a 'Canal Zone' straddling the channel. Opened to traffic just as the Great War began, the waterway was overlooked by US troops.

By this juncture America had become the dominant maritime power in this region of the world, with a strong navy and a web of supporting bases, one of which was on Cuba. Situated at the confluence of the Gulf of Mexico, the Atlantic and the Caribbean, that island dominated several major sea lanes, including the northern approaches to the Panama Canal. Her sugar industry especially had long made her a valuable trade partner for the USA. In the Spanish-American War's aftermath, Cuba, though nominally independent, remained occupied by US soldiers and her fate was largely determined by, initially, the Platt Amendment of 1901 and, thereafter, treaty provisions agreed to by Roosevelt. Among these was the stipulation that, whereas the armed forces of third parties were to be excluded from her territory, Cuba was to lease sites to Washington for use as naval outposts. America thereby gained a naval base at Guantánamo Bay.[40]

Proclaimed in 1823, the Monroe Doctrine had – until the American Civil War began transforming the military balance in the Western Hemisphere – been passively enforced by Britain's Royal Navy. By 1900 the USA was actively upholding it herself in what was still essentially a Eurocentric world. Preoccupied with a precarious situation on several fronts, notably South Africa, the North Sea and India's frontier, Britain was seeking to concentrate on her core concerns by sharing her peripheral anxieties with those emerging powers whose interests coincided, if only for the time being, with her own. Whilst there was something of an Anglo-American entente between 1890 and 1916, this had to remain an informal partnership, not least because of traditional American suspicions of entangling alliances.[41] In her courtship of Japan, Britain could and did go appreciably further.

In 1902, in an effort to alleviate the burden on the Royal Navy, Britain accepted an alliance with Japan that was to endure until the Washington Conference of 1922. Tokyo had initially contemplated forging an accord with Russia, only to conclude that disputes with the latter over their respective designs on Chinese territory would ineluctably turn deadly. Sure enough, when the tsar leased Port Arthur (Lüshun) from China, the Japanese, having tried unsuccessfully to resolve their differences diplomatically, turned on the Russians in 1904,

---

40 Morison et al., *American Republic*, p. 494.
41 See: John Arbuthnot Fisher, *Records by Admiral of the Fleet Lord Fisher* (London, 1919), pp. 38–9.

inflicting a startling defeat in a war that arose from these nations' conflicting attempts to penetrate into Manchuria and Korea.

The Anglo-German Agreement of 1890, by contrast, sought to defuse growing tensions over, above all, the colonization and division of Africa, but a conspicuous trade-off within this accord was the ceding to Germany of Heligoland, the tiny island in the North Sea that the British had first seized from Denmark during the Napoleonic Wars. Heligoland straddled the approaches to harbours that were destined to accommodate a key component of Germany's *Weltpolitik* of 1897–1914, her *Hochseeflotte*.[42] But, in British eyes, Heligoland was as untenable as, prior to the initiation of *Weltpolitik*, the Germans were unthreatening. Preoccupied with consolidating the (Second) *Reich*, which had only been founded as recently as 1871, Kaisers Wilhelm I and Frederick III and their Chancellor, Otto von Bismarck, had neither the wish nor the ships to compete significantly with the other major powers for influence and possessions overseas. The wars that had brought about German unification had destroyed the old balance of power on the European continent, and Berlin was anxious to avoid provoking Germany's resentful neighbours into fresh hostilities. Foreign policy was overwhelmingly cautious if not conciliatory: while, closer to home, his elaborate network of alliances kept France especially in check, Bismarck was content with the establishment, in 1884, of German protectorates in south-western Africa and Cameroon. Under the Anglo-German agreement of 1890 – the year Bismarck was ousted by Wilhelm II – Germany gained Heligoland and a strip of land that gave German South-West Africa (now Namibia) access to the Zambezi. In return, she acknowledged London's claims to enormous swathes of territory in East Africa as well as recognizing the island of Zanzibar, off Africa's eastern coast, as a British protectorate.[43]

Zanzibar was a key node for the British, lying roughly equidistant from the Suez Canal, the Cape and Bombay (Mumbai). In September 1914 the *Königsberg* – one of a scattering of cruisers the *Kaiserliche Marine* had beyond its home waters – was to surprise and sink HMS *Pegasus* within the island's anchorage. Eventually some two dozen British warships were to be drawn into the hunt for the raider before she was finally tracked down and destroyed in the Rufji delta in July 1915.[44] In the interim, in a bid to deny *Königsberg* bases for her sorties into the Indian Ocean, an amphibious attack was also mounted against Tanga, the cardinal port of German East Africa (now Tanzania). The assault was, however, thwarted. In fact East Africa was the one fragment of Berlin's overseas empire that was not to be overrun during World War I.

42 High Seas Fleet.

43 For an analysis of the rise of German naval power during the late 1800s and early 1900s, see: Michael Epkenhans, 'Imperial Germany and the Importance of Sea Power', in N.A.M. Rodger (ed.), *Naval Power in the Twentieth Century* (Annapolis, MD, 1996), pp. 27–35.

44 H.A. Jones, *The War in the Air: Being the Story of the Part Played in the Great War by the Royal Air Force: Volume Three* (Oxford, 1931), pp. 1–13.

By that conflict's advent, Germany's colonial possessions had grown to over four times the size of the Fatherland itself and, besides chunks of the Dark Continent, included parts of Micronesia and Samoa in the Pacific. Her most important maritime base in the Far East was, however, Tsingtao (Qingdao) on the Yellow Sea. This had been seized in 1897 as part of the 'Scramble for Concessions' and became home to the Asiatic squadron commanded by Admiral Maximilian von Spee. Shortly after the First World War's outbreak, Tsingtao was besieged by an Allied army that overwhelmingly consisted of Japanese troops. It fell in November 1914, obliging Spee to seek succour elsewhere.

Detaching the light cruiser *Emden*, which sailed into the Indian Ocean, attacking hostile shipping and bombarding Madras (Chennai), Spee headed for neutral Chile, if only in search of coal. Whilst, on paper, some of air power's inherent qualities would have rendered it of immeasurable assistance in tracking down and eradicating the German raiders, in practice aviation offered few solutions to the problems involved. Firstly, far too few aircraft and bases were available to maintain patrols over even the tiniest fraction of the area within which the hostile warships might, in theory, be located. Secondly, air power as such was just too inchoate. Short on reach, punch and navigational and communicative capabilities, those machines that were at hand were unlikely to do much more than incommode relatively formidable adversaries, particularly in such a large and otherwise complex environment. Even if a plane were to catch sight of one or more of the interlopers, the aircraft's own weaponry – if it had any at all – would, at best, pose a negligible threat to an opponent that was both armed and armoured. Interception by platforms with comparable firepower and protection was the deadliest risk being run by Spee and his colleagues.

Although aviation might help coordinate such an encounter, it could, at this juncture, do little more. Hunted by other surface vessels across vast swathes of water, *Emden* easily eluded her pursuers, causing utter consternation. But when she put a landing party ashore on the Cocos Islands with a view to disabling the wireless station, the operators managed to get off a message that alerted a nearby Australian cruiser, HMAS *Sydney*. *Emden* was finally overtaken and sunk. Spee himself proved luckier, if only for a few more weeks. His odyssey was to include his shocking victory over a weak British flotilla that, somewhat rashly, confronted him off Cape Coronel, Chile, in November 1914 and was to culminate in his death and his squadron's virtual annihilation near the Falkland Islands that December; only the light cruiser *Dresden* – sister ship of the *Emden* – survived this second clash with the Royal Navy, and she was to be scuttled by her crew when HMS *Glasgow* and *Kent* found her off the Juan Ferdinand Islands in March 1915. Aviation played no more of a part in either the Battle of Coronel or that of the Falklands than it had in the destruction of the *Emden*. In all these maritime engagements the victorious side used far more familiar means to subdue their opponents, notably superior speed, manoeuvrability and long-range gunnery.

These were essentially the same capabilities that had led to Japan's triumph over the Russians in the Tsushima Strait in May 1905. This crushing defeat of one of the established great powers, together with her alliance with another, Britain, heralded Japan's advance from the periphery of the international stage to its heart. After taking Tsingtao, the Japanese went on to acquire all of Germany's possessions in the Pacific to the north of the equator, notably the Palau Islands, the Carolines, the Marshall Islands and the Marianas; forces from New Zealand and Australia occupied those to the south. In the interim, with the exception of beleaguered East Africa, all of Germany's outposts on that continent were also taken from her, Cameroon, Togoland and South-West Africa all being lost within twelve months of the war starting. Thereby, the German navy's global network of coaling-stations and repair and maintenance facilities was, together with the wireless transmitters that gave any remaining, roving vessels some contact with Berlin, all but extirpated.

By this juncture, however, German confidence in the starkest symbol and instrument of *Weltpolitik*, the *Hochseeflotte*, had already been waning for some years, if only behind the scenes. In March 1890 the British magazine *Punch* had mischievously highlighted Kaiser Wilhelm's dismissal of Bismarck with a cartoon that had a nautical flavour: 'Dropping the Pilot', by Sir John Tenniel. In response to Germany's *Weltpolitik* in general and her Naval Acts of 1898 and 1900 in particular, Britain had started designing and building a new generation of battleships, the super-dreadnoughts. She had also formally allied herself with Japan and, in 1904 and 1907 respectively, had concluded the Anglo-French and Anglo-Russian Ententes, making useful friends out of putative foes. All of this allowed her to retain more of her own best warships in waters closer to home. The desire to avoid squandering resources through the duplication of effort also impacted on the planning and operational purview of the Royal and French Navies especially. In a memorandum drawn up between Sir Edward Grey and Paul Cambon – Paris' ambassador to London – in 1912, the British undertook to shield France's Atlantic coastline from attack by German shipping, allowing the French fleet to concentrate on the security of the Mediterranean. This, in turn, reduced the strain on the Royal Navy in that quarter, allowing it to focus on its principal concern, that of mustering sufficient strength to contain and, if necessary, defeat the *Hochseeflotte* in the English Channel and North Sea.

Whilst this secret, peacetime agreement between gentlemen did go some considerable way in helping to alleviate the material burdens on Britain's armed forces especially, it could only exacerbate the moral and political pressure that Grey and his critics and supporters within the Cabinet would be subject to if, as was to occur in August 1914, Germany and France were to come to blows. By 1912 the former's strategic priority was the fighting of, not a global war, but one on the European continent; the proportion of the military budget allotted to the army was rising, while that going to the navy was, after being boosted in 1906 and 1908, tapering off.[45] The *Kaiserliche Marine* was no longer to seek

45 See: Epkenhans, 'Imperial Germany', pp. 29–33, 35.

parity with the Royal Navy but, rather, to act as something of a counterpoise to it by remaining in being and seeking to deny it absolute maritime mastery. Certainly, the notion of pitting the *Hochseeflotte* in a set piece battle against Britain's Grand Fleet – which dominated the North Sea from its bases at Scapa Flow in the Orkneys, Cromarty on the Moray Firth and Rosyth on the Forth – seemed fanciful if not foolhardy. The best that could be hoped for was that a portion of the Grand Fleet might be ensnared and destroyed in hit-and-run attacks. Meanwhile, a *guerre de course*, which could be waged by a few enterprising cruisers and U-boats, might prove cost-effective in that this might oblige London to divert substantial resources to protecting the empire and sea lanes.

Nevertheless, a combination of strategic reasoning in Berlin and the characteristics of familiar and emerging weaponry placed Britain in something of a dilemma. Grey went so far as to say as much to the German government in February 1908, when the *Kaiserliche Marine* was enjoying another increase in funding:

> The independence and very existence of the British Empire depend on the preservation of its supremacy at sea, and the British Government is bound to . . . keep up such naval forces as are essential for that purpose. It would be futile to pretend that the increase in the German fleet is not one of those factors which have to be taken into account in any calculation of that strength at which the British Navy must be maintained.[46]

Grey's concerns in this regard were elaborated on in confidential memoranda he produced during summer 1908. He warned the king ahead of a meeting with Kaiser Wilhelm that:

> If the Germans continue to execute their Naval programme . . . we shall certainly have to ask Parliament to vote a considerable increase to our expenditure: no Government . . . could avoid doing so. The justification and necessity for this increase . . . would be the German expenditure. We have to take into account not only the German Navy, but also the German Army. If the German Fleet ever becomes superior to ours, the German Army can conquer this country. There is no corresponding risk of this kind to Germany: for however superior our Fleet was, no Naval victory would bring us any nearer to Berlin. . . .[47]

The waters surrounding Britain were, Grey and the Admiralty feared, in danger of being bridged by steamships that might in theory ferry an army across the divide in a matter of hours. In August 1908 he emphasized that:

46   Foreign Office, *British Documents on the Origins of the War, 1898–1914: Anglo-German Tension: Armaments and Negotiations 1907–1912*, G.P. Gooch and H. Temperley (eds), (London, 1928), pp. 134–5.
47   Ibid., p. 779.

Whereas, if the German Navy became superior or even attained such a relative proportion to the British as to enable it at an untoward moment to secure command of the sea for a few days, Great Britain would be not only defeated but occupied and conquered; Germany does not run so great a risk as this from any superiority of the British fleet, for the British army is so inferior to the German in size that occupation and conquest are out of the question.[48]

Nor was this the only potential threat that was looming on the horizon. Just three years after the Wright brothers pioneered powered, controlled flight in a heavier-than-air machine, the Brazilian engineer and aviator Alberto Santos-Dumont, using a plane he had designed and built in France, the '*Oiseau de Proie*', executed the first such flights to be documented in Europe. On witnessing this, the English press mogul Alfred Harmsworth, later Viscount Northcliffe, commented that Britain 'was no longer an island'.[49] Sure enough, in July 1909 Louis Blériot was to cross the English Channel in a heavier-than-air machine, covering the 22 miles from Calais to Dover in just 37 minutes. To maintain their security, it was no longer sufficient for the British to dominate the waves; they evidently had to control much of the adjacent air, too. Indeed as Edwardian fears of threats from the sea were transcended by dangers that emanated from the skies, Britain's heartland, not least the capital, London, began to appear rather more vulnerable than the sea lanes between it and the empire.

Just months before Blériot's celebrated flight, the government – anxious to avoid a costly arms-race but equally determined to maintain the Royal Navy's quantitative and qualitative advantages – was confronted with a censure motion in the Commons for pursuing a defence policy that, critics held, did not adequately safeguard Britain and her overseas possessions. Grey insisted that:

There is no comparison of the importance of the Germans' navy to them, and the importance of our navy to us. Our navy to us is what their army is to them. To have a strong navy would increase their prestige, their diplomatic influence, their power of protecting their commerce; but . . . it is not a matter of life and death to them, as it is to us. . . . [O]ur army is not maintained on a scale which, unaided, could do anything on German territory. But if the German navy were superior to ours, . . . for us it would not be a question of defeat. Our independence, our very existence would be at stake. . . .[50]

---

48 Ibid., p. 174. Also see Gilbert Murray, *The Foreign Policy of Sir Edward Grey, 1906–1915* (Oxford, 1915), pp. 27, 109.

49 Quoted in Alfred Gollin, *No Longer an Island: Britain and the Wright Brothers, 1902–1909* (London, 1984), p. 193.

50 Parliament of the United Kingdom (*Hansard*) *Parliamentary Debates, House of Commons*, 29 March 1909, Series 5, vol. 3, cols 60–61.

Grey and his colleagues won the vote comfortably and pressed ahead with measures aimed at maintaining the Royal Navy's lead in terms of the number and firepower of its dreadnoughts. By the end of 1910 the fleet was accounting for a quarter of all state expenditure. Germany, by contrast, now faced with the prospect of a continental war on two fronts, devoted a dwindling proportion of her military budget to bolstering her nautical strength. Nevertheless, the security of the British Isles remained precarious. Whereas Winston Churchill – First Lord of the Admiralty from 1911 to 1915 – was to opine that John Jellicoe, the Grand Fleet's commander, was the only man on either side who was capable of losing any war against Germany in an afternoon,[51] Jellicoe's sailors were unable to win any such conflict without complementary efforts by other forces. As both Grey and Richard Haldane – the war secretary – realized, naval supremacy was a necessary but inadequate condition for victory, just as it had been in Nelson's day. In fact wider reforms to Britain's armed services included the provision of a discrete army corps for deployment on overseas expeditions. Should, then, Britain invest in an even stronger fleet and entrust major land operations to her allies, or should she enlarge her own army through conscription? Could she bear the political and financial costs of either of these policies?[52]

Endeavouring to hold down military expenditure overall, Haldane already faced some thorny decisions regarding the allocation of resources when the advent of air power compounded his woes. Any investment in this nascent technology might prove premature at best. Yet the incorporation of aviation into both of Britain's armed forces seemed desirable if not essential. Aircraft already had much to recommend them for reconnaissance missions and, once platforms and weaponry improved, their employment in other tasks could only become more feasible. But what form, precisely, should air power take? Should it comprise dirigibles, fixed-wing planes or a mixture of the two? Should planes be mounted on wheels – and, thus, based ashore – or be essentially flyable boats that could use customized ships as harbours? Much depended upon exactly what strategic and tactical functions Britain expected this fledgling technology to fulfil.

Exploiting the third dimension: aircraft and submarines

As *the* maritime power, the precise details of Britain's requirements and hopes differed somewhat from those of some other states. Still, she had to try to keep abreast, if not ahead, in the race to exploit the third dimension. This wider competition engendered others, however, as established organizations vied with one another and with new ones for roles and resources. For instance, not least

---

51 Winston S. Churchill, *The World Crisis 1911–1918: Volume 2* (London, 1939), p. 1015.

52 See: A. J. A. Morris, *The Scaremongers: The Advocacy of War and Rearmament, 1896–1914* (London, 1984), pp. 98–110.

because of its superior procurement systems – which arose largely from its close collaboration with private companies, notably Shorts, Sopwith and Handley Page – the RNAS thrived initially in comparison with the RFC.[53] Conversely, production of airframes and aero-engines in Germany was dominated by the army, which begrudged any expansion of maritime aviation.[54] Naval air power was, likewise, far less important to France than soldiers were, particularly in the opening phase of the Great War, during which her capital almost fell and much of her industrial heartland was overrun. Meanwhile, although the Austro-Hungarian Empire managed to turn out sufficient aircraft for its relatively small navy, its army remained reliant on imported German machines because of a lack of indigenous industrial capacity.[55] Everywhere, the war created unprecedented demands for all manner of commodities and manufactures, the latter including the products of what at this juncture were fledgling enterprises founded on the internal-combustion engine.[56] Even in Britain the procurement of the biggest of aero-engines, for instance, encountered bottlenecks. Handmade, the Rolls Royce 'Eagle', for example, was as dependable as it was uncommon. Only 875 of these motors were turned out in the whole of 1917 and, overall, output was to peak at just 38 units per week.[57]

Although the RNAS was granted first call on large aero-engines essentially so that it could augment its fleet of seaplanes, it actually scattered these rarities across several projects, among them the production, by Handley Page, of heavy bombers, the O/100 and O/400, that might permit disabling strikes against distant, fixed targets, such as dockyards and Zeppelin sheds.[58] (A raid executed against Friedrichshafen Zeppelin base on 21 November 1914 by four, land-based Avro 504s aeroplanes had had only disappointing results.[59]) Out of kilter with the priorities of the army, which lacked any strategic air power of its own, this preoccupation alone illustrates the impact of inter-service rivalry on the development of British military aviation. The division of labour between the fledgling RFC and RNAS was still nebulous when the Great War erupted. In addition to its maritime commitments, however, the latter force was to be burdened with responsibility for defending the homeland's airspace until 1916

---

53 Report by J.C. Nerney, 'Aircraft Design and Production, 1914–1918', TNA, AIR 1/678/21/13/2186, p. 9.

54 John H. Morrow Jr, *The Great War in the Air: Military Aviation from 1909 to 1922* (Washington, 1993), p. 75.

55 Ibid., pp. 84–5.

56 See, for instance: *Documents Relating to the Naval Air Service: Volume I, 1908–1918*, S.W. Roskill (ed.), (London, 1969), p. 565.

57 Report by J.C. Nerney, 'Development of Aircraft Production, 1917–1918', TNA, AIR 1/678/21/13/2138, pp. 162–3.

58 The need for what became the Handley Page 0/100 was first identified at a meeting held by Churchill in April 1915. See: 'Minutes of meeting held by First Lord of Admiralty', 3 April 1915, TNA, ADM 1/8433/270B.

59 See: Francis K. Mason, *The British Bomber Since 1914* (London, 1994), p. 21.

and enthusiastically championed strategic bombing until 1917. The Admiralty – persuaded in any case that the best form of defence was attack – suggested that the army's opposition to the Royal Navy undertaking strategic bombing was rooted more in envy than reason, while the RFC dismissed the senior service as a bit too supercilious. Certainly, over time, damning suspicions that the Admiralty was deliberately seeking to obstruct attempts to co-ordinate aircraft production did little to further its cause.

Only one aircraft – the spotter plane launched from HMS *Engadine* and piloted by 'Rutland of Jutland' – was to make a noteworthy contribution to that battle, the greatest naval engagement of the First World War. On taking command of the Grand Fleet some months later, Beatty was aghast at what a small proportion of the RNAS was actually devoted to maritime missions and grumbled that it 'hitherto has been totally inadequate for the work we should be able to rely on it to perform'.[60] Under his leadership, maritime operations became the paramount concern of the RNAS: efforts were made to get as many aircraft to sea as possible; a tailor-made torpedo plane – the Sopwith *Cuckoo* – was pressed for; the Royal Navy's first squadron of aviation vessels was formed; and better shore facilities were created at Rosyth and Scapa Flow.

By this juncture the Germans had long recognized the utility of aircraft, not least as a cost-effective substitute for certain types of naval vessel. Although Zeppelins took a lot of time and more than a little money to build, if only because the trimmed *Kaiserliche Marine* ultimately lacked sufficient fast cruisers for scouting missions it was to pioneer the use of dirigibles as long-range reconnaissance platforms, an allocation of twelve operational airships being approved for the *Hochseeflotte* in June 1915.[61] Whereas the long reach of these dirigibles made them highly suitable for reconnaissance missions, like sailing vessels they were very susceptible to adverse weather conditions and extremely vulnerable to enemy fire. Although the *Kaiserliche Marine* occasionally employed them in raids against the British mainland, airships were mostly deployed nearer home, especially in the monitoring of coastal waters for signs of British mine-laying operations or threats to exposed German vessels, notably minesweepers.

Indeed whilst few, if any, new forms of technology have had more of a strategic impact than the aircraft, mines and submarines also exerted a substantial influence on the operations of naval surface units from the Great War's outset. On 5 September 1914 Germany's *U-21* obliterated the cruiser *Pathfinder* near St Abb's Head with a single torpedo, while, eight days later, the British submarine *E-9* sent the German cruiser *Hela* to the bottom near Heligoland. Patrolling the Dutch coast on 15 September, the *U-9* sank three ageing warships – HMS *Aboukir*, *Hogue* and *Cressy* – within an hour before going on to claim another victim, HMS *Hawke*, off Aberdeen just a month later. Torpedoed by the *U-26*,

60 Letter from Beatty to Secretary of the Admiralty, 21 January 1917, TNA, ADM 1/8478/10.
61 Morrow, *The Great War in the Air*, p. 107.

the Russian cruiser *Pallada* exploded off Hanko (Hangö), Finland, on 11 October, while HMS *Hermes* – the Royal Navy's first seaplane-carrier – was sunk by the *U-27* off Calais on 31 October. During November, the *U-18* actually penetrated Scapa Flow. 'At the present time', Admiral Fisher, the Royal Navy's First Sea Lord, lamented, 'no means exist of preventing hostile submarines emerging from their own ports and cruising more or less at will'.[62]

The Orkneys' very remoteness from the Germans' principal submarine bases (on Heligoland) would, it had been reasoned, afford the anchorage appreciable protection. Certainly, at the time of the *U-18*'s intrusion Scapa Flow's man-made defences were incomplete. Had it not already left for the relative safety of distant Lough Swilly in Ireland, the Grand Fleet might have been riddled with torpedoes. (The *U-18* was damaged trying to flee the scene and subsequently scuttled by her crew, who were imprisoned.) However, Jellicoe's precautionary move to Irish waters proved sadly ironic: days later, the new super-dreadnought *Audacious* struck a mine, capsized and blew up, an event witnessed by the *Olympic*, sister ship to *Titanic*.[63] The mine had been sown by what appeared to be another passenger vessel, the *Berlin*, an auxiliary warship of the *Kaiserliche Marine*.

A solitary mine had accounted for one of the Royal Navy's greatest behemoths, reducing the odds against the *Hochseeflotte*. Elsewhere, submersible weaponry made other significant adjustments to the military balance. The German cruiser *Friedrich Karl* was wrecked by a mine off Memel (Klaipeda) shortly after *Audacious* foundered, while the British submarine *E-11* plagued the Turkish fleet, sinking the battleships *Messudieh* and *Heireddin Barbarossa* in 1914 and 1915, respectively. Operations in the Dardanelles during spring 1915 saw the destruction, by mines, of three pre-dreadnought Allied battleships – *Irresistible*, *Ocean* and *Bouvet* – and, by torpedoes discharged from the *U-21*, of two more, *Triumph* and *Majestic*. These losses had serious ramifications for the conduct of the whole Dardanelles campaign, both by land and sea. Later that year, British submarines having infiltrated the Baltic, the cruiser *Moltke* was damaged by the *E-1* near Riga, while, off Sweden's southern coastline, the *E-19* claimed the cruiser *Undine*. When the *E-8* torpedoed the *Prinz Adalbert*, the sister ship of the *Friedrich Karl*, west of Libau (Liepaja) in October 1915, the *Kaiserliche Marine* suffered its greatest loss of life in a single incident in the entire Baltic campaign; *Prinz Adalbert* exploded, killing 672 of her 675 crew.[64]

The fate of the Austro-Hungarian dreadnought *Szent István*, which was fatally holed by a humble torpedo boat north of the Otranto Straits in 1918, was, alongside the loss of HMS *Audacious*, arguably the starkest illustration of the asymmetry that the mine and the torpedo had made possible in naval warfare. The sheer insidiousness of these underwater armaments had both psychological and

62 Fisher, *Records*, p. 183.
63 Letters from Commanding Officer, HMS *Audacious*, to Commander-in-Chief, Home Fleets, 28 October and 9 November 1914, TNA, ADM 137/1012.
64 Erich Gröner, *German Warships, 1815–1945* (Annapolis, MD, 1990), p. 51.

practical effects. In May 1916 the Grand Fleet and the *Hochseeflotte* finally met off Jutland in the largest naval battle seen since Trafalgar. By this juncture, the Germans had been trying to provoke Jellicoe for over a year, hoping to isolate elements of his force outside their havens. Borrowing a leaf from the writings of the *Jeune Ecole*, the *Kaiserliche Marine* had intermittently staged opportunistic raids on commercial shipping and ports on the North Sea, the most flagrant of which being the bombardment, by cruisers, of Scarborough, Hartlepool and Whitby just days after Spee's defeat off the Falklands. On 31 May 1916 the *Hochseeflotte* was persuaded that it had at long last managed to entice a few of Jellicoe's capital ships – Beatty's battlecruiser squadron – into an unequal fight. However, this turned out to be merely the vanguard of the entire Grand Fleet, which, alerted to the enemy's movements by decoded signal intercepts, had put to sea in anticipation of a decisive clash. The Germans quickly found the tables had been turned. Yet, just as the British were gaining the upper hand, Jellicoe, fearing that his retreating adversaries were trying to lure his dreadnoughts onto mines and U-boats, ordered his vessels to turn about.[65] Though badly damaged, the *Hochseeflotte* survived and even claimed to have been victorious.[66]

Any success, however, was purely tactical. In mid-August, after licking the wounds they themselves had received, the battleships and cruisers of the *Hochseeflotte* that remained serviceable bore down on the English port of Sunderland in a renewed bid to lure out, corner and destroy a portion of the Grand Fleet. But the British, alerted once more to the enemy's approach through signal intercepts, sallied forth in force, hoping to save their coastal towns and finish what they had started at Jutland. Conscious of the reconnaissance failings that had led to matters going awry in May, the German ships, divided into a 'fixing' and striking force, were preceded by Zeppelin scouts and a large screen of U-boats. Although the light cruisers *Falmouth* and *Nottingham* fell foul of the latter and the dirigibles did glimpse other elements of the closing British pincers, the Zeppelins' reports were too vague and few to be relied upon. Fearing that it was being outmanoeuvred and encircled, the *Hochseeflotte* aborted the planned bombardment of Sunderland and headed for home, harried by submarines. The Grand Fleet, wary of further torpedo attacks and that its adversaries might have covered their retreat with freshly sown minefields, likewise returned to its bases.[67]

The action of the 18–19 August 1916 proved to be the climax of the last major sortie into western waters by the *Hochseeflotte*. This was strategically momentous, for, as the naval theorist Julian Corbett had emphasized a few years earlier, command of the sea meant 'nothing but the control of maritime communications, whether for commercial or military purposes'.[68] If warships sought to

65　See: Churchill, *World Crisis: Volume 2*, p. 1015.
66　But see, for instance: Andrew Gordon, *The Rules of the Game: Jutland and British Naval Command* (London, 1996), p. 514; and Epkenhans, 'Imperial Germany', p. 34.
67　Ibid., pp. 515–16.
68　Julian S. Corbett, *Some Principles of Maritime Strategy* (London, 1911), p. 94.

avoid an adversary by taking refuge behind man-made or natural defences, they relinquished 'to the enemy . . . the control of sea communications, . . . permitting . . . operations which tend to exhaust the resources of . . . [their] own country. . . . [A] naval defensive means . . . keeping the fleet actively in being – not merely in existence, but in active and vigorous life'.[69] However, always reluctant to risk a truly conclusive encounter with their counterparts of the Royal Navy, by the winter of 1916–17 Germany's capital ships had effectively conceded that they could not loosen the Allies' maritime blockade.[70] The only real hope in this regard resided in the long-range submarine, a furtive platform that, if authorized to try, might ravage Britain's own trade through the stealthiest of attacks.[71] Experience to date seemed promising, such craft having inflicted rather more physical and psychological damage on the enemy than had the big guns of the *Kaiserliche Marine*. Indeed among the former's many victims was one of Jellicoe's most senior colleagues, Field Marshal Earl Kitchener of Khartoum and Broome, Secretary of State for War. Just days after Jutland, this national hero and his staff had embarked on a fact-finding mission to Russia, only to perish aboard HMS *Hampshire* when she struck a mine, laid by a U-boat, off the Orkneys.

By early 1917, bent on prosecuting unrestricted submarine warfare, the *Kaiserliche Marine* was diverting more and more of its U-boats into the *guerre de course*, further reducing the capacity of the *Hochseeflotte* for forays against the Grand Fleet. 'And what is it that the coming of the submarine really means?' mused Fisher:

> It means that . . . our traditional naval strategy, which served us so well in the past, has been broken down. The foundation . . . was blockade. The fleet did not exist merely to win battles – that was the means, not the end. The ultimate purpose . . . was to make blockade possible for us and impossible for the enemy. Where that situation was set up, we could do what we liked . . . on the sea. . . . But with the advent of the long-range ocean-going submarine that has all gone! Surface ships can no longer either maintain or prevent blockade. . . . All our old ideas of strategy are simmering in the melting pot.[72]

Certainly, through increases in the spectrum and potency of technological countermeasures, the close blockade of sophisticated opponents' coastlines had become steadily less viable until, by the 1900s, the British especially were looking to the notion of distant blockade, whereby shipping travelling to and from a given region would be intercepted much further out to sea.[73] Whilst this

---

69 Ibid., p. 212.
70 Epkenhans, 'Imperial Germany', p. 35.
71 See: Gordon, *Rules*, pp. 514–15.
72 Fisher, *Records*, pp. 182–3.
73 See, for instance, Corbett, *Principles*, p. 97.

approach often necessitated the allocation of more (and faster) vessels to any such undertaking, those participating were, on the other hand, less exposed to enemy action and that much nearer their own bases. Although the easiest and quickest way of securing naval mastery remained confronting one battle-fleet with another, there had been only a handful of such titanic clashes in many decades. This was largely because concentration is a relative concept: most navies spent most of their time dispersed on patrol rather than massed in anticipation of a decisive confrontation. Moreover, many of the larger engagements that had occurred had done so by mutual consent. For whereas it was difficult for an army to avoid being attacked by another that was in striking distance and bent on taking the offensive, warships might shun battle simply by taking refuge within waters where they could not easily be assailed, such as behind the passive and active defences of ports. 'And so', noted Charles Callwell, 'even the navy which is paramount upon the ocean may have to trust to land operations, if maritime control is to be abiding and assured'.[74]

Indubitably, supremacy at sea had seldom stemmed from discrete, crushing victories, far more frequently proving to be the cumulative effect of numerous minor actions, not least the patient, unglamorous vigil of blockade. This had always proved a somewhat blunt instrument against adversaries who were willing and able to sacrifice their seaborne trade and coastal settlements. On the eve of the Great War, however, changes in the geostrategic environment were suggesting the need for adjustments to both the concept of blockade and the characteristics of platforms entrusted with its enforcement. In the early 1900s, analysts such as Callwell and Mackinder pointed to the rise of new, faster modes of overland transport that might better compete with maritime power, notably trans-continental railways such as the Siberian line. Connecting Russia's interior with the Far East, this was not only a major conduit for trade and migration but also enabled reinforcements to be moved speedily to military bases on the Pacific rim and to do so without running the gauntlet of hostile fleets. (In fact, though not fully completed until 1917, enough of this track was finished for it to prove very useful in the Russo–Japanese conflict of 1904–05.) Besides such railways, armies also had at their disposal ever more motorized rather than horse-drawn vehicles. 'They have, too', noted Mackinder, 'the Aeroplane, which is of a boomerang nature, a weapon of land power as against sea power'.[75]

By the second decade of the 1900s, powered, controlled flight, especially in machines that were heavier than air, seemed destined to help make or break distant blockades. That, between 1914 and 1918, the submarine was to have a far greater impact than the plane in this regard can be accounted for by several factors, some of which overlap.

74 Callwell, *Maritime Preponderance*, p. 165.
75 Halford John Mackinder, *Democratic Ideals and Reality: A Study of the Politics of Reconstruction* (New York, NY, 1919), p. 157.

Firstly, at this juncture aviation was in practice more of an indirect than direct threat to most vessels in general and military ones in particular. Not only were there relatively few hostile aircraft for naval forces to contend with but also none of them could equal, let alone surpass, the submarine in several key respects, crude though it was. To this day, few aircraft can match the range of diesel-powered submarines and no plane can stay aloft for as long as any mainstream submarine can remain at sea. The machines best suited to maritime operations in the Great War tended to be among the largest airships and winged aircraft – notably seaplanes and flying boats – then in production. These were harder and more expensive to manufacture than the aeroplanes commonly used to support undertakings on dry land and were commensurably rare.

Secondly, whereas potential targets for aircraft were also comparatively few in number, the chances of finding them varied inversely with the dimensions and complexity of the theatre concerned. These conspired with the concomitant problems of navigation to make prey elusive. Thirdly, even if aircraft came upon a potential target, attacking it efficaciously was seldom viable, as illustrated by the case of the *Yavuz Sultan Selim*. Formerly the *Goeben* – one of two warships that had been transferred from the *Kaiserliche Marine* to the Turkish Navy in August 1914 and renamed – this battlecruiser was beached by her crew after being holed by mines near the entrance to the Dardanelles in January 1918. A sitting duck, she was attacked by planes from Britain's Number Two Naval Air Wing. However, so few were the hits that were scored and so ineffective did the aircrafts' small bombs prove against the ship's armour that the vessel was repaired and retrieved.[76] (She survived until 1974, when she was sent for scrap.)

Striking a moving object was that much harder than hitting a stationary one, and the intended victim might dodge bombs or torpedoes through evasive manoeuvres or sheer speed. In any case, whereas the armour on many warships was impervious to the small pieces of ordnance that could be accommodated aboard most planes, dropping torpedoes and bombs exposed the attacker to defensive fire that could easily prove fatal for ungainly, rickety aircraft. For instance, the Short 184 seaplane used by the Royal Navy had only a single engine that, in prime condition, generated just 225 horsepower. When burdened with a small (14") torpedo, this aircraft could only get aloft at all in the calmest of seas and with a take-off run of some 3,000 yards. It then had to penetrate to within 1,000 yards of its quarry, the maximum range of its weapon.[77] Early in 1915, Churchill pressed for both the elaboration of procedures for mounting torpedo attacks from the air and the production of platforms better suited to this mission.[78] Shorts duly started developing a seaplane that, fitted with an engine

---

76 Layman, *Naval Aviation*, p. 58.
77 'Torpedo Carrying Seaplanes', by Superintendent of Aircraft Construction, 29 April 1916, TNA, AIR 1/149/15/104.
78 Minutes of meeting held by the First Lord of the Admiralty, 3 April 1915, TNA, ADM 1/8433/270B.

that generated 310 horsepower, was capable of carrying an 18" torpedo. It was not until the autumn of 1916, however, that the first such machine was ready to start trials.[79]

There were in any event the perennial problems of how to transport significant numbers of seaplanes to within striking distance of potential targets and then launch and recover them in a safe and timely fashion. On taking command of the Grand Fleet, Beatty dismissed all of Britain's existing carriers as being 'of very doubtful value'.[80] Wheeled aircraft had far more to recommend them than seaplanes for most combat missions, but employing them would require both customized vessels with suitably lengthy, unobstructed flight-decks and planes tailored for use on such ships. Moreover, if the aircraft were expected to return to the carrier, the flight-deck would have to be fitted with arresting gear and signalling systems, while pilots would need training in the tricky manoeuvre of landing on a moving platform. Meteorological conditions were also a key consideration where carrier operations were concerned. To retrieve or launch planes, the ship needed to turn into the wind in order to help generate sufficient aerodynamic lift for the aircraft to remain controllable while landing or taking off. Consequently, as in the days of sail, too much or too little wind could cause crippling difficulties. The carrier had, at worst, to be able to accelerate to speeds that offset any dearth of headwind. Precipitation could also give rise to problems and risks. Snow or ice deposits – from frozen spray or rain – could render the flight-deck inutile and have detrimental effects on machines and personnel alike. Indeed besides chilling people and equipment, freezing conditions could jeopardize a vessel's very seaworthiness; its upper works encrusted with ice, a ship could become hard to manoeuvre and even fatally top-heavy.

In October 1917 Roger Keyes, the British Admiralty's Director of Plans, suggested towing some Curtiss *Large America* seaplanes far into the North Sea on massive lighters so as to bring them within range of Wilhelmshaven and Hamburg's wharves and shipbuilding yards.[81] Nothing came of this rather eccentric scheme, but a year later the establishment of the Royal Air Force (RAF) and the maturation of various programmes initiated by the Royal Navy had made the concept of maritime strike more viable. A raid on the airship base at Tondern (Tønder) was planned and, on 19 July 1918, after being postponed twice because of adverse weather, was finally enacted. Wheeled planes were used: launched from HMS *Furious*, seven Sopwith TF1s – a marinized version of the RAF's *Camel* – destroyed the Zeppelins *L54* and *L60* in their berths.[82] A grand attack on Wilhelmshaven was also suggested in May 1918, whereby 80 torpedo aircraft

79 Minute by Wing Commander, Royal Navy, 26 January 1917, TNA, AIR 1/145/15/66.

80 Letter from Beatty to Secretary of the Admiralty, 21 January 1917, TNA, ADM 1/8478/10.

81 'Long Distance Bombing Operations': Memorandum by Director of Plans, 9 October 1917, TNA, ADM 137/2706.

82 Report of attack on Zeppelin sheds at Tondern by aeroplanes from HMS *Furious*, 19 July 1918, TNA, AIR 1/344/15/226/285.

would strike the *Hochseeflotte* at anchor.[83] However, the most suitable plane for such a mission was the Sopwith T.1 *Cuckoo*, which had been specifically designed with carrier operations in mind. Barely a hundred such machines had entered service by the war's end. Neither, in spring 1918, was there sufficient carrier tonnage to support all of these.

In view of all of this, the value of aircraft as weapon platforms during the Great War was never likely to prove anything but marginal. In fact the tally of vessels destroyed exclusively by air power comprised just two destroyers, two torpedo-boats, six motor torpedo-boats, four naval auxiliaries and a dozen merchant ships.[84] These figures contrast starkly with the number of 'kills' achieved by other platforms, not least submarines.

Combat tactics are – or at least ought to be – largely a function of technology. The differing categories of craft that make up a fleet are, as Julian Corbett had commented in 1911, 'the expression in material of the strategical and tactical ideas that prevail at any given time, and consequently they have varied not only with the ideas, but also with the material in vogue'.[85] At the First World War's conclusion, battleships remained the principal icon of maritime might. For different reasons, neither submarines nor aircraft could entirely supplant them at this juncture. Change was afoot, however. As a concept, the battleship was nearing its zenith, whereas aviation and the submarine were in their infancy. The advent of the aircraft carrier especially was to transform naval tactics rather than strategy. Essentially gun platforms that relied upon firepower and armour plate for their survival, battleships possessed great strength when pitted against similar platforms. Yet, by the same token, their tactical flexibility was comparatively limited and, increasingly, they lacked strategic reach, endurance and utility. They were, moreover, poorly suited to the changing demands of either blockade or the *guerre de course* and, especially in coastal waters, were vulnerable to new and maturing weapons, notably submersible mines.

Such relatively inexpensive devices were a cheap, simple way of barring stretches of water to an opponent's submarines and surface vessels alike. The British, for instance, hampered U-boats seeking to penetrate the English Channel and harass shipping en route to Flanders by sowing minefields in the Straits of Dover. This barrage was patrolled by aircraft and proved very effective. But although mines and submarines did do appreciable damage to warships in World War I, generally speaking they were more of a potential than an actual threat. The belligerents' navies could normally predict where, if at all, mines might be laid, and they both could and did take steps to avoid or otherwise neutralize them. The scope of submarine operations was similarly constrained, largely by the amount of fuel such craft could accommodate; 'distant' blockade sought to

83 Letter from Commander-in-Chief, the Nore, to Secretary of Admiralty, 4 May 1918, TNA, ADM 1/8525/136.
84 Layman, *Naval Aviation*, p. 61.
85 Corbett, *Principles*, p. 107.

keep capital ships especially beyond their reach. If only because of the resulting paucity of viable targets, submarines often had to prey on merchantmen.

The *Kaiserliche Marine*, moreover, chose to do just that. Whereas the Royal Navy's blockade was irksome to neutral powers, Germany's policies proved increasingly lethal for them. Her U-boats sank around twelve million tons of shipping, much of it in the unrestricted campaigns that they embarked on in 1915 and 1917, respectively. As appreciable numbers of submarines converged on British coastal waters, where potential targets were at their most plentiful, swashbuckling but indulgent commerce-raiding like that once practiced by the *Emden* gave way to an increasingly merciless *guerre de course*. Especially as a consequence of the torpedoing of the passenger liners *Lusitania* and *Arabic* in 1915 and the attack on the ferry *Sussex* in March 1916, the first of these campaigns aroused world-wide condemnation. Berlin was obliged to adopt the '*Sussex* Pledge', an undertaking that, henceforth, its submarines would adhere to international law. Culminating in the Hague Conventions of 1907, the codification of maritime warfare had been endeavouring to keep abreast of changing technology and tactics ever since the Paris Congress of 1856. However, the Germans' second unlimited campaign completely abrogated the rights of even neutral vessels, notably those of the USA. Intended to strangle Britain's commerce and starve her population, thereby winning the war before an alienated America could mobilize her manpower and intervene on the deadlocked Western Front, this desperate undertaking initially yielded some tactical success yet failed to meet its strategic objectives. For all the U-boats' efforts and despite the collapse of tsarist Russia, the weary European Allies grew stronger, receiving infusions of fresh manpower from the USA. Meanwhile, their shipbuilding yards churned out new merchantmen to replace those that were lost and their own blockade hamstrung the Central Powers, inflicting widespread hunger and other hardships. For Austria-Hungary, Germany and Turkey alike, the resulting economic and social disintegration was to prove as debilitating as any battlefield defeat.

Germany's submariners had menaced British shipping increasingly since their first unrestricted campaign in 1915. Initially, the Royal Navy's response comprised regular patrols in which aviation came to play a major role. Non-rigid airships that scoured England's eastern coastal waters and the southern reaches of the North Sea especially were supplemented with F.2A *Felixstowe* flying boats in late 1917. These were powerful planes that had an operating radius of some 400 miles and an endurance of more than nine hours.[86] In December 1916, however, the Anti-Submarine Division of the Admiralty was established and instigated fresh endeavours to neutralize the U-boats: efforts in the Irish Sea, along Britain's eastern coasts and in the Channel and Western Approaches were redoubled, with numerous aircraft flying 'spider web' sweeps of 60 miles in

86 Thetford, *British Naval Aircraft since 1912*, pp. 197–202.

diameter.[87] Rigid airships were first introduced in spring 1917, while a second generation of non-rigid dirigibles, such as the North Sea type, which had better endurance than their predecessors, executed sorties over the North Sea for between 20 and 48 hours at a time.[88]

The F.2A *Felixstowe* was a British re-engineered version of the Curtiss H.4 *America*, which suffered from several defects, not least poor hydrodynamics and unreliable motors. (H.4s often had to be towed into a harbour by 'annoyed' destroyers![89]) With its vastly superior hull and engines, the F.2A had better endurance and could accommodate more firepower.[90] By the war's end a new anti-submarine plane, the Blackburn *Kangaroo*, which had a still longer reach and a bigger punch than the *Felixstowe*, was also making its debut.[91] Indeed in August 1918, in concert with the destroyer HMS *Ouse*, a *Kangaroo* destroyed the *UC-70*, which it espied lurking beneath the waves off Whitby.

A wheeled aircraft that was a spin-off from a faltering project to devise a new seaplane, the case of the *Kangaroo* helps illustrate how the need to create a balanced range of capabilities complicated research and procurement decisions. These in turn affected operational and strategic considerations. The RNAS, as anxious to thwart the U-boat as it was to truncate the Germans' capacity for reconnaissance and bombing forays into British airspace, relied overwhelmingly and somewhat incongruously upon land-based aviation for the fulfilment of these diverse missions. As far as countering submarines was concerned, the strategy included, among other measures, the molestation of U-boat anchorages at Bruges, Zeebrugge and Ostend, largely with Allied aeroplanes based just along the littoral at Dunkirk. In May 1917 fears arose that a combination of French losses and a reduction in the overstretched RNAS contingent in this quarter was contributing to a loss of control over the southern part of the North Sea and adjacent parts of the Channel.[92] In fact so threatening did German seaplane and submarine operations appear at this juncture that Britain's ability to sustain the war for another year was being cast into doubt. Ousting the *Kaiserliche Marine* from Belgian ports that had long proved a thorn in Britain's side became an imperative. The army, hoping it might succeed where others had failed, duly launched the ill-fated Passchendaele offensive, the ultimate objectives of which were Zeebrugge and Ostend.

87 Historical Section, Admiralty, *The Defeat of an Enemy Attack on Shipping: A Study of Policy and Operations, Volume 1A* (London, 1957), pp. 6–7.

88 Ces Mowthorpe, *Battlebags: British Airships of the First World War* (Stroud, 1998), pp. 67–8.

89 P.I.X. [Douglas Hallam], *The Spider Web: The Romance of a Flying-Boat War Flight* (London, 1919), pp. 21–2.

90 Thetford, *British Naval Aircraft since 1912*, pp. 197–202.

91 H.A. Jones, *The War in the Air: Being the Story of the Part Played in the Great War by the Royal Air Force: Volume Six* (Oxford, 1937), p. 340.

92 Minute from Commodore, Dunkirk, to Vice Admiral, Dover Patrol, 29 May 1917, TNA, AIR 1/631/17/122/44.

Another key component of the response to unrestricted submarine warfare also had ramifications for the military balance in the North Sea. The institution of convoys in May 1917 necessitated the diversion of destroyers and light cruisers to escort duties, reducing the capabilities of the Grand Fleet, the counterpoise to the *Hochseeflotte*. On the other hand, the U-boats were left with few alternative targets. In attacking convoys, they ran greater risks of being detected and engaged by the escorts which, where practicable, might include aircraft, notably airships. Although only one submarine was sunk exclusively by aviation and a further four by combined aerial and surface forces, air power evidently proved a significant deterrent: in the entire war, only five Allied merchant ships were lost while in convoys that had air cover, two of them being sunk in a single incident off Falmouth on Boxing Day, 1917.[93]

The following year, 'scarecrow' patrols by sluggish DH.6 biplanes were initiated over British coastal waters, the mere sight of an aircraft usually being sufficient to send a submarine diving for cover.[94] If only because using their deck guns necessitated exposing themselves to observation (and possibly counter-fire), by this stage in the *guerre de course* U-boats were normally remaining submerged when attacking, relying on unheralded strikes with torpedoes. The viability of this tactic relied on the only windows submariners had on their environment from their sub-aquatic world, namely the relatively tiny apertures of periscopes.

This highlights an intricacy that was as commonplace as it was inhibiting. In an era when the theoretical range of battleships' guns – the supreme symbol of martial might – was continuing to grow, the efficacious exploitation of fire from many platforms was constrained by the human eye's innate capabilities. In the absence of dependable, remotely-guided weaponry, it was in principle possible only to shoot with any accuracy at what could be seen. During March 1918 Paris was intermittently (and indiscriminately) shelled from all of 75 miles away by colossal German artillery that could use triangulation to approximate the vast city's location. But particularly amidst the immensity of the seas, very precise reconnaissance and navigation were prerequisites for almost everything and anything. Because of its reach, air power promised both to discern what might otherwise not be spotted and to hit what might otherwise not be struck. Moreover, because of its ubiquity and relative speed, it might suddenly appear from virtually any direction, catching victims unawares; a few mountain peaks aside, the skies were free of obstructions, certainly of minefields and barbed-wire entanglements. Obvious potential targets included ports, vessels – both military and commercial – and states that had long derived a degree of immunity from being encircled by natural moats.

93 John J. Abbateillo, *Anti-Submarine Warfare in World War I: British Naval Aviation and the Defeat of the U-Boats* (London, 2006), p. 108.

94 H.A. Jones, *The War in the Air, Volume Six*, pp. 330–32.

In 1907 Halford Mackinder had warned that: 'Mere insularity gives no inde-feasible title to marine sovereignty'.[95] By the Great War's end he was depicting the face of the Earth as having insular lineaments, postulating that there was a 'World-Island' comprising Europe, Asia and Africa, with offshore islets, notably Britain and Japan, and outlying islands, including the Australian and American continents. The 'World-Island' had a 'Heartland' that was essentially the region that had constituted tsarist Russia in 1917 and was to form the core of the Soviet Union for almost eight decades thereafter. Whoever dominated that Heartland and, thereby, the World-Island was, Mackinder insisted, best placed to achieve success in any competition for global hegemony.[96] Relegated to the fringes of this geostrategic model, the great maritime nations – ineluctably preoccupied with sea rather than land power and with Mahanian notions about the former's utility – were at a disadvantage, he argued, not least because 'the utmost limit of sea power' was an opponent's coastline.[97]

By contrast, Corbett, who published his *Principles* before the First World War, had a more nuanced view. Consistently exhibiting a greater awareness of sea power's limitations than Mahan (who died the very year the Great War began), Corbett had stressed the importance of maritime – as opposed to purely naval – might. Britain for one, he reasoned, needed military doctrines and forces that could tap strategic opportunities arising from mastery of the waves. The army should, as Sir Edward Grey had put it, be a projectile fired by the Royal Navy; allies who lacked maritime options but who had the requisite manpower should furnish the majority of soldiers for grand campaigns on land, just as they had in the Napoleonic and Crimean Wars. Charles Callwell, a soldier himself, con-curred, concluding that: 'Maritime preponderance more correctly defines the conditions which ordinarily arise in warfare between States laying claim to a measure of sea-power'.[98]

For sure, actual experience in the First World War provided little incontrovert-ible evidence in support of Mackinder's thesis. Indeed two prominent advocates of aviation felt that the conflict's greatest single lesson was that: 'In the future a nation which dominates the aerial highways will dominate also those of the land and sea; that a dominion of the air must mean, ultimately, the dominion of the world'.[99] Yet a prerequisite for air or sea power was control of pivotal points on the Earth's surface, not least the sources of raw materials and components without which ships, submarines and planes could neither be constructed nor operated. Aviation was thus ineluctably rooted in, not just aerodromes, be they ashore or afloat, but also in a web of trade routes, petroleum refineries and other industrial plants that were vulnerable to a variety of threats. Between 1914 and

95 Mackinder, *British Seas*, p. 358.
96 Mackinder, *Democratic Ideals*, p. 186.
97 Mackinder, *British Seas*, p. 310.
98 Callwell, *Maritime Preponderance*, p. 1.
99 Claude Grahame-White and Harry Harper, *Air Power: Naval, Military, Commercial* (London, 1917), p. v.

1918 the British were only able to continue mounting aerial and amphibious sorties from their island homeland – which, unlike the neighbouring states of continental Europe, was relatively secure from conquest by hostile armies – because they prevailed against Germany's maritime blockade. Air power made an important contribution to this victory, but primarily as a means of reconnaissance; when it came to actually destroying targets other than its own kind, the aircraft was too often of doubtful utility. Nevertheless, much like the tank, a still more recent innovation, it clearly had great potential, if only as a cog in bigger military machines. By 1918 the requirements for organic aviation for Britain's Grand Fleet alone were set at all of 127 fighters, 80 two-seater reconnaissance aircraft and 100 torpedo planes, the last two types to be housed aboard carriers, with the fighters squeezed onto cruisers and battleships.[100]

Certainly, during the Great War their maritime preponderance enabled the Allies to blockade the Central Powers efficaciously enough and to safely transport colossal amounts of manpower to the fronts in the West and Middle East especially. (In fact Kitchener was one of just a relative handful of Allied soldiers who were lost at sea as a result of enemy action.) These troops came, not only from the USA, but also the British and French empires. Scores of thousands of them arrived in Europe from the rims of the Pacific and Indian Oceans via the Suez Canal. Tellingly, as early as February 1915 the Turks strove to staunch this flow, dispatching an army across the Sinai from Palestine to occupy what Mackinder himself described as 'the cross-ways of the world'.[101] The attack was repulsed by the waterway's defenders, nine warships and 30,000 soldiers.[102]

The Dardanelles campaign

The Turks' domination of the Dardanelles and their attempt to sever the Suez Canal were practicable propositions only because of the topography of the Levant's seaboard. Neither here nor elsewhere has the Earth's atmosphere a geography that is truly comparable. This simple fact rendered some of the analogies drawn by early theorists between aerial and naval warfare of dubious value. Since the skies and the oceans themselves are indestructible, however, air and sea operations alike have sought to constrain opponents in their exploitation of either one or both of these media, often as a prelude to intervention on adjacent terra firma. As Corbett suggested with regard to control of the sea:

> Maritime communications . . . correspond in strategical values . . . to those internal lines of communication by which the flow of national life is maintained ashore. . . . At sea, the communications are, for the most part, common

---

100 'North Sea Air Forces', from Admiralty War Staff (Air Division) to Deputy Chief of Naval Staff, 13 February 1918, TNA, AIR 1/308/15/226/188.

101 Halford, *Democratic Ideals*, p. 157.

102 See Strachan, *Call To Arms*, pp. 695, 700, 729, 734–42.

to . . . belligerents, whereas ashore each possesses his own. . . . [S]trategical offence and defence tend to merge in a way that is unknown ashore. Since maritime communications are common, we as a rule cannot attack those of the enemy without defending our own. . . . By . . . control . . . we do not mean that the enemy can do nothing, but that he cannot interfere with our . . . operations so seriously as to affect the issue of the war, and that he cannot carry on his own . . . except at such risk . . . as to remove them from the field of practical strategy. In other words . . . that the enemy can no longer attack our lines of . . . communication effectively, and that he cannot use or defend his own.[103]

The grandest joint operation of the Great War arose from the Allies' attempts to establish a line of communication and supply with Russia through the Dardanelles, a notorious nautical choke point. It was also hoped that this undertaking would encourage Bulgaria, Greece and Romania to side against the Central Powers and ultimately knock Turkey out of the conflict altogether.

The mission was initially entrusted to naval forces alone: in February and early in March 1915 warships twice attempted to force the straits, the western end of which was guarded by artillery emplacements on both shores and by thick belts of submerged mines. Progress in subduing resistance was slow and intermittent, partly because of changeable, wintry weather. Meanwhile, the Ottomans, alerted to the threat, were able to bolster the defences with comparative ease. When, on 18 March, 18 Allied battleships and heavy cruisers tried to push up the channel for a third time, three were sunk by mines and another three damaged. With the weather deteriorating again, the offensive was suspended once more. Concluding that the Turkish forts could not be subdued by sea power alone, the Allies renewed the attack on 25 April, landing some 75,000 troops along the Gallipoli peninsula. Three footholds were established on its rugged shoreline, but its spine proved insurmountable and the front became as static as that in Flanders. A phased, meticulous evacuation of the Allied army commenced in December, the last troops embarking on 9 January 1916. The spectre of this failed amphibious assault was, however, to haunt military planners, not least those of the US Marine Corps, for the next two decades.

Allied aviation – in the form of kite balloons, aeroplanes and seaplanes – made a significant contribution to the Dardanelles campaign. A cardinal task for all three types was identifying targets for, and monitoring the effects of, fire from ships and the army's howitzers, many of the Turks' positions lying well out of the gunners' sight. The kite balloons – anchored to tenders such as HMS *Manica* and *Hector* and linked to them by telephone – achieved considerable success in this respect, despite being fired on from the ground and, occasionally, by hostile (German-made) aeroplanes.[104] These were countered by similar

103 Corbett, *Principles*, pp. 100–105.
104 Report from Colonel Sykes, RFC, to Secretary of Admiralty, 9 July 1915, TNA, AIR 1/625/17/12.

aircraft based at makeshift strips on islands in the Aegean. Wheeled machines also assisted in observing the enemy's trenches. The Farman HF.20, for instance, was too slow for aerial combat, but its engine's very puniness minimized fuel consumption, enabling the aircraft to loiter that much longer. The seaplanes' performance proved mixed as well, partly because of maintenance difficulties; among the 11 machines aboard HMS *Ark Royal* there were five different designs using three varieties of engine. Moreover, when carrying their normal crew of two – a pilot and observer – these ponderous aircraft struggled to climb above 3,000 feet, which precluded them from seeing much besides exposing them to fire from the surface.[105] Although the Dardanelles campaign also witnessed the first use of torpedoes by planes, the available aircraft proved of limited value in this respect, too. Even with a reduced fuel stock, which cut its reach, the Short 184 could only carry a 14" torpedo, which lacked punch. Whilst it was believed that three Ottoman merchantmen were sunk by such weapons, these claims remain controversial.

Air-defence

During the First World War command of the air yielded mixed and marginal results. Whereas it helped make land offensives that much more viable, in maritime settings aviation's accomplishments were largely less tangible. Yet, in the years before the conflict, it had been anticipated by some visionaries that air power would exert a decisive influence: hostilities would start and end – perhaps only hours later – with a crushing aerial offensive, a knock-out blow that would shatter an opponent's political, psychological and social cohesion. Such ideas were exemplified in H.G. Wells' *War in the Air* published in 1908. Abandoning any attempt to distinguish between combatants and non-combatants, attacks would be directed, not so much against the military per se, as against the roots of military power, namely the country's population and economic infrastructure.

Certainly, as the Great War loomed, it appeared that the Germans were poised to disrupt Britain's maritime communications in a novel way. Already her most ominous competitors on and below the waves, they were also emerging as the primary challengers to her so far as war from the skies was concerned. In their dirigibles, they seemed to possess a capability for strategic bombing that outstripped those of any other power. Dread of the Royal Navy being overwhelmed in a decisive battle – and, thereby, abruptly losing any war – persisted, yet fear of hostile air power gradually began to eclipse the Edwardian preoccupation with numbers of dreadnoughts, partly because aviation promised to distort the balance of maritime might.[106] The Zeppelin appeared a still bigger threat than the *Hochseeflotte*, insofar that the former might be used to cripple efforts to mobilize

---

105 Ibid.
106 See, for instance, Francis Wrigley Hirst, *The Six Panics and Other Essays* (London, 1913), pp. 103–18.

Britain's potential for war, not least her fleet, through pre-emptive attacks on key infrastructure, notably arsenals and naval dockyards.

Rushed through Parliament, the Aerial Navigation Act of 1913 designated parts of Britain's airspace as prohibited zones and authorized the shooting down of any aircraft that violated them. However, questions remained as to who exactly should police these cordons and with precisely what assets. Air-defences at the time were at best porous, not least because they lacked any robust mechanism for recognizing and tracking intruders. Several acoustic mirrors were erected along England's eastern and southern coastlines from 1916 onwards, but these audiological sensors lacked the sophistication of their electromagnetic off-spring, the radar of the 1930s. Moreover, the usefulness of these listening systems declined further as the average speed of planes increased, reducing the time in which warnings might be issued. The human eye remained the principal means not only of spotting potential targets but also of focussing fire upon them, be it from either combat aircraft – which had to be kept aloft and sufficiently close to intercept any raiders – or from guns on the ground, which had to be maintained in a similarly suitable state of readiness. In any event bespoke anti-aircraft (AA) weapons were ineluctably as novel as the things they were supposed to counter. Most cannon (and their projectiles) were designed with very different targets in mind. Customizing them took time and resources. Existing armaments, above all coaxial machine guns, had to be turned to.

Such weaponry could be positioned on the Earth's surface to protect spe-cific points from, among other intruders, low-flying aircraft. Whereas photo-reconnaissance planes were particularly vulnerable as they had to fly straight and level in order to take an interlocking sequence of pictures, combat aircraft that manoeuvred at high and medium altitudes were that much harder to molest. Traversable cannon with adequate reach and, ideally, a capacity for exceptionally rapid fire were called for. However, just as ringing every potential, individual target with a mixture of contrasting armaments was virtually impossible, so too was achieving a theatre-wide blend and density of defences that would prove appropriate and adequate in every conceivable scenario. Whilst some air raid precautions inevitably turned out to be insufficient when actually put to the test, others were effectively wasted investments as the need to activate them never arose at all. Ineluctably, the demand for substantial quantities of AA weaponry to protect hinterlands siphoned resources away from the trenches of the various fronts, where, above all, copious machine guns and artillery were indispensable. Such competing demands could only exacerbate the shortages of suitable weap-onry and munitions that, particularly between 1914 and 1916, plagued the land operations of the Allies especially.

As the Great War progressed, machine guns were also added to fighter and, for defensive firepower, to other planes, too. However, the incorporation of guns – even light, air-cooled weapons such as the Lewis – and their requisite ammuni-tion added substantially to the burden on an aircraft's propulsion system and had to take account of the configuration of its airframe. Arcs of fire were ineluctably limited by the plane's own architecture, not least the location and dimensions of propellers. So-called 'pusher' biplanes – such as the Vickers F.B.9 and the Airco DH.2 that were used by the RFC in the war's early stages – were thrust along by propellers mounted on the tail, leaving the front of the fuselage free for gun emplacements. However, the vast majority of designs that promised to prove far superior fighters than ungainly 'pushers' had a solitary 'tractor' engine housed in the nose. Until a synchronization mechanism could be perfected that allowed bullets to be discharged safely between the whirling blades of a propeller, forward-firing armaments could only be employed on such biplanes if they were affixed to the upper wings, sufficiently clear of the engine. Either the pilot had to try simultaneously to fly the aircraft and operate its weaponry, or, alternatively, observers had to take on the role of gunners, too. In the latter case, pilots had to seek to anticipate manoeuvres by hostile machines – which might achieve closing speeds of 180 mph or more – if their colleagues were to have any chance of training fire on such fleeting targets.

Pioneered by the RFC, the so-called Foster Mounting alleviated this ergonomical nightmare by permitting guns on the upper wings of biplanes to be retracted to the level of the cockpit; jammed or empty weapons could thereby be attended to relatively quickly and safely. The Germans, however, won the race to introduce a synchronization mechanism, an innovation that helped make the single-seater fighter a much more viable proposition. Pilots now had only to align their aircraft with targets and squeeze the trigger. This modus operandi was superior to that of 'pusher' fighters and ultimately precipitated the demise of such designs. Indeed the advent of the Fokker *Eindecker* – a rather unwieldy monoplane, the principal merit of which was its forward-firing guns – temporarily swung the aerial war in Germany's favour. However, spurred on by the 'Fokker Scourge' of 1915–16, the Allies, too, devised dependable synchronization devices, notably the Constantinescu gear, a hydraulic mechanism that appeared in 1917. The first British plane equipped with synchronized guns, a Sopwith *Strutter*, made its debut in 1916, as did several other new aircraft, many of them fighter variants. Prominent among these were three French machines, the Nieuport 11 and the Nieuport 17, which were used by both the RNAS and RFC as well as by French squadrons, and the *SPAD* S.XIII, which was utilized by British, French and American units; the *Bristol* F.2; the Sopwith *Pup*, which, light and nimble enough to take off from a platform atop a gun-turret, was widely used by the RNAS as a scout and fighter; and Germany's *Albatros* D.II. These were joined in the course of 1917 and 1918 by the Sopwith *Camel* and *Dolphin*; the Royal Aircraft Factory's S.E.5 biplane; and by Germany's formidable *Albatros* D.III.

The elements of air-defence can be as mutually competitive as they are reinforcing. This in turn exacerbates force-design and strategic planning difficulties. During 1917, alongside Germany's unrestricted submarine campaign, *Gotha* G.IV bombers based in Belgium commenced daylight raids against Britain's heartland. The principal ramification of these attacks was the establishment of the Smuts Committee to review air-defence. The panel recommended the formation of a single air service, the RAF. Britain's military aviation forces were duly unified under its auspices from 1 April 1918, some seven months before hostilities ended.

The RAF was the world's first autonomous air force. Indeed Britain was unique among the great powers in amalgamating her air arms and the decision to do so reflected the peculiar dynamics within and between her political and military institutions and her paramount geostrategic concerns. Whereas in other European states the army was predominant, for Britain, maritime interests were pre-eminent. Yet where air power was concerned, the senior service found itself embroiled in operations, such as strategic bombing, that lay on the periphery of its purview and had a detrimental impact on its ability to fulfil what were intuitively its core functions. The RNAS predictably resented government interference and resisted attempts by politicians to co-ordinate procurement programmes between it and the RFC. Consequently, the RAF's creation initially enjoyed considerable backing from within the Royal Navy, not least from those officers who felt that maritime air power had not been allotted the attention and resources it warranted. Although, under the RAF's aegis, there was in fact to be little change in emphasis with regard to maritime aviation, the effect of the re-organization was to become apparent soon after the war's end.

2 The interwar years and their legacy

Battleships or bombers?

The Naval Defence Act of 1898 acted as an aid to self-discipline for the British and simultaneously signalled their resolve to preserve the Royal Navy's quantitative superiority over prospective challengers. In spite of this and the subsequent warnings and entreaties of, among others, Sir Edward Grey, the Germans pursued the creation of a surface fleet with which they hoped to jeopardize the security of Britain's maritime communications and, thereby, the roots of her economic and military might. That this ambition ultimately went unfulfilled was partly because Germany lacked the wherewithal and partly because of Britain's response to what was predictably seen as a security crisis. Immense though the resources devoted to the *Hochseeflotte* were, Germany, burdened with other commitments, notably the need to maintain an army large enough to confront those of Russia and France simultaneously, could not afford a naval arms-race, too, particularly one against the world's leading maritime power. In fact after Jutland the *Hochseeflotte* spent most of its time rusting away in its home ports, where its sailors turned so despondent as to become mutinous. Its constituent squadrons and Germany's other naval assets were destined to be divided between the victorious allies as part of the few tangible spoils of the First World War.[1] Impounded at Scapa Flow, most of the mighty cruisers and battleships that had once caused Britain especially such anxiety were eventually to be scuttled by their own crews under their captors' very noses.

The capital vessels of the *Hochseeflotte* had never come close to achieving strategic success in the conflict. Once available in significant numbers, German ocean-going submarines had had, by contrast, a major if insufficient impact. Indeed to some analysts the submarine's maturation and the growing capabilities of aviation seemed to herald a new age of naval warfare in which the battleship would be eclipsed if not rendered entirely obsolescent. Aviation, it was argued, promised to supplant the long-range gun as the primary means of bringing firepower to bear and might more cheaply and easily fulfil many of the other

---

1 See: Epkenhans, 'Imperial Germany', pp. 36–7.

missions that in the past had been entrusted to fleets. Naval gunnery, with its maximum range of 27,000 yards, might still be useful, but it could scarcely compete with bombers and torpedo planes that were limited only by fuel, meteorological conditions and the physical and psychological endurance of their crews. Even the largest of shells could not, moreover, match the destructive force of the biggest of bombs or torpedoes. The capability to project air power into maritime environments could only become more important, making the aircraft carrier, not the battleship, the capital vessel of the future.

For the few states that possessed both air forces and fleets, such hypotheses were the focal points of rivalries between the armed services and their respective advocates. Among those who contributed to the incipient public debate in Britain were the likes of 'Jacky' Fisher, Bernard Acworth, the naval architect and novelist E.F. Spanner – who might also have written under the pseudonym 'Neon' – Claude Grahame-White and Harry Harper. Meanwhile, in Italy, Giulio Douhet was gradually establishing himself as air power's pre-eminent theorist and champion.[2]

The Kingdom of Italy was a constitutional monarchy under the House of Savoy. From 1861 until the rise of the fascist dictatorship of Benito Mussolini in the 1920s, it was a parliamentary democracy. Although a relatively new polity, Italy had been the first state to use powered, controlled flight in anger. Eager to establish themselves as major actors on the international stage, the Italians had first sought to join the other European powers in the 'Scramble for Africa' by trying to conquer Ethiopia. This had ended in calamity at Adowa in 1896. However, between 1911 and 1912 they successfully wrested the Libyan provinces of Tripolitania and Cyrenaica from the tottering Ottoman Empire in the so-called Tripolitan War, during which they employed a handful of aeroplanes to observe and harass Turkish troops. Indeed there was a widespread reverence for trendy, novel forms of technology – notably aircraft and fast motor cars – among the Italian population, not least within the proto-fascist movement that encouraged and exploited it. Mussolini himself was as much a part of the modernist cult that surrounded the aircraft and flight in Italy as was the celebrated pilot Italo Balbo, who became under-secretary and, subsequently, secretary for aviation following the creation in 1923 of a discrete air ministry similar to that founded in Britain some years earlier.[3] The *Duce* also organized Italy's air forces into an independent corps, much like the RAF. Regarded as the 'fascist service', the *Regia Aeronautica Italiana* (RAI) took pride of place among the armed forces and quickly received a sevenfold increase in its budgetary allocation.

2 See, for instance: Fisher, *Records*, pp. 143–4; Bernard Acworth, *The Navy and the Next War: A Vindication of Sea Power* (London, 1934); E.F. Spanner, *The Broken Trident* (London, 1929); Neon, *The Great Delusion: A Study of Aircraft in Peace and War* (London, 1927). Also see: Barry D. Powers, *Strategy Without Slide-Rule: British Air Strategy, 1914–1939* (London, 1976), pp. 138–42, 174–7.

3 Azar Gat, *Fascist and Liberal Visions of War: Fuller, Liddell Hart, Douhet, and other Modernists* (Oxford, 1998), pp. 63–70.

During the First World War, environmental factors had severely curtailed the scope for aerial combat along much of the frontier between Austro-Hungary and Italy. Largely mountainous, huge swathes of the borderlands, notably the Tyrol, were inimical to planes, if only at this juncture. Not least because they had cockpits that were open to the elements and engines that were as vulnerable to falling temperatures and thinning air as their crews were, few machines could ascend sufficiently to surmount the higher ranges, the tallest mountains in Europe. This funnelled flight paths into the larger passes, wherein lurked more than a few natural dangers: whenever cliffs and isolated peaks were not obscured by clouds and mist, bouts of snow blindness or dazzling shafts of direct sunlight could conceal them from the human eye equally well. There were invariably man-made obstacles, too. Any aircraft seeking to enter or leave such a valley would almost certainly have to run the gauntlet of waiting AA guns and fighters, for Vienna had taken the precaution of bolstering this region's natural barriers with elaborate fortifications and a formidable garrison, rendering it virtually impassable for surface forces, too.

Although there were ferocious clashes along the Alpine border, geography pushed and pulled the Italian front's epicentre onto the foreland of the mountains, the narrow plain at the Adriatic's northern tip. From here, the Italians repeatedly attempted to thrust into the Danube basin and threaten Vienna. Uniquely for this front, there was considerable scope for the tactical use of air power over the trenches and other earthworks that developed along the Isonzo and Piave especially. However, as casualties mounted and the conflict dragged on, outflanking these fortifications with aircraft that had sufficient reach to cross the Adriatic became an ever more seductive idea. Air power in the form of strategic bombers could, some perceived, offer an alternative to the ghastly, attritional battles – there were a dozen on the Isonzo alone between 1915 and 1917 – that were costing the lives of hundreds of thousands of soldiers yet yielding few, if any, tangible gains.

So it was that the protagonists' lines of communication and nodes of supply became a focal point of the aerial war in this quarter. Although both sides embraced the concept of strategic and interdictory bombardment, it was above all those Italian aviators who, equipped with impressive, tri-motor bombers designed by Giovanni Caproni, came to see such grandiose raids as their forte. The Austrians struck at industrial and population centres such as Milan, Venice and Padua, while, for the Italians' fleet of Ca.40 and other large planes, much of the Adriatic's eastern edge lay within reach and was a plainly more promising environment than the Tyrol. Several of the sorties executed by Italy's bombers were directed at the principal anchorage of the Austro-Hungarian navy, Pola (Pula), and at its shipyards and seaplane base at Trieste. Others were mounted against the Whitehead torpedo plant at Fiume (Rijeka) and against the submarine base at Cattaro (Kotor), where Austro-Hungarian boats were supplemented with German ones brought in by rail so as to circumvent the Allied blockade of the Mediterranean.

Douhet was the most vociferous advocate of strategic bombing. Like the USA's William Mitchell, through his somewhat cavalier agitation on behalf of air power he antagonized many within his country's military and political hierarchies. After advocating that Italy participate in the First World War, he had rapidly become a vehement critic of the way in which the enterprise was conducted. Not least because of the region's topography, Italy's frontier with Austria-Hungary witnessed some of the most extreme and futile instances of trench warfare seen in the whole of the Great War. (Indeed the Battle of Caporetto – the twelfth major clash on the Isonzo – in autumn 1917 culminated in the Italians losing many of the few gains they had made since joining the conflict in May 1915.) Disillusioned and embittered by the unavailing offensives and sanguinary, siege-like combat, Douhet urged that air power be employed imaginatively to end the stalemate. The army, he insisted in a series of memoranda to the top brass, should stay firmly on the defensive while resources were switched to expanding Italy's capacity for aerial offensives. An armada of 1,000 heavy bombers could, he asserted, simply fly over or around any defences and, venturing up to 300 miles beyond the front line, might end the war by striking at the Central Powers' very core.

What Douhet envisaged appeared to be the quintessence of what strategy was supposed to be, the art of distributing and applying military means for political ends. His reasoning partly mirrored that of, for one, Rear Admiral Mark Kerr of the Royal Navy, who, in a memorandum of October 1917 entitled 'Air Policy', surmised that the Germans' *Gotha* raids were the start of a campaign that would deprive Britain of her ability to wage war on land, sea and in the air. 'It is a race between them and us', Kerr insisted, calling for 2,000 large bombers with which the British might flatten Germany's factories.[4] Douhet likewise proposed undermining the enemy's armed forces indirectly through a strategic bombing offensive. Along with ports and other communication and supply conduits, arsenals and industrial centres, the initial targets would comprise banks, governmental departments and similar nodes without which any wider war effort could not be sustained. Only once the Austrians' morale and means of production had been severely weakened would their armed forces be assailed directly. Even then the initial attacks would comprise an aerial interdiction campaign that would deprive front-line units of supplies, reinforcements, information and political guidance.[5] The aeroplane in general and the bomber in particular, Douhet insisted, constituted 'the best weapon to strike a fatal blow . . . [for it] attacks not only the fist but the heart, and cuts the nerves and veins of the arm'.[6]

To the likes of Douhet and Caproni, just as war had become the science and industry of destruction, so too had aviation supplanted the battleship as the supreme form of martial might. Published in English as *The Command of the Air*, Douhet's magnum opus first appeared in 1921 and, enlarged upon in 1926,

4 See: *Documents Relating to the Naval Air Service: Volume I*, (ed.) S.W. Roskill, pp. 562–4.
5 See: Giulio Douhet, *Diario critico di Guerra* (2 vols, Rome, 1921–22) vol. 2, pp. 21–2 and *Scritti inediti*, (ed.) A. Monti, (Genoa, 1951), pp. 14 ff.
6 Douhet, *Diario*, vol. 2, pp. 16, 19–21.

attracted international attention, primarily because of its sheer radicalism.[7] His subsequent book on future warfare, *Probabili aspetti della Guerra futura*, continued in the same vein, portraying a conflict that was prosecuted and, thereby, swiftly decided almost exclusively through strategic, aerial bombardment.[8] Indubitably, during the 1920s and early 1930s, a widespread, popular concept of war was essentially that of 'Douhetism' and many people's notions of air power duly boiled down to fleets of strategic bombers. However, within professional military circles, Douhet's vision got a very mixed reception, even within Italy, not least because of its very radicalism.

This was partly because Douhet, basing his reasoning on the limited experience of the Great War, underestimated the improvements to air-defences that the science he so venerated would, in time, render possible. He was far from alone in this respect. Stanley Baldwin – who served as Britain's prime minister three times during the interwar period – openly despaired over the sheer size and complexity of the problems inherent in air-defence. He once told the House of Commons that:

> The speed of air attack, compared with the attack of an army, [is such that] . . . in the next war you will find that any town which is within reach of an aerodrome can be bombed within the first five minutes of war from the air, to an extent which was inconceivable in the [First World War]. . . . Take any large town you like . . . within such reach. For the defence of that town . . . you have to split up the air into sectors. . . . Calculate that the bombing aeroplanes will be at least 20,000 feet high in the air, and perhaps higher, and it is a matter of simple mathematical calculation that you will have sectors of . . . tens or hundreds of cubic miles. Now imagine 100 cubic miles covered with cloud and fog, and you can calculate how many aeroplanes you would have to throw into that to have much chance of catching odd aeroplanes as they fly through it. It cannot be done. . . . The only defence is in offence. . . .[9]

In combat, aircraft had always derived a degree of protection from the innate limitations of their opponents' eyesight, particularly when these were exacerbated by darkness, cloud, glare or precipitation. Writing at the start of the 1930s, the devisers of American strategic bombing tactics pinned their hopes on an opponents' inability to seal off air space, drawing analogies with the partial success the mighty Royal Navy had enjoyed in containing Germany's surface ships during the Great War. Exploiting ever higher altitudes and faster speeds, bombers, they maintained, would fly beyond the reach of most ground defences and get to their targets before they could be harassed by fighters. Grouped together

7 Giulio Douhet, *Il Dominio dell'aria* (Rome, 1921 and 1926).
8 Giulio Douhet, *Probabili aspetti della Guerra future* (Palermo, 1928).
9 *House of Commons Debates*, 10 November 1932, series 5, vol. 270, col. 632.

in defensive formations and bristling with machine guns, they would in any event be able to repel interceptors with interlocking fields of fire.[10] However, as Baldwin himself once observed, aerial warfare was still in its infancy, and its potentialities seemed, if only at this juncture, 'incalculable and inconceivable'.[11] Foremost among the innovations of the 1930s that were destined to influence the conduct of such warfare was the perfection and introduction of Radio Detection and Ranging (radar).

Within six years of Douhet's death in 1930, eight countries – Britain, France, Germany, the Netherlands, Italy, Japan, the USA and the Soviet Union – were all independently developing radar for military purposes. Although this technology did not itself destroy things, it did permit combatants to perceive what could otherwise not be seen. An electromagnetic variation on the principle underlying the acoustic mirror, large, ground-based systems were far more capable of detecting the presence of planes in a timely manner than any audiological device could ever be. Provided with adequate warning of oncoming hostile aircraft, fighters could be scrambled specifically to head them off. The subsequent movements of both sides might then continue to be tracked by a web of monitors as well as radar. Equipped with optical instruments and connected to command nodes by telephone or radio, such observers could help vector interceptors towards their unsuspecting quarry. In the meantime, ground-based defences could be activated as and when necessary.

This methodology afforded far more flexibility and economy of effort than could be derived from the type of defensive shields erected during the First World War and encapsulated in the suggestions of, among others, Noel Pemberton-Billing, a former RNAS pilot who had participated in the Friedrichshafen raid and who, at the time of the Blitz nearly two decades later, was to produce a book on the challenge of countering nocturnal aerial attacks.[12] The defences with which he had had to contend during the Great War were very much more porous, essentially because of an inability to focus sufficient firepower in time and space. Concentration being a relative concept, an attacker might, exploiting air power's reach and comparative speed, bring a superior force to bear against any one of perhaps thousands of potential targets. How, then, were all these to be adequately protected against a threat that, in principle, was ubiquitous?

Until the advent of radar, air-defence turned on the domination of the airspace immediately adjacent to certain points on the surface through, above all, AA weaponry that was essentially static; mobile firepower in the form of fighters might occasionally be to hand, but this occurred as often by accident as design. With radar, the defence might be promptly and systematically, if not always

---

10 See: Tami Davis Biddle, *Rhetoric and Reality in Air Warfare: The Evolution of British and American Ideas About Strategic Bombing, 1914–1945* (Princeton, 2004), p. 167.

11 *Commons Debates*, 10 November 1932, series 5, vol. 270, col. 633.

12 See: Noel Pemberton-Billing, *Air War: How to Wage It: With Some Suggestions for the Defence of the Great Cities* (London, 1916), pp. 22–32; and *Defence Against the Night Bomber* (London, 1941).

proportionately, strengthened. Air-defence, furthermore, became more a matter of control of the aerial lines of communication – the whole of a country's
airspace – rather than the domination of mere bubbles within it. Henceforth,
intruders would run a far greater risk of interception on their way to and from
targets as well as over them.

It is worth recalling that a quarter of the Avro 504s the RNAS dispatched to
bomb the Zeppelin factory at Friedrichshafen in November 1914 did not return
and that the last raids by German *Gotha* bombers on Britain in May 1918 saw
several of the attacking planes brought down. Nevertheless, in 1932 and for
some years thereafter, Baldwin, Britain's once and future prime minister, could
intone with some confidence that: 'The bomber will always get through'.[13]
Combat experience in the Sino-Japanese and Spanish Civil Wars suggested,
however, that change was afoot. Certainly, by 1940 the situation had altered
substantially, essentially because of technological advances. When, in the Battle
of Britain, Germany's *Luftwaffe* sought to circumvent Britain's maritime strength
by gaining control of the skies, its endeavours foundered on an integrated air-
defence network – a state-of-the-art system that, constructed around radar,
comprehensively monitored the airspace throughout the United Kingdom
(UK). Furthermore, when the British developed the cavity magnetron – an
electronic vacuum valve that generated microwaves – radar came to enhance
the capabilities of individual air and surface components within such defences;
for this innovation was to make 'tactical' centimetric sets that might be accommodated aboard suitably spacious planes and other mobile platforms a viable
proposition. In fact so effective did their respective air-defences become that
both Germany and Britain, having suffered prohibitive losses, were to abandon
diurnal, strategic raids against one another. Increasingly, they exploited the cover
of darkness to conceal their bombers from ground-based defences as much as
possible. But whereas this could not hide them from growing numbers of night
fighters guided by centimetric radar, it did intensify the problems of navigation
and, thereby, those of striking targets with precision.

Douhet misjudged the effectiveness of the few episodes of strategic bombing that had occurred in the Great War. Certainly, its repercussions paled into
insignificance when compared with those of the Allied blockade of the Central
Powers. Killing just 1,413 people, the Zeppelin and *Gotha* raids on Britain, for
instance, exerted an impact that was far more psychological than physical, elicit-
ing a knee-jerk reaction from her political leaders. Tragic though it was for the
relative handful of civilians who lost friends and relations in it, this sporadic and
essentially haphazard bombardment proved, like the shelling of Paris in 1918,
much more of a nuisance than a decisive stratagem. This was so not least because
the whole notion of delivering crippling blows from the skies went hand in hand
with that of picking out an opponent's Achilles heel from amidst a plethora of

13 *Commons Debates*, 10 November 1932, vol. 270, col. 632.

potential targets. Dropping ordnance that missed its mark was, at best, a contradiction of such a strategy. Then as now, precision bombing called for extremely accurate navigation. By assisting in achieving the latter, the application of Radio Detection and Ranging was destined to help with the former. However, this was not to prove possible until a decade and more after Douhet's death.

During the 1930s, radio direction finders were fitted to many commercial and military aircraft. These picked up signals from beacons, enabling planes to home in on aerodromes, even in poor visibility. Since the accuracy of the bearings obtained from these emissions declined with distance, this was a rather crude navigational aid that was of negligible use beyond 200 miles from one of the beacons. However, as the Second World War loomed, the *Luftwaffe* adapted this concept, fitting its larger aeroplanes with *Knickebein*, a device that detected Lorenz beams. Produced from adjustable transmitters, these narrow streams of Morse code could be made to intersect one another at a known point over the Earth's surface. Simply by following one such beam until it crossed a second, an aircraft could, at worst, get close to the highlighted spot.

This electronic navigation system enabled the Germans to carry out strategic bombing that was consistently more accurate than that seen in the First World War and considerably more precise than that executed by the RAF in the early years of the Second. Using *Knickebein*, *Luftwaffe* bomber crews could consistently drop ordnance on or very close to a target without even needing to establish visual contact with it. The system made it possible to launch attacks with a circular error probable (CEP) of about a thousand yards, while two derivatives – the *X-Gerät* and the *Y-Gerät* – promised to improve on this performance.[14] Lorenz beams were, however, very susceptible to deliberate jamming – another possibility that arose as human faculties and mechanical engineering were increasingly supplemented with electrical technology. Far harder to interfere with was *Oboe*, a device that was to be introduced by the RAF in 1943. By means of radar emissions from ground stations, *Oboe* tracked the movement of a bomber towards a selected point, where it automatically received a signal to release its payload. Although the Earth's curvature restricted the 'envelope' within which such a system could work, *Oboe* was very precise and could steer planes towards selected points as far as 280 miles beyond the British coast. A centimetric surface-scanning radar, the H2S, was also to be developed by the British at this juncture. By displaying the radar's varying echoes on a cathode-ray tube, where they could be compared with maps, it was possible to obtain a guide to the topography below; stretches of water, for instance, produced a weak echo, whereas built-up areas had the strongest of signatures.

This, however, highlights an important point about such electrical navigation and targeting media. Extracting truly useful intelligence from these 'infographics' called for skill and practise. What phenomena, exactly, did squiggles and blips

---

14 The Circular Error Probable is the radius of a circle around the target within which a bomb has a 50 per cent chance of landing.

on a screen represent? What was important and what was mere 'clutter'? Such questions had to be answered by human operators, and the more information sensors gathered, the harder it became to separate the wheat from the chaff.

A similar caveat should be raised regarding aerial photography, which, along with the interception and scrutinization of signal traffic, was the commonest mechanism for gleaning information about an adversary. The British took the lead in enhancing the potential of photographic intelligence by applying photogrammetry and related techniques to the analysis of images. Successions of pictures caught by lenses aboard planes travelling at either high or (often extraordinarily) low altitudes were thereby manipulated in ways that revealed hitherto indiscernible details. As the aircraft moved along its flight-path, the camera would take a sequence of overlapping shots, the perspective of each frame differing ever so slightly from its predecessor. It was found that, if two such prints of a given subject were laid side by side and viewed through a stereoscopic magnifier, these simple, two-dimensional photographs would appear as having fused into a solitary but three-dimensional image.

Acquiring such pictures at all thus called for skilled, courageous pilots, suitably manoeuvrable aircraft, and sophisticated cameras and processing equipment. But if potentially useful data was to be extracted from these images, considerable contextual knowledge and imagination would be essential, too. Through camouflage and other deception measures, opponents would routinely endeavour to hide what would otherwise be in plain view. To be effective, photographic interpreters had to be able to spot what were essentially anomalies in the environment they were examining and make apposite deductions about them. Not least in maritime operations, such intelligence could frequently be of ephemeral value. This, together with some of the other problems encountered in trying to exploit reconnaissance reports, is adequately illustrated by an episode of the Norwegian campaign of 1940. On 11 April, advised by the RAF that there were German cruisers at Trondheim, HMS *Furious* launched aircraft against these vessels, only to find that the port contained no enemy warships. The captain of *Furious* later surmised that a vessel found anchored further inshore was 'probably a small destroyer' and that his planes only succeeded in striking a second such ship, 'since later intelligence reports spoke of an enemy cruiser ashore . . . and it is more than probable that . . . destroyers were originally reported . . . [by the RAF] as cruisers'. It was noteworthy, he continued, that 'experienced naval observers found difficulty in identifying the . . . ships attacked and were quite unable to form a united opinion as to the approximate displacement of either of them'.[15]

If bombing by electrical instrument could prove difficult, bombing by eye was, if less controversial, frequently more demanding. In principle, air power's

---

15 Document 39, 'Report . . . from Commanding Officer, HMS *Furious* to Vice Admiral Commanding, Battle Cruiser Squadron', 30 April 1940, in *The Fleet Air Arm in the Second World War: Volume One, 1939–1941*, Ben Jones (ed.), (Farnham, 2012), p. 97. It is possible that, in this setting, atmospheric phenomena distorted the ships' appearance.

cardinal attribute is its ability to exploit the third dimension, thereby gaining vantage points that are superior in every sense to those available on the surface. Yet in the absence of electrical aids, the extent of aviators' awareness of their surroundings is dictated by, above all, the strictures of their visual acuity. Height can thus be a curse as well as a blessing. If visible at all, objects located on the Earth appear minuscule from even moderately high altitudes. Lower down, visibility might be better, but here aircraft are commensurably more exposed to observation and fire from the ground; in searching for a target, they might all too easily become one. Furthermore, in order to strike a discrete spot on the Earth with an unguided object dropped from a moving aircraft, an appropriate trajectory must be imparted to that object. Environmental factors – notably the strength and direction of any wind – are among the particulars that have to be taken into consideration when determining the exact point in time and space at which the projectile should be released. The very earliest attempts at bombing depended upon luck more than design: aviators lobbed suitably small (and fused) munitions from their cockpits if and when they chugged over potential targets. With improving AA defences, larger projectiles and increasing aircraft operating speeds and ceilings, such haphazard techniques had perforce to give way to more sophisticated ones. By the same token, however, achieving precision in bombing also became a more convoluted undertaking.

Mechanical bombsights that sought to exploit the certainties of Euclidean geometry and vector arithmetic formed the cornerstone of such endeavours. The most seminal of the instruments that emerged in the Great War was the Course Setting Bomb Sight (CSBS) devised for the RNAS in 1917. This employed a calculator that was similar to a slide-rule in order to gauge the effect of the prevailing wind on a projectile's trajectory. The other fundamental problem surrounding optical bomb sights – the imperative of keeping them stable in relation to the target – was addressed through the incorporation of gyroscopes.

Several attempts to improve the sophistication of these targeting media proved self-defeating, however. Some promising systems, such as the American Mark XI bomb sight, were too labyrinthine to be used well in the heat of battle. Indeed their innate limitations aside, CSBS and its cousins could only be as accurate as environmental variables, not least time constraints, and the quality of their operators permitted them to be. Visual contact with the objective had to be established and maintained, despite the human eye's limitations. Even on diurnal raids, cloud, haze, smoke, precipitation or industrial smog could – and often did – prove no less impenetrable than the darkness of night. Try as they might to fly straight and level towards a target, bombers, if not actually brought down by AA fire, were often driven off course by it and, during nocturnal raids, by the glare of searchlights. Even if it successfully completed a bombing run, an attacking plane might still find that its payload missed its mark or did not prove efficacious for other reasons. Once released, projectiles were at the mercy of the prevailing aerodynamics. In the event that the bombardier's guesstimate turned out to be a touch too imprecise, bombs approaching all but the largest of targets from their leeward or windward side would either overshoot or fall short. In any event the extent to which ordnance might affect a target depended upon the peculiar characteristics of both. The Douhetian conviction that a tenfold increase in bomb tonnage would necessarily cause ten times as much physical and psychological damage was, in this and other respects, as deceptive as it was alluring.[16]

The Second World War was to see the application of aspects of Douhet's reasoning in both the Pacific and European theatres. In the latter, the bombing offensive that was to be undertaken by, initially, the RAF's Bomber Command and, later, by both it and the United States Army Air Forces (USAAF)[17] comprised raids in which planes unleashed their payloads while flying horizontally at high speeds and levels. The targets were almost invariably fixed points on the Earth's surface and were to include key nautical facilities, among them German U-boat pens in the occupied territories of France and Norway. However, whilst the rationale for the Allies' strategic bombing campaign was clear enough, the fulfilment of its objectives turned on the ability to focus sufficient firepower in time and space. This was lacking for much of the conflict. As late as the summer of 1941, the RAF was far from matching the bombing accuracy achieved by *Luftwaffe* planes equipped with *Knickebein*. According to the Butt Report – which analysed the damage inflicted on 28 targets within Germany by 100 nocturnal raids carried out by the RAF that June and July – only 4,065 of the 6,013 sorties came close to their designated aiming mark at all. Of this latter proportion, moreover, only a third of the planes disgorged their payloads within five miles of it.[18] In the course of the campaign, the Allies' fleets of heavy bombers were duly compelled to adjust their approach, redefining their concepts of exactitude: although most attacks were nominally precise in that they focussed on strategic cogs within a country's war-machine – such as communication networks, key industrial labourers, ball-bearing or petroleum production – the tactics employed were increasingly those of 'area' or 'pattern' bombing, whereby targets were plastered with projectiles in the expectation that a sufficient proportion would actually hit something of value. Where destroying buildings was the object, a combination of high-explosive and incendiary warheads often proved the most efficacious, the fires caused by the latter completing the destruction started by the former.

The trading of precision for mass was to reduce swathes of many German and Japanese cities to ashes and was to culminate in the use of atomic weapons against Hiroshima and Nagasaki. Against targets on water, however, there were some very early indications that horizontal bombing from on high was likely to prove too inaccurate a methodology to be of much use. In April 1939 the RAF

---

16 See: Michael Sherry, *The Rise of American Air Power* (New Haven, 1987), p. 27.

17 It should be noted that the US Army Air Service was reformed in 1926 as the US Army Air Corps. It underwent further reorganization, becoming the US Army Air Forces in 1941. This body survived until 1947, when the autonomous US Air Force was established.

18 The Butt Report on Strategic Bombing, TNA, AIR 14/1218.

drew up a contingency plan to 'Copenhagen' the German fleet in its berths at Wilhelmshaven. Six squadrons of *Blenheim*, *Halifax* and *Wellington* aircraft were to be committed to the bombardment, which was to be conducted in daylight and from high altitudes, it being thought that these were the only circumstances in which any bombs at all might not only hit their targets but also penetrate the warships' armoured decks.[19] Whereas most of the units earmarked to carry out this projected mission had no particular expertise in such undertakings, seven squadrons of medium bombers were given special training in the bombing of ships. This programme culminated, in July 1939, in dummy runs against HMS *Centurion*, the outcome of the trials confirming that very large amounts of ordnance would have to be dropped if a single hit was to be scored from on high.[20]

Nevertheless, several large and medium bombers that were conceived and built with this *modus operandi* in mind were to be used extensively in maritime settings during the Second World War. Foremost among these were the American-made B.17 *Flying Fortress*, the B.29 *Superfortress* and the versatile B.24 *Liberator*, which, if only because they were products of a country flanked by vast oceans, were designed to have exceptional reach. The most superficial of comparisons between the B.17, which came on stream in 1935, and strategic bombers dating from just a few years earlier, such as the Vickers *Virginia* – first introduced by the RAF in 1922 and withdrawn in 1938 – highlights the sheer pace of technological change during the interwar period. This was just one facet of a wider conundrum faced by all air forces especially, the twin fears of, on the one hand, obsolescence and, on the other, premature investment in emerging technology proving as influential in many instances as the actuality. The growing exploitation of mechatronics, the streamlining and strengthening of airframes, improvements to propeller designs and the development of powerful radial engines were the major trends that, together, transformed aviation in the space of a few years. Although small biplanes that could be operated from battleships and cruisers were to remain popular with most navies throughout the Second World War, generally speaking monoplanes supplanted tri- and biplane designs. Thicker, sturdier, cambered and sloped wings not only improved aerodynamic performance but could also be used to accommodate retractable undercarriages, fuel, weapons, munitions and engines. Aircraft could fly faster, higher and further than ever as well as being more dependable and resilient than their quite recent ancestors.

Devised with combat survivability in mind, the B.17, for example, was a large, all-metal, four-engined monoplane, the fuselage of which could be completely enclosed to the elements. Its gun-turrets were mechanized and its fuel tanks were, in the event of punctures, self-sealing. The B.17 could travel at speeds of up

19 Letter from Air Officer Commander-in-Chief, Bomber Command, to Under-Secretary of State, Air Ministry, 3 April 1939, TNA, AIR 2/3018.
20 Minute from Director, Staff Duties, to Director of Operations (Home), 10 November 1939, TNA, AIR 2/2608.

to 287 mph and climb to 35,000 feet. However, bristling with machine guns and weighed down with armour plate, on most missions it could carry a payload of bombs no greater than that of the much smaller De Havilland *Mosquito*, which, although it had a wooden airframe, was very much a product of the 1940s, not the 1930s or earlier. For its survival in battle, the *Mosquito* depended essentially upon its ingenious monocoque design and two powerful engines, which gave it remarkable speed and agility. Originally envisaged as a fast bomber, it proved an exceptionally versatile aircraft and was to serve in a variety of roles throughout the Second World War. So, too, did the Fairey *Swordfish*, a small bomber that dated from 1934 but looked as antiquated as it was slow. Affectionately known as the 'string bag' – more for its ability to accommodate appreciable amounts of differing weaponry, such as depth-charges and rockets as well as torpedoes than for its rickety appearance – the *Swordfish* was as wieldy as it was adaptable and was to prove useful against both surface vessels and submarines. If only because its exceptionally low take-off and stall speeds enabled it to operate from the smallest of flight-decks, it outlived the Fairey *Albacore*, for instance, the very machine that, introduced in 1939, was supposed to supplant it.

A plane that lay at the far end of the technology spectrum from the *Swordfish* for one was the awesome Boeing B.29. The big brother of the B.17, it could fly at speeds of up to 350 mph and, thanks to its fully pressurized fuselage and very efficient engines, at exceptionally high altitudes. Astonishingly, the B.29 was to be envisaged, created and fielded in the space of just a couple of years, essentially as a means by which the USA might assail distant Japan directly. The rationale that was to underpin much of the fighting across the Pacific in 1944 comprised the Americans' need to seize islands, notably the Marianas, from which B.29s might reach as far as the Japanese homeland. Indeed this strategic bombing campaign was destined to decide the conflict's outcome, insofar that it was to culminate in the dropping, by B.29s, of atomic weapons on Hiroshima and Nagasaki. The B.17, by contrast, was to see service in both the European and Pacific theatres until the war's end. In the former especially, many bombers were to have relatively short careers, notwithstanding being contemporaries of the B.17. For example, of three British medium bombers, the *Wellington*, *Whitley* and *Hampden*, two were to be quickly overtaken by events. Only the *Wellington* – which had a geodesic, alloy airframe designed by Barnes Wallis – was retained for front-line service throughout the conflict of 1939–45, fulfilling, among others, anti-submarine missions, where it operated alongside a new generation of flying boats, notably the Consolidated PBY *Catalina* and the Short S.25 *Sunderland*, both of which were monoplanes.

The *Sunderland*, which was first introduced in 1938 and subsequently underwent several refinements, was loosely based on the S.23 *Empire*, one of several large flying boats utilized by civil airlines on long-haul flights in the interwar years. A riveted metal construction with four large engines, the *Sunderland* was a roomy, robust machine that, even when burdened with large quantities of fuel, defensive weaponry and navigational and targeting devices, could accommodate 2,000 lbs of bombs and depth-charges, which it unleashed from retractable – and thus reloadable – racks. Able to stay aloft for a dozen hours and more, it could cover great distances at fairly high speeds and, with a spacious hydrodynamic hull, was useful in air-sea rescue as well as maritime reconnaissance and combat operations. So, too, was the American-made *Catalina*. Dating from 1933, this plane had a steel frame that was mostly riveted together and snugly covered in stitched fabric. Although its wing-tip floats could be retracted to improve aerodynamic performance, powered by just two engines, it was rather more ponderous than its British cousin, while its amphibian sibling, the Mark 5A, was heavier and slower still. Nevertheless, what the *Catalina* lacked in speed it made up for with reach, having a range of around four thousand miles.

Not least because of her peculiar geopolitical circumstances, interwar America was eager to acquire the wherewithal for precision bombing. The USA shared Italy's fascination for emerging technology and surpassed most countries in needing planes that could contribute to coastal defence and maritime reconnaissance operations.[21] But besides facing the challenge of keeping abreast of rapid technological change, particularly in the sphere of aeronautics, the competing states also grappled for much of this era with an acute shortage of money. This hampered research, development and procurement programmes. Indeed 'Billy' Mitchell – who had risen to prominence in the Army Air Service during the Great War and who became America's most vociferous advocate of aviation – was, like Douhet, persuaded that the bomber had supplanted the battleship if only because of the costs associated with the latter. Certainly, for all the developed powers the interwar period was a time of austerity, complicating decisions about expenditure on defence, for one. The indebtedness and other economic turbulence that had resulted from the ruinously expensive conflict of 1914–18 manifested themselves in a depression that persisted until 1922. Worse was to follow: a brief boom, particularly in the USA, gave way to the 'Wall Street Crash' of 1929 and a global slump of unprecedented ferocity, the effects of which dragged on throughout the 1930s, exacerbated by the vicious circle of war reparations and loan repayments that increased tensions between former friends and foes alike.

A rather backward country, with a maladroitly structured economy and high levels of poverty and unemployment, Italy suffered more than most as a result of the Great Depression, which, moreover, coincided with some lavish, overseas entanglements. Her intervention in the Spanish Civil War of 1936–39; the ongoing costs of developing – and, where necessary, pacifying – her territories in Eritrea, Somalia and Libya; her conquest of Abyssinia during 1935–36 and of Albania in 1939: all of these exacerbated Italy's financial woes, absorbing some three quarters of the budget allocated to her armed forces and colonies between 1935 and 1940. At a juncture when many leading states were continuing to devote money and other resources to navies and their supporting infrastructure,

21 See: Biddle, *Rhetoric and Reality*, p. 161.

Douhet, like Mitchell, believed that spending on aeroplanes would prove far more cost-effective, if only because of the greater flexibility afforded by air power. Through a combination of shore-based guns and aviation, the USA's immense littoral could be adequately protected, Mitchell argued, while aircraft carriers were clearly superior to battleships when it came to projecting military might abroad.

Italy, Douhet was persuaded, could go even further, doing virtually everything with strategic bombers alone, including dominating the Mediterranean basin with planes based in both her various possessions around its rim and in the homeland peninsula, Sardinia and Sicily. Many servicemen ventured to disagree with him, if only insofar that they called for a range of assets, among them interceptors for air-defence and fighter-bombers that could furnish close support to the army and fleet. The navy especially was as fiercely independent as the RAI and nearly as myopic. The latter justified its existence with, above all, its capacity for 'strategic' bombing that, it was widely assumed, could easily be adapted to the tactical level of operations. Although its claims of pinpoint accuracy were essentially spurious, the RAI certainly looked impressive and was duly overrated by other states and many Italians alike. By contrast, the *Regia Marina Italiana* (RMI) deluded itself more than others for much of the interwar era, not least by mistakenly taking its French counterpart as its sole prospective foe. Anticipating, moreover, that Jutland would remain the paradigm for sea warfare, Italian naval construction focussed on maintaining parity with a fleet that had not built a carrier since the *Béarn* was commissioned in 1927.[22]

Rivalry between the RMI and the RAI helped to kill off any thoughts of adding aircraft carriers to the former's order-of-battle. Indeed according to Robert Mallett, a decision taken in August 1925 by a 'conservatively minded admirals' committee' that Italy did not require vessels of this kind set the pattern for the rest of the interwar period. In the early 1930s, the Italian Navy's Treaty Office concluded that carriers would, on the one hand, be vulnerable and would therefore require numerous naval escorts while, on the other, would themselves not pack much of a punch. If enough strike aircraft – dive- or torpedo-bombers – to conduct a successful attack were ever going to be amassed, the backing of some shore-based planes would in any case be essential. When presented with a report by the Naval Construction Board on the feasibility of building a carrier in August 1935, Admiral Cavagnari, the Chief of the Naval Staff, dismissed the heretical notion with disdain, furiously scrawling 'No!!!' across the document.[23]

The Italians were preparing for, as Mallett terms it, a 'Mediterranean Jutland' conducted by battleships, cruisers and submarines with land-based air support.[24] Yet there were those who had grave doubts about the viability of this approach.

22 James J. Sadkovich, *The Italian Navy in World War II* (Westport, CT, 1994), pp. 8–9.
23 Robert Mallett, *The Italian Navy and Fascist Expansionism, 1935–1940* (London, 1998), p. 62.
24 Ibid., p. 3.

In September 1937 a report for Mussolini on future policy from the Naval War Plans Office underscored the risks of embarking on a war without carriers. It noted that: 'There are regions within the Mediterranean where any intervention by our land-based air force should be considered as far too onerous an undertaking; there are strategic situations where only carrier-based aircraft would have any chance of success'.[25] Even Cavagnari evinced concerns. On 16 May 1939 he wrote to Marshal Badoglio, head of the Italian supreme command, and to Guiseppe Valle, the Under-Secretary of the Air Ministry, emphasizing that, if the RMI was to dominate the Mediterranean, it was essential that the RAI gain total air superiority on the outbreak of war and attack the foreign naval bases at Malta and Bizerte in order to disable their ships and facilities.[26] Valle vaingloriously assured him that the ten squadrons based on Sicily, Sardinia and in Libya, which included seven bomber units, would definitely prove capable of putting the bases out of action. But, he cautioned, the RAI could not be certain it could protect its Sicilian and Sardinian airfields and could only 'guarantee control of a specific theatre of operations for a limited period of time'.[27]

Having left the question of air support to the RAI, the navy was, furthermore, at the mercy of the former's choice of aircraft designs. If the shortcomings of the RAI's Breda Ba.85 dive-bomber were alarming, the failure to develop customized torpedo-bombers proved crippling. In June 1939 Valle had informed Mussolini that he did not believe in the utility of such planes. Neither was the navy willing to fund either their development or that of their armaments, which were so *sui generis* in nature. In fact the RAI did not form its first torpedo-bomber squadron until December 1940 and, even then, aerial torpedoes were only to be produced in dribs and drabs.[28] Although, during 1941 and 1942, ponderous Savoia-Marchetti SM.79 and SM.84 aircraft were to be used with some success as torpedo planes off Malta especially – where they sank or damaged several merchantmen and warships, including the battleship HMS *Nelson* – on the whole torpedoes launched from Italy's numerous submarines were to pose more of a threat to enemy shipping than those dropped from aircraft.

Whilst, throughout this period, Italy's pilots continued to bask in a formidable reputation as far as sport and spectacle were concerned, when, in June 1940, she declared war on not only France but also on Britain, the shortcomings in her aviation were quickly revealed, not least in maritime settings: too many machines were ageing, if not antiquated, wooden designs; most aero-engines were either German imports or crude imitations of foreign motors; some squadrons had, in the motherland of Guglielmo Marconi, yet to be equipped with surface-to-air radio, let alone the wherewithal for intercommunication, while even primitive radar systems were still in gestation; bespoke torpedo-bombers had not been

25 Ibid., pp. 102–3.
26 Ibid., p. 146.
27 Ibid., p. 147.
28 Sadkovich, *The Italian Navy in World War II*, pp. 11–12.

developed and makeshift ones were to prove slow in coming on stream; few aircraft were capable of operating at night; and little thought had been given to the practicalities of hitting moving – as opposed to fixed – targets, notably warships. Indeed when, in July 1940, some five hundred Italian planes fell upon vessels of the Royal Navy in the Mediterranean, they experienced the selfsame problems the British themselves had encountered in attacking the stranded *Yavuz Sultan Selim* near the Dardanelles in 1918; very few bombs actually struck their targets and those that did were too puny to penetrate armour plate.[29]

Generally speaking, the performance of Italy's air power was to prove lacklustre at best. Germany's apart, hers was the only major navy in the world that did not control its own aviation. Likewise, aspects of air-defence per se were, rather incongruously, shunned by the air force itself. Left to a separate command structure and local authorities, it was a rather haphazard undertaking, largely founded upon the tactics and technology of yesteryear. Just as the British torpedo raid of November 1940 on warships moored in Taranto harbour was to underline the deficiencies in these arrangements, so too would Malta's failure to succumb to aerial bombardment *à la* Douhet highlight the gaps between Italian doctrines and capabilities.

Until the Battle of Crete in May 1941, the Taranto raid and the Norwegian campaign of spring, 1940, were arguably the first great tests under genuine combat conditions of the ability of a modern battle-fleet to defend itself against aerial assault. If, during the interwar period, the capabilities of many aircraft types were improving, then so too, it was reasoned in some quarters, were those of systems intended to actively repel their attacks. There were officers within the Royal Navy, for instance, who were increasingly confident about the future effectiveness of AA fire. The outcome of bombing trials conducted in 1922 against the pre-dreadnought battleship *Agamemnon* – which had been adapted as a radio-controlled target vessel – had been worrying.[30] Captain Collard, the Director of the Gunnery Division, noted in the experiments' aftermath that: 'It is not considered that these attacks could have been in any way seriously interfered with by the AA armament of the fleet in its present state of development'.[31] However, in 1919 a Naval Anti-Aircraft Gunnery Committee had been formed specifically to push through technical enhancements that might help counter the evolving air threat. Innovations such as the introduction of the new High Angle Control System Mark 1 and of the Vickers *Predictor* were designed to

29 See the reports on the Italian air raids that took place during the second week of July submitted by Admiral Andrew Cunningham, Commander-in-Chief, Mediterranean: Documents 64 and 65 in Jones, *FAA in Second World War: Volume One*, pp. 196–8. Some comments regarding the fallout from the failure of these attacks can be found in Jack Greene and Alessandro Massignani, *The Naval War in the Mediterranean 1940–1943* (London, 2002), pp. 80–81.

30 See: R.A. Burt, *British Battleships 1889–1904* (Annapolis, MD, 1988), p. 295.

31 Quoted in Geoffrey Till, 'Airpower and the Battleship in the 1920s', in Bryan Ranft (ed.), *Technical Change and British Naval Policy 1860–1939* (London, 1977), p. 115.

improve the accuracy of AA armaments, while the Mark M multiple pom-pom gun accelerated the rate of fire for defence at close range. The calibre of large AA weaponry also increased with, for instance, the introduction of the 4.7" High Angle Gun. By 1936, Admiral Chatfield, the First Sea Lord, was anticipating that the Royal Navy's battle-fleet would ultimately be furnished with AA weaponry to such an extent that he did not believe it would be 'a profitable thing for aircraft to approach it'.[32] Chatfield's views were echoed by Vice Admiral James, the Deputy Chief of the Naval Staff, who perceived a tilt in the balance between attack and defence:

> The decision in all matters connected with the air is complicated today by the prevailing uncertainty with regard to the potency of air attack in face of the steadily increasing efficiency of the defence. Five years ago torpedo droppers and bombers enjoyed a dominating position. Today it is questionable whether launching a torpedo attack by planes is justifiable when the target is well equipped with short range anti-aircraft weapons.[33]

Nevertheless, there were those who had lingering doubts. Reginald Henderson, the Rear Admiral, Air, remarked in 1932 that: 'No realistic firing against aircraft has taken place since the last war and, in my opinion, the value of our own High Angle Control System Mk 1 is rated too high. In common with others we are apt to over-rate the capabilities of our weapons in peacetime . . .'.[34] Certainly, under combat conditions, the new systems turned out to be less efficacious than many within the Admiralty had expected and did not in any event compensate for the rather low complement of AA armaments with which British ships entered the Second World War.[35]

That the allocation of such weaponry to shipping often turned out to be rather meagre was largely a consequence of the inability to reconcile limited means with almost infinite demand and to insure adequately, if at all, against potential threats that defied delineation. Any concentration is, by definition, relative. Although quantity does have a quality of its own, security founded on superior numbers can prove illusionary, particularly when subject to the vagaries of actual combat, not least morale. Almost every corner of Britain's enormous empire and virtually every unit within her armed forces could lodge a persuasive claim for being furnished with at least some scope for AA defence. Even at the best of times, demand ineluctably outran supplies. But within a year of hostilities breaking out, a genuine crisis was developing in this respect: the army's core – the British Expeditionary Force (BEF) – had had to abandon virtually all of its heavier equipment, including its AA artillery, at Dunkirk; Malta was on course

---

32 Quoted in Till, 'Airpower', p. 116.
33 Minute by Deputy Chief of the Naval Staff, 15 June 1936, TNA, ADM 1/11971.
34 Quoted in Till, 'Airpower', p. 116.
35 See: ibid., p. 115.

to becoming the most heavily bombed place on Earth yet was protected by just 42 AA guns;[36] and so much of Europe's northern and western fringe had already fallen under Axis control, exposing the British Isles and the surrounding sea lanes to aerial assault from almost every quarter. Indeed metropolitan Britain had become the focus of the most famous, self-contained aerial campaign in history.

The most flexible instrument by far for protecting things from attack by planes was another aircraft. Yet fighters – and pilots who were sufficiently trained and experienced to operate them well – were particularly precious commodities during the war's first two years especially, the Battles of France and Britain alone absorbing so many of the available personnel and machines. Parts of the Mediterranean, too, also had a pressing need for interceptors. For instance, all that was initially to hand to guard the Maltese skies against hundreds of Italian raiders were three *Gladiator* biplanes – dubbed 'Faith', 'Hope' and 'Charity' – that, fortuitously, had been discovered in crates aboard the carrier *Glorious*. As far as shielding vessels on the high seas was concerned, in September 1939 the FAA had only six carriers from which it could function,[37] of which two of the most capable – *Courageous* and *Glorious* – were to be sunk within nine months of the war's outbreak.[38] HMS *Ark Royal* was also to be lost before 1941 was out; she was destined to be torpedoed by *U-81* after delivering fighter planes to beleaguered Malta.[39] If only because production efforts had to be concentrated on the mainstream designs employed by the RAF's Fighter and Bomber Commands, during this phase of the conflict in particular Britain's remaining carriers were intermittently hampered by a shortage of replacement machines, notably *Fulmars* and other interceptors.[40]

Such aircraft were indispensable in regions where enemy planes might seek to monitor and harass maritime undertakings. Since in 1939 neither Germany nor Italy had possessed serviceable carriers of their own, London had anticipated that there would be little need for customized fighters over the high seas stretching northwards and westwards from Europe. By contrast, the centre of the Mediterranean, ringed as it was with Italian aerodromes, threatened to prove a very different operating environment in the event that Rome turned actively belligerent, while Japan's powerful carrier fleet posed a looming danger to some of Britain's possessions, allies and trade in parts of the Pacific and Indian Oceans.

36 Correlli Barnett, *Engage the Enemy More Closely: The Royal Navy and the Second World War* (London, 2000), p. 225.

37 Namely: *Ark Royal, Glorious, Courageous, Furious, Eagle* and *Hermes*, plus the seaplane carrier *Albatross*. See: Document 15, 'Memorandum by Fifth Sea Lord', 4 September 1939, in *FAA in Second World War: Volume One*, p. 35.

38 Regarding the loss of *Courageous*, see: Documents 18 and 18a, ibid., pp. 42–9 and for the loss of *Glorious*, see: Documents 55, 55a and 58, ibid., pp. 169–71 and 179–83.

39 On the loss of *Ark Royal* see: Documents 164, 167 and 173, ibid., pp. 552–3, 540–41 and 554–9.

40 For example in May 1940 the production of *Hurricanes, Spitfires, Blenheims, Whitleys* and *Wellingtons* was prioritized. See: Document 45, 'Paper by Director of Air Materiel: Priority for naval aircraft production', 20 May 1940, ibid., pp. 115–17.

The conquest by the Axis of so much of the European and Oriental coastlines during the war's opening months and years dashed many of these assumptions, however, insofar that it gave hostile, land-based aircraft access to far more than just the central Mediterranean. Geared to protecting shipping primarily from surface raiders and submarines, the Royal Navy abruptly found itself compelled to equip more and more vessels to ward off attacks from the skies as well.

This development exacerbated the already urgent need for heavy machine guns and artillery of small or medium bore. Although dual-purpose guns – namely those that, with appropriate ammunition and focussing mechanisms, could be employed against both aerial and other targets – offered a promising solution to this dilemma, it could only be a partial one. Their very exoticism rendered these weapons that much more difficult, expensive and slow to produce at a juncture when industrial capacity, money and time were all short. The Royal Navy could not in any case afford to have many of its existing vessels tied down in extensive refits, particularly after France had fallen and the Italians had sided with the Germans, transforming the balance of maritime power in the Mediterranean especially. In any event comprehensive air cover called for a blend of weaponry with differing ranges and arcs and volumes of fire. This, in turn, had implications for the architecture and manning of ships, among them the need to decide just how much of a given vessel's limited space should be allocated to AA command and control cells, ordnance and ammunition stockpiles.

Among Britain's stop-gap efforts to tackle this and the burgeoning, wider quandary of AA defence was the conversion of several older cruisers into specialist platforms. Begun in the midst of the frugal 1930s, this programme was yielding some fruit by the war's outbreak. Configured almost exclusively to countering hostile planes, these AA ships, however, were vulnerable to other threats and were best deployed as part of a balanced flotilla.[41] Moreover, to have much chance of shielding any accompanying vessel from aerial attacks, they needed to stay within a half a mile or so of their charge. This could prove immensely hazardous, particularly when the ships were manoeuvring at speed. In October 1942, while steaming side by side on zigzag courses to evade interception by lurking U-boats, HMS *Curaçoa* and the huge liner RMS *Queen Mary* collided with one another off the coast of Northern Ireland. Literally sliced in two, the AA cruiser began sinking almost instantaneously, taking 338 of her crew with her.[42] *Queen Mary*, which was laden with American soldiers bound for Britain, was left severely damaged and with no active defences. Fortuitously, she did not come under attack as she limped on to her destination.

This horrifying incident highlights, among other things, the risk in concentrating a large proportion of the AA assets available to a flotilla within a solitary hull. Distributing systems across several platforms spread the danger as well as

---

41 For details of AA cruiser conversions, see: David K. Brown, *Nelson to Vanguard: Warship Design and Development, 1923–1945* (London, 2000), pp. 155–6.

42 See: Loss of HMS *Curaçoa* in collision with SS *Queen Mary*, TNA, ADM 116/6158.

the defence, often proving tactically advantageous, too. But identifying suitable AA armaments and sensors and then procuring and fitting them to numerous ships were difficult undertakings, particularly when supplies of tried and tested weapons were limited. So-called Unrotated Projectile systems (UP) were but one of a variety of novel AA armaments experimented with by the British especially. Others included flamethrowers and steam and compressed air mechanisms for launching explosive devices.[43] The only one of these alternatives that promised to rival the gun was, however, the basis of the UP system, the rocket, the exploitation of which posed some thorny technical problems.

So as not to obscure targets from observation and further fire, the propellant within any AA rocket had not only to combust in a controlled way but also do so without generating much smoke. Extrudable in variable patterns, a form of pliable cordite developed by Britain's military scientists in the mid-1930s proved ideal in these regards. Mortar-like, UP systems had tiers of tubes from which were launched fin-stabilized rockets. These in turn dispensed small mines suspended from miniature parachutes at an altitude of around five thousand feet, the aim being to create an aerial barrage around the vessel that would impede if not destroy oncoming planes, particularly any flying at relatively low levels, such as torpedo-bombers. Being easier and cheaper to manufacture than guns, UP and similar systems were comparatively plentiful and were to be seized on as a means of bolstering AA defences both ashore and afloat at a time when the demand for artillery was essentially insatiable. The protection given to, for instance, the new battleship *Prince of Wales* included a UP network until this was replaced by additional pom-pom guns a few months after she first saw action. Likewise, what became known as Z batteries – large banks of rocket launchers – assisted in the guarding of nodal points in the UK, most notably harbours.[44]

The quest to make these all too haphazard rockets more efficacious had a wider significance, however, since it demanded technological innovations that went on to have tremendous ramifications for air-defence as a whole. It was plainly necessary to choose a trigger mechanism for the warheads mounted on the rockets. Whereas any reliance on simple percussion would call for the projectile's trajectory to be calculated with phenomenal exactness, time fuses were hardly better insofar that they required presetting in keeping with the estimated duration of the rocket's flight to the target. The linking of detonators to sensors – as was being done concurrently in nautical mines especially – offered the most promising solution to the problem and, owing to the miniaturization of electrical components, was, by 1940, becoming a viable one. Initially, electro-optical devices that could detect the shadows cast by aircraft were employed, but, if only because these were useless in poor light, proximity fuses were soon being devised,

43 See: Minute from the First Sea Lord to the Prime Minister: 'Defence against suicide aircraft', 22 January 1945, TNA, ADM 205/43.
44 See: Minute by Director of Gunnery and Anti-Aircraft Warfare: 'Type "K" Rockets', 17 January 1945, TNA, ADM 205/43.

too. Such instruments comprised an emitter, the signals of which were picked up by a receiver as they bounced off the rocket's quarry. When the echo frequency peaked, the warhead was detonated, showering the target with shrapnel.

The development, refinement and mass production of Britain's prototype proximity fuse was to become one of the most important projects within the scientific and industrial cooperation between Washington and London that ensued after the mission led by Sir Henry Tizard to the USA in September 1940. Code-named 'VT' by the Americans, a device of such compactness was eventually perfected that it could be squeezed into any warhead with a nose no narrower than 2.9" in diameter. This encompassed not just rockets but also the AA artillery shells fired from most ships' guns. First used operationally in 1943, the proximity fuse was to have an immense effect on aerial warfare in maritime environments especially. Indeed, for fear that telltale remnants of this top-secret technology might fall into the enemy's hands, its employment was to be largely restricted to operations over water, most notably in the immensity of the Pacific. Mark Peattie's insightful study of the aerial warfare in that theatre suggests that, by the time the Japanese surrendered in August 1945, 'VT' had formed the key element in the destruction of over three hundred of their planes by AA fire from Allied vessels.[45] Certainly, together with other improvements to AA defences, this precision weapon was to help blunt attacks from even the most fanatical of adversaries, the *Kamikaze*, making the heavy cruiser HMAS *Australia*, for example, a tougher opponent in 1944 than either HMS *Cornwall* or *Dorsetshire* had proved only two and half years earlier; harassed by swarms of *Kamikaze* planes while covering the Lingayen Gulf landings in the Philippines, *Australia* downed several of her attackers and, although damaged, was able to complete her mission and withdraw to safety.[46]

Arguably the most eye-catching instance of the early successes notched up by air over sea power was to be the sinking of HMS *Repulse* and *Prince of Wales* by Japanese planes in December 1941. At the time, however, these capital ships had no alternative but to rely on AA defences that were appreciably inferior to sensor and weapon systems that were to come into service with ever more of the Allies' maritime forces during the concluding half of the Second World War. In the final analysis the Imperial Japanese Navy (IJN) was to suffer more than any other as a result of the accumulating shortcomings in its air-defence capabilities. Among the numerous Japanese vessels that were fated to fall victim to aviation were the two largest battleships ever built, *Yamato* and *Musashi*.

45 Mark R. Peattie, *Sunburst: The Rise of Japanese Naval Air Power, 1909–1941* (Annapolis, MD, 2001), p. 199. Also see: Ralph B. Baldwin, *The Deadly Fuze: The Secret Weapon of World War II* (Princeton, NJ, 1947), pp. 233–49; Eric M. Bergerud, *Fire in the Sky: The Air War in the South Pacific* (Boulder, CO, 1999), pp. 562–8; Norman Friedman, *US Naval Weapons: Every Gun, Missile, Mine and Torpedo Used by the US Navy from 1883 to the Present Day* (Annapolis, MD, 1982), pp. 88–9.

46 See: Minute from First Sea Lord to Prime Minister: 'Defence against suicide aircraft', 22 January 1945, TNA, ADM 205/43.

Much of Mitchell's reasoning about the respective merits of battleships and bombers stemmed from experiments that were carried out by the US Navy and the Army Air Service during the summer of 1921. These were intended to establish the effectiveness of aerial attacks upon differing types of nautical craft, ranging from lightly protected destroyers and submarines to dreadnoughts sheathed in heavy armour.[47] A variety of vessels – mostly relics of the defunct *Kaiserliche Marine* – were struck from altitudes as low as 200 feet with ordnance that weighed up to 2,000 pounds. Ultimately, all of them were sunk, among them the modern German cruiser *Frankfurt*, the old American battleship *Alabama* and the German dreadnought *Ostfriesland*. 'Thus ended the first great air and battleship test that the world has ever seen', wrote Mitchell. 'It conclusively proved the ability of aircraft to destroy ships of all classes on the surface of the water'.[48]

Impressive though this sounded, what had actually been demonstrated was the self-evident: things could be damaged *if* adequate firepower were to be focussed on them in time and space. The *Ostfriesland* herself had sought to achieve as much when she participated in the Battle of Jutland some years earlier. Indeed it was generally believed during the interwar era that, in a surface engagement, twelve to sixteen direct hits from the largest shells could destroy any warship afloat.[49] The experiments, such as they were, that Mitchell witnessed tested only the most passive defences of the rusting vessels involved, notably their armour plate above and below the waterline. The submarine – the old *U-117* – had none whatsoever and, lying motionless on the surface at a known location, was a sitting duck. *Frankfurt*, *Alabama* and *Ostfriesland* had more in the way of armoured protection, but were in other respects more vulnerable than the *Yavuz Sultan Selim* had been when she was assailed back in 1918. Unmanned, all the vessels attacked in the tests had no active defences at all, including damage-control teams. The various craft could neither jink – either to throw off the bombardiers' aim or to dodge incoming projectiles – nor destroy or deflect approaching planes with defensive fire. In fact for the purpose of these experiments, it was assumed that a preliminary sweep by fighter-bombers would scatter and destroy any aircraft or escort vessels seeking to cover the target ship. Keeping aerial attackers at bay with AA guns would thus be rendered 'impossible', allowing larger bombers to administer the *coup de grâce* 'with little danger'.[50]

Mitchell speculated that these tests 'would lead people to believe that the navy should be entirely scrapped, as a thousand airplanes could be built for the price of one battleship'.[51] However, in professional circles at least it was appreciated that replicating the outcome of these experiments in authentic combat conditions

47 See: William Mitchell, *Winged Defense: The Development and Possibilities of Modern Air Power, Economic and Military* (New York and London, 1925), pp. 56–76.

48 Ibid., p. 73.

49 Peattie, *Sunburst*, p. 139.

50 Ibid., pp. 59–60, 61–2.

51 Ibid., p. 71.

would be easier said than done. Views on the issue varied in Britain as well as in the USA. By the Great War's end 'Jacky' Fisher had already concluded that the dreadnought had had its day: 'All you want is the present naval side of the air force!' he wrote. 'That's the future Navy!'[52] In 1921, at the Bonar Law Enquiry, which was established to assess the Royal Navy's request for new battleships in response to the US Navy's latest acquisitions, Hugh Trenchard, Chief of the Air Staff, remarked that: 'We are ineffective against an enemy capital ship even now. . . . I want to say that none of the people I speak for are against the Capital Ship'.[53] On 21 March 1922, in his statement on the Air Estimates, Captain Frederick Guest, the Secretary of State for Air, went so far as to tell the Commons:

> It is already proved that one bomb can sink the most powerful battleship in a few minutes. A battleship may survive a direct surface hit, but you cannot protect it from the explosion of a bomb underneath its water-line. It is merely necessary to perfect the bomb sight, which is purely a matter of practice and experiment. . . . In ten years' time I believe that a combat between the forces of the air and the forces of the sea will have become a grotesque and pathetically one sided affair.[54]

Guest was an old soldier. Another MP from the same stable, Lieutenant-Colonel Moore-Brabazon, suggested that it was no longer necessary for aircraft to actually hit a ship with a bomb in order to hole it: dropping a depth-charge within 600 feet of the target would surely suffice. This, he asserted, was not a very difficult thing to do, even from 10,000 feet. Such developments had rendered the Royal Navy's battleships obsolescent, he surmised.[55] On 30 March the Marquess of Linlithgow asked Lord Lee of Fareham, the First Lord of the Admiralty, whether battleships would 'be defenceless against the under-water explosion of bombs dropped by aircraft' and whether the growing capabilities of bombers 'had substantially reduced the fighting value of the capital ship'. 'Are our airmen', Linlithgow also enquired, 'to stand – or, I suppose I ought to say, fly – idly by while enemy machines hover at low elevation over our battleships and drop bombs on them or near them?'[56]

In response, Lee stated: 'I cannot . . . permit public confidence in the Navy . . . to be undermined – perhaps I should say in this case to be bombed – without making some reply exposing the absurdity of some of the claims which have been advanced. If these claims are well founded . . . then the maintenance and, still more, the building of . . . [battleships] would be quite inexcusable . . .'. He

---

52 Quoted in Robert L. O'Connell, *Sacred Vessels: The Cult of the Battleship and the Rise of the U.S. Navy* (Oxford, 1991), p. 250.

53 Till, 'Airpower', p. 111.

54 *Commons Debates*, 21 March 1922, vol. 152, cols 309–10.

55 Ibid., col. 320.

56 (*Hansard*) *House of Lords Debates*, 30 March 1922, vol. 49, cols 1,036–9.

went on to summarize the main findings of the American military's report on the sinking of the *Ostfriesland*, namely, that, had the ship been manoeuvring at speed and protected by effective AA defences, including fighter cover, the threat to her from bombers would have been negligible.[57] It should not be supposed that Britain's admirals were blind to possible developments with regard to aerial attacks upon ships, Lee insisted: 'They are devoting a great deal of time, thought and experiment to . . . these matters, but at present . . . the battleship holds its own, [if only] . . . because of the inaccuracy under existing conditions of any attack from the air'. Striking battleships with suitably big bombs called, he pointed out, for commensurately capacious planes, which could only operate from land bases. Such aircraft also had a relatively short reach. This stratagem was, therefore, only really viable in coastal defence. In any event littoral operations were, he continued, of minor importance to the Royal Navy, the principal function of which was to protect far-flung trade routes over wide oceans. Whilst, in this setting, carrier-borne aircraft might, Lee acknowledged, pose a threat, he was reassured by the fact that:

> no . . . carriers, either actual or . . . permitted by the [Washington] Naval Treaty . . . can carry these huge . . . [bombers] at all; and even if they could, and these machines could be launched from them, the machines could not . . . land on their mother ship. . . . [W]e are hopeful that, by [AA] gun fire alone, . . . it may be possible to make warships . . . practically immune against air attack of any description. But, simultaneously, we are developing the defence in the air . . . carried out from . . . carriers, which are going to play a most vital part in the fleet of the future. . . . [N]othing whatever has occurred in connection with recent developments in aircraft or methods of [aerial] . . . attack substantially to reduce the fighting value of the capital ship. The attacks to which it may be exposed from the air are being provided against . . . by improvements in its passive and active defence. . . .[58]

Lee concluded that the battleship was still 'the backbone of the fleet and the bulwark of the nation's sea defence'. The aeroplane like the submarine, destroyer and mine, had added to the dangers to which all shipping was exposed, but had yet to make the battleship obsolete. Still, although as late as the end of the 1920s the General Board of the US Navy was to remain insistent that 'the battleship is the ultimate measure of the strength of the Navy',[59] there were nagging doubts as to how, in the face of the challenge posed by fast-evolving aircraft and submarines, such warships might remain viable components of battle-fleets. In May 1936 Winston Churchill demanded the impossible of Admiral Chatfield, the First Sea

57 Ibid., cols 1,041–5.
58 Ibid., cols 1,045–8.
59 S.W. Roskill, *Naval Policy Between the Wars, Volume 2: The Period of Reluctant Rearmament, 1930–1939* (London, 1976), p. 23.

Lord, when he told him that: 'What you have got to prove is that the Admiralty of the future will be able to construct vessels so immune from these risks as to enable their being used freely at sea for all tactical and strategical purposes'.[60]

At this juncture, however, electrical engineering was promising to reduce if not eradicate the submarine threat for one. In 1917 an Allied Submarine Detection Investigation Committee (ASDIC) had been established. This Franco-British body had begun developing a system that, a cousin of the acoustic mirror seen in aerial warfare, used sound waves to locate submerged U-boats. By the late 1930s the Royal Navy especially had honed this technology – known to the Americans and subsequently to the British as 'Sonar' – to a point whereby it was believed warships fitted with it could be confident of establishing the bearing and range of any submarine within striking distance. The interloper could then be assailed by escorts dropping patterns of ordnance set to explode at differing depths.

Convoys accompanied by escorts equipped with Asdic would, in the Royal Navy's view, thus be able to blunt the threat posed by submarines. Since, in the First World War, aviation had also proved very successful in deterring attacks, if not actually destroying U-boats, it was also expected that aggressive patrolling by planes would compel submarines to remain below the surface, where they were short-sighted and far slower.[61] However, between 1919 and 1939, no exercises involving aircraft and commercial shipping were conducted by the British at all. Indeed for an air force that was anxious to survive as an autonomous entity, anti-submarine operations were scarcely a priority. The RAF allotted resources sparingly, spawning critical shortcomings: there was a lack of appropriate planes with sufficient range and adequate weapons and bomb sights. Commander Ellis, the Director of the Naval Air Division, observed in September 1938 that the RAF lacked a proper methodology for engaging, rather than deterring, submarines, for nobody in that organization fully appreciated the complexities involved.[62] In the opinion of the Royal Navy's tacticians, planes of small or medium size were best for such missions; large flying boats were less suitable when it came to shadowing, hunting or attacking. It was adjudged that, under favourable environmental conditions, an aircraft had an appreciable chance of executing a successful attack, particularly if it could catch its prey above the waves or while submerging; quarry on the surface would most likely spot a big, unwieldy flying boat and crash-dive. The likelihood of sighting submarines operating at peri-scope depth off the UK was dismissed as negligible, unless they were travelling at high speed. While above the water, on the other hand, they might be discerned from all of six miles away by planes flying at 1,500 feet.[63]

60 Till, 'Airpower', p. 110.
61 See Marc Milner, *Battle of the Atlantic* (Stroud, 2005), pp. 13–14.
62 Report of meeting held by the Fifth Sea Lord, 13 September 1938, TNA, ADM 116/4038.
63 'Anti-Submarine Striking Forces', Tactical Division of Naval Staff, September 1939, TNA, ADM 199/124.

In the event the performance of the RAF's shore-based aircraft against the U-boat menace was to fall far short of many people's expectations, certainly during the Second World War's early years. The Coastal Area – the division of the air force that supposedly specialized in maritime security – became Coastal Command in 1936. By the outbreak of hostilities the mainstay of this organization was the Avro *Anson*, a plane that was verging on obsolescence and that had such a small operating radius that it could scarcely monitor the eastern sea lanes of the North Sea; it was to be quickly supplanted by, initially, *Whitley* and Lockheed *Hudson* aircraft. After an investigation in 1931, Admiral Chatfield, commander of Britain's fleet in the Mediterranean, had recommended that flying boats be allocated to that quarter for trade protection as well as reconnaissance, arguing that 'if aircraft accompany convoys the danger of submarine attack will be reduced'. He requested that naval cooperation should henceforth form the cardinal function of the RAF's flying boat squadrons. However, it was not until as late as October 1935 that all the details of the remit of the newly-established Coastal Command were agreed upon.[64] Only then could capabilities start being matched to strategic requirements and procurement schedules fashioned accordingly.

At this juncture, the Admiralty in any case regarded the surface raider as much more of a threat to merchant shipping than the submarine. Even before the Nazis came to power in 1933, Germany had embarked on a programme of nautical re-armament. In contrast to the old *Kaiserliche Marine*, the *Kriegsmarine* was essentially designed for a *guerre de course*. Between 1931 and 1939, the Germans launched three so-called 'pocket battleships' as well as two heavy cruisers, two battlecruisers and the *Bismarck*, the biggest battleship then afloat. Work on a second such colossus, *Tirpitz*, was already underway, too, and she was to be commissioned in February 1941. By 1934, attempts to restrain Japan through arms-limitation and arms-control treaties had also failed; she had established a massive and sophisticated shipbuilding industry and, when she entered the Second World War, her naval tonnage was twice what it had been in 1922 and included some formidable capital ships. Furthermore, whereas the Germans had formally ceded their dependencies at the conclusion of the First World War, the Japanese had since acquired further outposts from which they might molest the trade and overseas possessions of the other European powers and the USA. As war again loomed in Europe, this expansion was continuing: early in 1939, Japan seized Hainan Island and the Spratly archipelago from the Chinese, adding to her existing conquests.

Only years before, Japan had been Britain's ally, as had the Italians, who now had a fleet that included 25 battleships and cruisers. By the mid-1930s both powers were turning rancorous. Indeed the geopolitical environment in general

---

64   Minute, Director of Staff Duties to Chief of Air Staff, 4 October 1935, TNA, AIR 2/8875. Roskill, *Reluctant Rearmament*, p. 200, suggests this occurred as late as 1936, the year in which the RAF was formally reconfigured into Bomber, Fighter, Coastal and Training Commands.

and the maritime situation in particular with which London was confronted had become far more convoluted by the decade's end than it had been at its beginning: the enforceability of the Versailles and Locarno Treaties of 1919 and 1925, respectively, had withered, culminating in the demise of the Stresa Front in the wake of the Anglo-German Naval Agreement and Italy's invasion of Ethiopia in 1935; the USA, the Versailles settlement's principal architect, was seeking to isolate herself from international events through the Neutrality Acts of 1937 and 1939, the last of which included provisions that impinged on the doctrine of the freedom of the seas; the USSR under Stalin, initially choosing collaboration over confrontation with Hitler, was to conclude a non-aggression pact with Germany in August 1939 and actively participated in the partition of territory adjoining the Baltic; Turkey – fearful of Soviet expansionism besides any threat from Germany and Italy – was prepared to forge ties that might enhance her own security but, by the same token, was unwilling to be seen to be taking sides; Portugal, a traditional ally of Britain, with precious harbours – Madeira, the Azores and the Cape Verde Islands – in the Atlantic's midst as well as several abutting its eastern edge, was struggling to remain unaligned, as was Spain, where there was appreciable sympathy for the fascist powers. Should another global conflict arise, India – one of several important sources of manpower, raw materials and manufactured items within the British Empire – would lie, not only in the path of any Japanese thrust westwards and any Axis drive eastwards, but also, like Australia, Canada, New Zealand and dependencies in the Caribbean, at the far end of immense sea lanes that, for much of their length, were more exposed to attack by surface vessels than aircraft or submarines.

There were difficulties looming closer to home, too. While the Locarno regime had prevailed, there had been no hostile bombers based within range of metropolitan Britain. With the crumbling of the Stresa Front and the proclamation of the Rome–Berlin 'axis', which was to be formalized by the Germano-Italian 'Pact of Steel' in 1939, this situation changed for the worse. Friendship with Italy was one of the keys to security in the Mediterranean, the cardinal route, via the Suez Canal, to and from Britain's possessions in the Orient. Future control over areas of the seas and skies immediately adjacent to the UK had also been slackened when, in 1938, London had agreed to relinquish several sovereign bases – Berehaven, Cobh and Lough Swilly – within Eire, a state that overlooked the Atlantic and Channel approaches and, like several other small countries, was to endeavour to stay out of the impending Second World War. As proved the case with the USA, however, Eire was soon obliged to acknowledge whom her neutrality ultimately benefitted: the Axis. In the event a blind eye was to be turned to infringements by Allied forces of her territorial skies and waters, not least hunts for U-boats. Dublin also habitually permitted RAF Coastal Command flying boats stationed on Lough Erne to take the most direct route to and from the Atlantic – the 'Donegal Corridor' – through Eire's air space.[65]

---

65 See: Robert Fisk, *In Time of War: Ireland, Ulster and the Price of Neutrality, 1939–1945* (London, 1996).

Whereas Eire could scarcely enforce her own claims of sovereignty, let alone challenge those of other states, the USA was a major player on the global stage. As underscored by her international treaty commitments, especially those regarding naval strength, she had far-flung interests and immense potential. Moreover, she evidently possessed the capacity to influence the course of affairs in distant parts of the globe, should she choose to do so, as she had demonstrated in the Philippines in 1898 and in Europe less than twenty years later. Her self-proclaimed neutrality was thus more problematic than that of Eire, Sweden or Switzerland, particularly for the other big maritime powers. After all, the USA had managed to stay out of the First World War until as late as April 1917, when the Germans had gambled that active American participation would prove too belated to save France from being overwhelmed and Britain blockaded into submission.

The USA's intervention, when it did finally come, resulted not so much from Germany's submarine warfare, which had claimed hundreds of American lives, as from Berlin's overtures to Mexico, a state that, in the 1800s, had lost thousands of square miles of territory to the neighbouring USA. Washington's relationship with London, although generally amicable enough, had also had its peaks and troughs, the worst of the latter being the so-called War of 1812, the ostensible cause of which was the principle of freedom of the seas but which really had far more to do with dominion over Canada. The 1920s and 1930s were also to witness intermittent disputes between Britain and the USA, especially over the limitation of naval armaments and over economic issues, notably the repayment of loans that the British had obtained on behalf of their allies in the Great War. In fact, like several other powers, during the mid-1930s the USA developed a contingency scheme for armed conflict with Britain. 'War Plan Red', as it was dubbed, envisaged a fresh attempt to conquer Canada, including amphibious landings on her eastern seaboard. Air support for this invasion was to be provided in part from new aerodromes that were to be built clandestinely along the USA's northern border, hitherto the longest unfortified frontier in the world.

By September 1939 squabbles and petty rivalries between Britain and the USA were being supplanted by a far more sober preoccupation with the Axis powers' behaviour. Although America was not to enter the Second World War formally until December 1941, her neutralism worked increasingly to Britain's benefit. An agreement of September 1940 – whereby the Royal Navy, desperately short of escort vessels, was to receive some old destroyers in return for the USA being granted leases on bases in Antigua, the Bahamas, Bermuda, British Guiana, Jamaica, Newfoundland, Trinidad and St. Lucia – was an implicit acknowledgement by Washington that, if only for the time being, the British armed forces formed America's first line of defence, certainly on her eastern flank.[66] Indeed in May 1941, shortly after President Franklin D. Roosevelt

66 See: David Gates, *Sky Wars: A History of Military Aerospace Power* (London and Chicago, IL, 2003), p. 61.

endorsed the Lend–Lease Act that was to grant, in lieu of loans and credit, mate-
rial aid to nations opposed to the Axis, the USA was to set up its Caribbean
Defense Command, much of which was dedicated to protecting the Panama
Canal and its approaches. A few weeks later Roosevelt and the British Prime
Minister, Winston Churchill, were to meet aboard warships moored in Placentia
Bay, Newfoundland, from where they issued not only an 'Atlantic Charter' but
also a warning to Japan of a joint response if she persisted in acting aggressively
in the Pacific.

The establishment of the USA's new bases and the expansion of existing ones
in Cuba, the Dominican Republic, Haiti and Puerto Rico were to prove timely,
for the Caribbean basin – Britain's principal source of oil and bauxite – was
destined to become a focal point for U-boat raids in 1942 especially. In the
North Atlantic, too, America's diplomacy on the eve of her formal entry into
the war was far more nationalistic than neutralist and did much to determine
the fighting's geometry. In July 1940 Roosevelt revived and recast the Monroe
Doctrine by proclaiming that the USA would take responsibility for the security of much of the western hemisphere: henceforth, any warships or aircraft
of belligerent states (other than those that enjoyed sovereignty over territory in
the region) that ventured into this area would do so at the risk of being seen as
having hostile intent. By mid-summer 1941, the 'hemisphere' – as defined by
the US Navy – had come to encompass Iceland. An autonomous state under the
Danish crown until it became an independent republic in 1944, this land mass,
commanding the sea and air corridors of the North Atlantic, was of immense
strategic importance. The hard-pressed British, who had quickly taken the precaution of occupying it after the Germans' conquest of Denmark in April 1940,
were more than content to accede to the indigenous government's request that
they withdraw in favour of an American garrison. The handover occurred in
July 1941, just five months before the USA was finally dragged into what was
fast degenerating into a global conflict. Iceland was destined to become a key
base for the Allies in the so-called Battle of the Atlantic, during the lengthy
course of which the Germans became increasingly dependent upon submarines
rather than on capital vessels. Not least because the *Kriegsmarine* lacked convenient, secure bases from which to support widespread sorties by the latter, aircraft
and even mines were to inflict rather more damage in the *guerre de course* than
the battleships and cruisers of the *Kriegsmarine*.

The perils for both sides inherent in surface raiders operating far from friendly
bases were to be highlighted very early in the Second World War by the fate of
the German pocket battleship *Graf Spee*. In December 1939, having sunk nine
merchant ships in the Indian and South Atlantic Oceans, she was confronted
off neutral Uruguay by one of the eight Allied flotillas that were searching
for her. Already plagued by engine troubles after such a protracted patrol, the
Graf Spee was damaged in the ensuing clash with the cruisers *Exeter*, *Ajax* and
Achilles and, seeking to make herself seaworthy for the lengthy journey back
to Germany, withdrew into Montevideo on the Rio de la Plata. Convinced he
was cornered by far superior forces, her commander subsequently scuttled her.

Just seventeen months later, the *Bismarck*, unable to implement crucial repairs or refuel at sea, was encircled and sunk by British ships, including aircraft carriers, while trying to intercept convoys in the North Atlantic. Likewise, fuel shortages and a dearth of sufficiently spacious dry docks were to have major ramifications for the operations of the *Tirpitz* and other big German vessels based along the Norwegian littoral.

Although aviation was to play an important part in locating and destroying both *Bismarck* and *Tirpitz*, *Graf Spee* did not have to contend with hostile air power other than a solitary *Seafox* spotter aircraft that was catapulted from HMS *Ajax*. (Both of the reconnaissance machines aboard HMS *Exeter*, the heaviest of the British cruisers, were damaged by shell splinters at the very outset of the Battle of the River Plate and could not be launched.) In any case, even at this juncture there were lingering doubts in many quarters about the ability of planes to inflict terminal damage on large warships, especially by horizontal bombing from high altitude. Whilst the 'lessons' drawn by one senior British admiral from this particular engagement were to include the conviction that: 'Naval forces which can arrange effective co-operation with aircraft will have a great advantage over those that do not', he identified the principal benefit stemming from such partnerships as aviation's ability to assist surface vessels in exchanging gunfire and in dodging torpedoes.[67] The field of vision afforded by a spy in the sky was palpably superior to that of battleships and cruisers, but the firepower of most planes – certainly of those machines small enough to be based at sea – was likely to prove more of an irritation than a real threat to such warships.

Indeed throughout the interwar period advocates of capital ships could still derive some comfort from the limitations of aircraft, their crews, navigational aids and armaments. As Eustace Tennyson d'Eyncourt, the Royal Navy's Director of Naval Construction, had observed in the aftermath of the *Ostfriesland* test, 'whenever new weapons are introduced, enthusiasts among their advocates invariably claim much greater power and results for . . . [them] than is actually shown by their use in war'.[68] Overawed by the selfsame experiment, Britain's Secretary of State for Air, Frederick Guest, had avowed in 1922 that, to consolidate the superiority of the bomber over the battleship, it was 'merely necessary to perfect the bomb sight, which is purely a matter of practice and experiment'.[69] Yet determining the location of what, amidst the immensity of the oceans, were relatively tiny objects and then disabling them from the air was in fact a sequential process that relied upon, among other things, exceptionally precise navigation. If any single link in the chain failed, then so would the system as a whole.

Chiming with popular notions of moral and ethical rectitude that had been honed by the indiscriminate aerial and submarine attacks seen in the Great War,

67 Document 74, 'Lessons from Battle of the River Plate': Letter from Commander-in-Chief, The Nore, to Secretary of the Admiralty, 9 September 1940, in *FAA in Second World War: Volume One*, pp. 226–7.

68 Minute by Director of Naval Construction, 24 March 1922, TNA, ADM 116/3477.

69 *Commons Debates*, 21 March 1922, vol. 152 cols 309–10.

the Americans' quest for targeting accuracy during this epoch was to climax in 1933 with the unveiling of a tachometric bomb sight, the Mark XV.[70] Designed by Carl Norden and perfected at immense expense, this employed an analogue computer to determine bomb trajectories in the light of prevailing conditions. Linked to the aircraft's autopilot, the system fine-tuned the plane's flight-path to allow for last-minute variations in the wind speed and, on paper, reduced the CEP to a few feet. In practice, however, such exactitude was seldom achieved. Peacetime testing suggested that, even from no higher than 12,000 feet, bombs would only hit their mark if released in excellent weather conditions and against a clearly designated, unprotected target that was sited in open terrain.[71]

Incorporated into American strategic bombers such as the B.17, this device was to form the very cornerstone of the doctrine of precision bombing that the USAAF sought to apply in the European theatre on joining the Second World War. Here, however, the realities of combat – not least atmospheric conditions that, even in broad daylight, could make the initial identification of targets very difficult – quickly made adjustments to the preferred tactic inescapable. 'Pattern' bombing was soon being resorted to, whereby all of the planes participating in an attack simultaneously released their payloads at a signal from the formation commander. Such expediency – which implicitly acknowledged that the planes' sophisticated targeting mechanism was all but inutile in poor visibility – jarred both with crews encouraged to believe in the superiority of their precepts and with public perceptions of the strategic bombing campaign. Anodyne titles, notably 'overcast bombing techniques', were duly affixed to the practice, which was accompanied by a growing reliance on a surface-scanning radar system derived from Britain's H2S. Generally not as adept as those of RAF Bomber Command at bombing by instrument, when reliant on such aids USAAF crews often failed to surpass the standards of exactitude than had been documented in the Butt Report of 1941 and that had compelled the British to turn increasingly to 'area' bombardment as the most dependable way of hitting something.[72]

The targets of the Allies' Combined Bomber Offensive were overwhelmingly fixed points, many of which appeared on readily available maps, and all of the navigational and bombing aids mentioned above were primarily designed with such objectives in mind. (The USAAF's first raid in Europe was against the French town of Rouen.) Finding and hitting moving targets, such as shipping, was that much more complicated and often called for preliminary intelligence reports – be they from visual reconnaissance units, radars, or from listening stations that monitored and intercepted the enemy's signal transmissions (and, where possible, deciphered them) – as well as precise navigation on the part of allotted strike platforms. An instance of this that occurred during the opening

---

70 See: Stephen McFarland, *America's Pursuit of Precision Bombing, 1910–1945* (Washington, DC, 1995), pp. 68–88.
71 Ibid., pp. 94–8.
72 See: Biddle, *Rhetoric and Reality*, pp. 228–9 and 243–5.

stages of the Battle of Midway in June 1942 was destined to be the very first test of the Mark XV Bomb Sight under combat conditions. Vectored towards the Japanese invasion flotilla – which had been stumbled across by *Catalina* maritime patrol aircraft beforehand – nine B.17s finally located the hostile vessels after a flight of more than four hours from Midway Island. The situation seemed very propitious: the Japanese lookouts were slow to notice the approaching bombers, which, unmolested, lined up and commenced their attack runs at altitudes of between just 8,000 and 12,000 feet. Indeed ordnance was already falling around the troop transporters before they realized anything was amiss, started evasive manoeuvres and began firing their AA weaponry. The bombardment lasted ten minutes without a single hit on a ship being scored.[73] During the Battle of Midway's closing phase, the destroyer *Tanikaza* was also to be engaged by five B.17s that bombarded her from 11,000 feet. Although no bombs found their mark, two of the attacking planes failed to regain Midway for lack of fuel.[74]

The following year was to see the first use in anger of a guided bomb that Germany had been developing specifically for employment against warships since 1938. Dropped from a suitably large aircraft, the so-called *Fritz-X* had radio-controlled spoilers and fins that enabled it to be steered towards its objective. A flare, mounted astern, helped the bombardier keep track of the projectile, which, from altitudes of around 18,000 feet, could travel three miles or more. Fitted with a powerful 3,100 lb warhead that was designed to penetrate deep into its quarry before exploding, the weapon was first used by the vengeful Germans in September 1943 in a bid to decimate Italy's fleet following her capitulation. The results were impressive: the battleship *Roma* was sunk and her sister ship *Italia* slightly damaged by a mere handful of bombs.[75] Thereafter, Allied ships elsewhere in the Mediterranean were struck in the same fashion, the light cruisers USS *Savannah* and HMS *Uganda* and the battleship HMS *Warspite* all sustaining appreciable damage.[76]

A lighter but faster relation of the *Fritz-X*, the Henschel HS-293 was a jet-propelled guided bomb intended for use against unarmoured vessels up to ten miles away from the plane that released and controlled it. Together, these armaments posed a novel and significant threat to Allied shipping. Such devices had their inherent weaknesses, too, however. They were not 'fire-and-forget' projectiles; operators had to retain visual contact with both them and the target, which dictated that, after unleashing the weapon, the bomber had to maintain an appropriate heading and distance until the bomb reached its mark. This rendered the plane that much more vulnerable to AA fire or interception by fighters. Radio-controlled, these armaments were, furthermore, ineluctably susceptible to

73 Jonathan Parshall and Anthony Tully, *Shattered Sword: The Untold Story of the Battle of Midway* (Dulles, VA, 2007), p. 106.
74 Ibid., pp. 364–6.
75 Greene and Massignani, *The Naval War in the Mediterranean*, pp. 304–5.
76 S.W. Roskill, *The War at Sea 1939–1945, Volume Three: The Offensive, Part One, June 1943–May 1944* (London, 1960), pp. 177–9.

electronic countermeasures, which the Allies were quick to devise in the form of ship-based jamming instruments, notably Britain's Type 650 and the American XCJ-1. The Americans also went on to produce a steerable bomb of their own, the *Azon*, which was initially used against bridges – lengthy, narrow targets, much like ships – during the Burma campaign in 1944. Airframes packed with explosives and controlled remotely via closed-circuit television links were also experimented with by the USA that same year, whereas the Japanese, increasingly lagging in the technology stakes, were ultimately to turn to the *Kamikaze* as a means of solving the interconnected problems of accurate navigation and bombing. Among the equipment employed by these suicidal pilots was a small, rocket-propelled aircraft, the so-called '*Baka*' – 'Idiot' – Bomb.

The British carriers *Illustrious*, *Indefatigable* and *Formidable* were among the numerous Allied ships destined to be assailed by *Kamikaze* planes during the war in the Pacific's last months. Their heavily armoured flight-decks proved their salvation, whereas many other vessels were to be sunk or very seriously damaged by opponents who presented AA defences with some novel psychological and technical challenges. In the European theatre, by contrast, largely through the Allies' electronic countermeasures and broader air superiority, the danger posed by Germany's radio-controlled projectiles was being contained if not eliminated by the time of D-Day in June 1944. The following November, RAF *Lancaster* aircraft, using immense 'Tallboy' bombs, struck at the German battleship *Tirpitz* that was languishing in Tromso Fjord, Norway. After sustaining two direct hits, she capsized.[77] She had long since been immobilized by earlier blows and, like the battleships berthed in Pearl Harbor on 7 December 1941, was stationary when attacked.

Here, several of the American battlewagons, most notably the *Arizona*, which blew up when a 1,750 lb bomb penetrated a magazine, were targeted by Japanese aircraft that disgorged their payloads at just 10,000 feet, the minimum height from which armour-piercing projectiles would gather sufficient momentum to slice through deck plating. So discouraging had its experiments in high-level horizontal bombing proved that the IJN had come very close to abandoning the practice altogether by the end of the 1930s, but, faced with some of the complexities inherent in attacking 'Battleship Row', had hastily implemented a programme of tactical and technological innovation. The crews who were to carry out this particular mission were exceptionally well prepared for it and comprised a fusion of outstanding bombardiers and pilots. Nevertheless, they knew from their intensive rehearsals that, even when attacking static targets, fewer than half of their bombs – which were specially fashioned from naval artillery shells and dropped from planes flying in tight patterns of three – would be likely to hit home.[78]

77 Interpretation Reports K.3361 and SA.2923 of attack on *Tirpitz* at Tromso, 12 November 1944, TNA, AIR 20/1309.

78 Peattie, *Sunburst*, pp. 138–9; Alan D. Zimm, *Attack on Pearl Harbor: Strategy, Combat, Myths, Deceptions* (Havertown, PA, 2011), pp. 92–4.

Attempts during the Second World War to bombard moving warships from planes flying horizontally at high altitudes were in the main to prove unsuccessful. No major surface vessel that was under way was ever sunk exclusively by such means. When, on 10 December 1941, the battleship HMS *Prince of Wales* and the battlecruiser HMS *Repulse* were attacked by dozens of torpedo and high-altitude bombers off Malaya, it was the Japanese torpedoes rather than the bombs that were to prove deadly. Whereas both ships were to be struck by just one bomb apiece, *Repulse* was hit by four or five torpedoes and *Prince of Wales* by four.[79] The bombs were in any case not the armour-piercing variety used at Pearl Harbor, which were as rare as they were potentially lethal; so the objective allotted to the bombers in this mission was that of ravaging the British ships' upper works and keeping their AA gun crews' attention away from the approaching torpedo planes.[80] The Japanese might well have overwhelmed any defending fighters, too, but in the event there were none. Owing to a sequence of miscalculations and misunderstandings, *Repulse* and *Prince of Wales* were not provided with what little air cover was available from aerodromes in the region.

The bombardment from on high of these ships was carried out by land-based Mitsubishi G4M medium bombers, part of a new generation of aircraft that were far more sophisticated than planes that had first appeared only a few years earlier. The IJN was unique in its development of a shore-based, long-range maritime strike aircraft, the twin-engined Mitsubishi G3M1. Championed by Rear Admiral Yamamoto Isoroku in his capacity as head of the Technology Bureau of Naval Aviation, the G3M1 entered service in 1936 and was intended to pare down any large fleet that threatened the Japanese home islands prior to it being confronted in a decisive sea battle.[81] Its successor, the G4M, was even more capable, as the British were to discover to their cost. Although aware of the G3M's existence, they mistakenly believed it to be just another large bomber or a coastal reconnaissance plane with a range of around seven hundred miles.[82] The G4M had an exceptionally long reach, but its range came largely at the cost of robustness. To safely penetrate defended air space, it had to either be escorted by fighters or bomb from high altitude, or both. Yet no Japanese fighter could match the maximum range of the G4M, while the precision of bombing varied inversely with height. Still, the G3M and G4M did confer some scope for 'strategic' strikes and the Japanese were to execute raids of this genre, firstly in the ongoing conflict in China and, later, against some of the Europeans' and the Americans' enclaves in the Pacific. They were even to bomb targets in distant Australia, notably on 19 February 1942, when 54 of these planes flew via

79 Report by First Sea Lord, 25 January 1942, TNA, PREM 3/163/2.
80 See: Peattie, *Sunburst*, pp. 168–70.
81 Peattie, *Sunburst*, pp. 80–81, 86–7.
82 Thomas C. Hone, Norman Friedman and Mark D. Mandeles, *American and British Aircraft Carrier Development 1919–1941* (Annapolis, MD, 1999), p. 110.

captured aerodromes in the Netherlands East Indies to join with carrier aircraft in ravaging Darwin, a key port.[83]

However, Japan's peculiar geostrategic circumstances dictated that her procurement priorities be centred on tactical air power, particularly of a kind that could function closely with her navy, among the vessels of which were counted the two mightiest battleships ever seen, *Yamato* and *Musashi*. By the middle of the 1930s her sailors were divided over the relative superiority of the aircraft and such capital ships. Some officers, most notably Onishi Takijiro and Genda Minoru,[84] held similar views to those of Mitchell on the tactical relevance and cost-effectiveness of what America's most vocal advocate of air power had dismissed as 'museum pieces' as early as 1919.[85] Yet, despite all the promises and warnings emanating from the various air power theorists, on the eve of the Second World War both the USA and Britain maintained fleets that struck a balance between aerial and surface units.

The IJN, too, proved reluctant to put all of its eggs into a basket of untried technology and doctrine. Indeed its basic strategy for neutralizing the USA's numerically stronger Pacific fleet was to call for a blend of aerial and surface forces. It was envisaged that the American armada would be worn down by attacks with G4M and other land-based planes as it tried to penetrate a defensive perimeter of outposts, notably the Mariana, Caroline and Marshall Islands. Cut down to more manageable proportions, it would then be confronted and decisively defeated by the Japanese fleet. A combination of the immensity of the seas and the reach of her shore-based aviation would, in the meantime, keep Japan's home islands safe. They were too remote to face grand, Douhetian counteroffensives by land-based bombers and, it was reasoned, modest AA defences would adequately secure them from lesser threats. The very puniness of the Doolittle Raid of April 1942 seemed to vindicate this calculation.[86]

The conundrum of targets and tactical techniques

The principal preoccupation of much of the RAI, of RAF Bomber Command and of the bombardment groups of the USAAF was the bombing of fixed targets by sizeable planes flying level at high altitudes. Far more common than horizontal attacks were two other varieties for which other forms of aircraft

---

83 Ministry of Defence (Navy), *War With Japan: Volume II: Defensive Phase* (London, 1995), pp. 80–81.

84 See: Peattie, *Sunburst*, pp. 82–5.

85 O'Connell, *Sacred Vessels*, p. 254.

86 Led by the celebrated aviator James Doolittle, this raid was executed by 16 USAAF B-25 *Mitchell* bombers. These had been squeezed onto the flight-deck of the carrier USS *Hornet* and brought within striking distance of Japan. Stripped of all unessential equipment, including their lower gun-turrets, to minimize their weight, these medium bombers, whose volunteer crews had been specially trained for the feat, managed to get aloft with the shortest of take-off runs. They bombed Tokyo and other cities before heading for China, where, for lack of fuel, their crews either crash-landed their machines or bailed out. See: Carroll V. Glines, *The Doolittle Raid* (New York, 1988).

were preferable if not essential, namely dive- and skip-bombing. Of these, the former – pioneered by the US Navy in the second quinquennium of the 1920s – was the more precise. The attacking plane would descend towards the target at an angle of between 60° and 90° from around 10,000 feet. The steeper the dive, the more accurate the aim tended to be, not least because, once following a vertical flight path, the bomber's pilot could concentrate more on keeping the quarry in his cross hairs than on the minutiae of maintaining an appropriate horizontal heading. However, besides operating the weapon systems, he did have to be careful not to exceed a safe speed and to leave sufficient sky and time to pull up after releasing the ordnance. This normally occurred at about 1,500 feet.

Machines employed in this technique had to be of a suitably robust construction to withstand the stresses and strains of such extreme manoeuvres. In skip-bombing, the angle of approach was much shallower and the payload was released as the aircraft pulled away, the object being to lob bombs from low altitude directly onto the target or to send them skimming across the surface of the water into the vessel's sides. Dive-bombing by accomplished aviators was typically four times more accurate than horizontal attacks, with a CEP of less than a hundred feet. Skip-bombing, like bombing from planes flying horizontally at high altitude, depended that much more upon the attacker's ability to gauge the exact point in time and space at which to release the projectile so as to avoid it overshooting or falling short. This was especially difficult where moving vessels were concerned, although, on the other hand, a ship targeted in this fashion would be left with insufficient time to actively dodge bombs that were on a collision course. Skip-bombing could also be resorted to where aerial torpedoes were unavailable or could not be relied upon, such as against craft that had a shallow draught or lay in water of insufficient depth. Indeed its attendant complexities notwithstanding, there were instances of this technique being used to great effect, notably by USAAF aircraft against Japanese shipping during the Battle of the Bismarck Sea in March 1943.[87]

Although, particularly in the Pacific theatre, many ships were to be lost to dive-bombers in the course of the Second World War, the destruction by the Japanese of the heavy cruisers HMS *Dorsetshire* and *Cornwall* in the Indian Ocean in April 1942 forms one of the most eye-catching instances of such an operation. Here, 90 per cent of the ordnance was to find its mark, with the *Dorsetshire* being put out of action in barely four minutes and sinking in seven; *Cornwall*, listing badly and robbed of all power and communications within moments of the attack starting, followed her to the bottom just five minutes later.[88] In European waters, the raid by British dive-bombers on the German light cruiser *Königsberg* during the Norwegian campaign was similarly successful if less spectacular. Operating at extreme range from an aerodrome on the Orkneys, 16 Blackburn

87 Peattie, *Sunburst*, p. 334.
88 'HMS *Cornwall* and HMS *Dorsetshire* Bomb Damage, 4 April 1942', Report (DNC 4B/R.158) by Director of Naval Construction, 13 November 1942, TNA, ADM 267/84.

Skua aircraft crossed the North Sea in poor weather and fell on this vessel as she lay anchored in Bergen harbour. Approaching at an angle of 60° with the sun behind them, the planes caught her crew and the surrounding AA batteries by surprise and, encountering little defensive fire, unleashed their payloads at just 2,000 feet. Three of these 500 lb armour-piercing bombs struck the ship, one causing a secondary explosion within her. A fourth detonated on the adjacent jetty and at least one other in the water directly alongside the cruiser, which subsequently sank. Two of the *Skuas* were slightly damaged by shell splinters and a third crashed on the return journey.[89]

All over Europe, the spearhead of Germany's *Blitzkrieg* was normally to consist of a dive-bomber, the Junkers Ju.87. As ungainly as it was ugly, this plane was easy prey for fighters, but, if it could penetrate sufficiently close, posed a grave threat to a wide variety of targets. The 'Stuka' and its Allied counterparts were not only capable of bombing with exceptional exactitude but, by virtue of their angle of attack and speed of descent, were also often able to thwart any AA weaponry that sought to ward them off. Besides furthering the Germans' operations by land, the Ju.87 also offered some scope for anti-shipping missions in littoral waters. Exploiting captured airstrips, large numbers of these planes were to harass the Allies as they strove to evacuate central Norway at the end of April and the beginning of May, 1940; the French destroyer *Bison* and the British destroyer HMS *Afridi* were both sunk during this undertaking by *Sturzkampf-flugzeuge*.[90] Although, a few weeks later, adverse weather (including darkness), palls of smoke, logistical shortcomings and numerous sorties by RAF fighters were to help constrain efforts by the *Luftwaffe* to impede the embarkation of the BEF and thousands of French troops from Dunkirk, evacuating the remnants of Crete's garrison from Sphakia and Heraklion in May 1941 was to prove a very different matter. Here, virtually bereft of either carrier- or shore-based air cover, the Royal Navy was to sustain serious losses as a result of aerial bombardment, much of it from *Stukas*; three cruisers and six destroyers were sunk and seventeen other vessels were damaged.[91] Likewise, there were to be instances of Ju.87s based in Sicily inflicting considerable losses on Allied shipping plying the waters around Malta especially.

For much of the Second World War, large swathes of the Black Sea and virtually all of the Baltic were to be overshadowed by German land-based aircraft. Acting either independently or in conjunction with sorties by U-boats and surface raiders, such as the *Tirpitz*, planes operating from Norway were also to harass British and American convoys seeking to replenish the USSR through Murmansk and Archangel (Arkhangelsk) on the Arctic. Nazi Germany, too, had important lines of communication in these regions, not least with neutral

89 Documents 38 and 38a 'Sinking of *Königsberg* at Bergen, 10 April 1940', in *FAA in Second World War: Volume One*, pp. 93–5.

90 Geoffrey Till, *Air Power and the Royal Navy, 1914–1945: A Historical Survey* (London, 1979), pp. 22–3.

91 See: 'Defence of Crete, 1941', TNA, CAB 121/537.

states. Iron ore, for instance, was shipped in from Sweden, while chrome ore was brought across the Black Sea from Turkey to Romania and Bulgaria, both of which had fallen increasingly under Berlin's sway as the Versailles settlement faded and fresh hostilities loomed. Bulgaria, rather reluctantly, was to side with the Germans in 1941, while the Romanians, impoverished and politically fragmented, were likewise to join the Axis powers, their forces participating in Hitler's invasion of the USSR that same year.

Rankled by the 'Winter War' of 1939–40 and the ensuing Peace of Moscow, the Finns, too, were to ally themselves with the Axis. Finnish and *Luftwaffe* aircraft based in northern Finland were able to molest Murmansk and Archangel and the sea lanes between them. Their partnership with the Finns was also to enable the Germans to bottle up the Soviet fleet in the Baltic, sealing off the Gulf of Finland with, not just patrols by aircraft, warships and U-boats, but also with an enormous barrier of minefields and anti-submarine nets.

By this juncture, in addition to the contact mine that had first been used in the Baltic in the 1850s, so-called 'influence' varieties had been introduced. Mines that detected the magnetic field of steel hulls had started to appear in the First World War and the Germans especially had worked on refining such weapons. Now, these devices were supplemented with types fitted with acoustic or pressure sensors that detected passing vessels. Between 1939 and 1945, many nautical mines were to be sown by parachute from suitably capacious bombers, although the viability of this practice was, again, largely determined by the exactitude with which aircraft could drop projectiles at specific points on the Earth's surface. German endeavours to seal off ports in the UK with magnetic mines commenced with the outbreak of hostilities and caused significant problems until one of the devices was retrieved, intact, from some mudflats off Shoeburyness in the Thames estuary. This find was to enable the British to devise countermeasures, including degaussing, whereby a ship's magnetic field was neutralized with gigantic copper coils carrying an electric current.[92] In time, ways of disarming acoustic and pressure fuses were also to be contrived.

Nevertheless, mine warfare was to become another sphere in which one innovation followed another amidst a welter of measures, countermeasures and adaptive tactics. When in February 1942 the German heavy cruiser *Prinz Eugen* and the battlecruisers *Scharnhorst* and *Gneisenau* dashed up the Channel from Brest to new berths in Norway, both of the battlecruisers were damaged by mines that British planes had dropped in areas that the Germans had just cleared.[93] From 1944 onwards, radar was to enable Allied bombers to sow mines from altitudes as high as 15,000 feet. The cardinal victims of this tactic were to be the Japanese, who had negligible sweeping capabilities. Mines dropped from

---

92 S.W. Roskill, *The War at Sea 1939–1945, Volume One: The Defensive* (London, 1954), pp. 78, 99–102.
93 'Escape of German battlecruisers *Scharnhorst* and *Gneisenau* and heavy cruiser *Prinz Eugen* up the Channel: Operation "Fuller" and the Board of Enquiry, 1942', TNA, ADM 116/4528, p. 12, paragraph 75; Barnett, *Engage the Enemy*, p. 453.

B.29s consolidated the blockade of Japan that was to be begun by Allied subma-rines and aircraft, all but eradicating her merchant shipping and constraining the movements of her battle-fleets. This, in turn, was to complete the isolation of the island bastions that were supposed to shield the Japanese homeland. Deprived of infusions of such essentials as fuel, munitions, personnel and spares, these outposts became increasingly unable to protect themselves, let alone dominate swathes of the adjacent oceans and skies.

In the Black and Baltic Seas, the Soviet Navy – which, until 1943, lacked both influence mines and the knowledge and means to deal with them – was also to be contained by a combination of minefields and land-based planes for much of the war. When mine-laying aircraft first attacked Sevastopol at the outset of Hitler's invasion of the USSR, the base's intelligence officer, Colonel Hamgaladze, seeing parachutes but hearing no explosions, was convinced that the Germans were seeking to seize the anchorage with airborne troops.[94] Together with this formidable fortress – which, between 1941 and 1944, was to be besieged, cap-tured and retaken – the Black Sea's western shore was destined to become the focus of Soviet maritime operations in this region. Here, an amphibious land-ing was to be mounted in a bid to relieve the beleaguered naval base at Odessa. When that failed, the garrison and a mass of equipment were evacuated in an operation that one authority has described as a 'small Dunkirk'.[95] Intermittently, Soviet submarines, aircraft and surface vessels were also to molest shipping in the navigable channels linking Turkey, Romania, Bulgaria and occupied sectors of the Ukrainian seaboard. As well as cargoes of chrome ore, troops and munitions, oil shipments were often targeted, Romanian petroleum production having by this juncture become indispensable to Germany's war-effort. Indeed the refineries at Ploesti were to end up among the most tempting and challenging targets for Allied strategic bombers. Launched by the Ninth USAAF from North Africa, one raid in August 1943 comprised 178 *Liberators*, of which 54 were lost, 41 of them to enemy action.[96] Meanwhile, mines sown in the Black Sea's extensive shallows were to prove a daunting obstacle to the Soviet Navy, which, on paper, was superior to the opposing fleet. Their land-based aviation, however, tipped the balance in the Germans' favour for much of the conflict; after the destroyers *Kharkov*, *Sposobny* and *Besposhchadny* were all sunk by *Stukas* in a single engage-ment off the Crimea in October 1943, all the remaining Soviet warships of any size were to be withdrawn out of harm's way. Likewise, harried by shore-based planes and hemmed in by vast minefields and anti-submarine nets, the Soviet flotilla in the Baltic was largely to be confined to the sea's most easterly reaches.

Besides bombs, ships might also face torpedoes. Intended to strike on or below the waterline, these had frequently proved lethal in the First World War,

---

94 John Erickson, *The Road to Stalingrad* (London, 1985), p. 161.
95 Ibid., pp. 293–4.
96 See: James Dugan and C. Carroll Stewart, *Ploesti: The Great Ground-Air Battle of 1 August 1943* (New York, NY, 1998), pp. 222, 233, 244.

particularly if fired from a lurking submarine without any warning. (Some subsequent varieties did not even leave a telltale wake on the surface.) Barely had the Second World War begun than HMS *Courageous*, one of Britain's few aircraft carriers, was to fall victim to this insidious threat while patrolling to the south-west of Ireland.[97] Just four weeks later, on the night of 14 October 1939, HMS *Royal Oak* was also lost when a U-boat crept into the Home Fleet's base at Scapa Flow and fired two salvoes of '*Aale*' towards the battleship as she lay at her moorings. She capsized with the loss of 833 of her crew.[98] It is also noteworthy that, to supplement their aerial raid in December 1941, the Japanese were to seek to penetrate Pearl Harbor with five midget submarines, at least one of which got into the anchorage and unleashed its two torpedoes.[99] The IJN was also to infiltrate Sydney's harbour with three miniature submarines when it struck at that port on 31 May 1942.

By this stage in their evolution, torpedoes typically mounted a warhead of some six hundred to nine hundred pounds and, for the sake of accuracy, were best launched no further than around a thousand yards from the target, although many had an appreciably longer reach. Whereas the Royal Navy continued to rely on a design that dated from 1928, during the 1930s the Japanese especially had forged ahead in this field, developing torpedoes that were virtually trackless and exceptionally fast, powerful and dependable. Again, however, the extent to which ordnance might affect a given target depended upon the peculiar characteristics of both. Whilst quite small weapons could cripple if not sink many merchantmen, battleships and many cruisers had bulging belts of steel straddling their waterlines so as to mitigate any damage sustained here. Yet even the most heavily armoured warships had their weak spots. Torpedoes fitted with magnetic-influence fuses might detonate underneath a vessel's keel, breaking its back. Rudders and propellers – which were, perforce, exposed – formed another Achilles heel. Indeed in May 1941 the mighty German battleship *Bismarck* was to be rendered very much more vulnerable when, while trying to dodge torpedoes dropped by *Swordfish* from the carrier HMS *Ark Royal*, she sustained a hit on her steering gear.[100] Her rudder jammed, she was condemned to sail in a circle and was subsequently overtaken and sunk by British battleships and cruisers – the only real match, some sailors remained persuaded, for an adversary of this calibre. Could air power alone have secured such an outcome? Could it be made capable of doing so?

Whereas the tests that climaxed in the sinking of the *Ostfriesland* focussed on the use of bombs and depth-charges, during the 1920s the Royal Navy carried out trials using torpedo-bombers. Mock attacks were staged against the

97 See: 'Loss of HMS *Courageous*: Board of Enquiry, 1939', TNA, ADM 156/195.

98 Roskill, *The War at Sea 1939–1945, Volume One: The Defensive*, pp. 73–4.

99 See: Burl Burlingame, *Advance Force Pearl Harbor* (Annapolis, MD, 2002).

100 Report from Commanding officer, HMS *Ark Royal*, to Secretary of the Admiralty, 6 June 1941, TNA, ADM 199/1187.

battleships *Malaya* and *Barham* and the new battlecruiser *Hood*. Superficially, these experiments demonstrated the great potential of such aircraft, with five hits claimed on the *Hood* out of six torpedoes fired. Nevertheless, the report of the attack by five aircraft on *Barham* emphasized that the only countermeasure available to the battleship in such a rehearsal was jinking and that, had there been a destroyer screen and a genuine AA barrage, the undertaking would have been very much more formidable.[101] By the end of the 1920s the Royal Navy was persuaded that the aerial torpedo was the most promising means of attack available, since its accuracy far surpassed that of horizontal bombing. On the other hand, it was recognized that, to achieve success, pilots would have to run the gauntlet of defensive fire and get in close to their quarry.[102] Air Vice Marshal Lambe, the Air Officer commanding the Coastal Area, warned that any such endeavour would have to 'warrant very severe losses'. Only two thirds of an attacking squadron would, he estimated, get so far as to launch their torpedoes and only one third of these weapons would actually hit home – an overall strike rate of 20 per cent. Still, with better armaments and more extensive training for the personnel, he anticipated that this performance could be improved upon. The statistics from horizontal bombing trials were, he reiterated, even more discouraging: if the weather conditions were good, it was estimated that there was no more than a 14 per cent chance of hitting a battleship from 10,000 feet with a 1,500 lb armour-piercing bomb.[103]

In spite or because of the outcome of its various trials, on the eve of the Second World War the British Admiralty was to remain persuaded that, of the available projectiles and bombing techniques, attacks by aircraft armed with torpedoes had the best prospects of success against surface vessels at sea. Even allowing for the likelihood of being engaged by enemy fighters and AA guns, the probability was that such planes would score more hits and that these would inflict relatively more critical damage. Furthermore, providing that the configuration of the coast and seabed allowed for the use of such armaments at all, the torpedo was also 'the best weapon for direct air attack on ships in harbour'.[104]

Besides the bombing tests involving the *Ostfriesland* and other vessels, in 1924 the US Navy also conducted dummy torpedo attacks against the battleship *Arkansas*, whereby half of the eighteen projectiles fired struck the ship, which was steaming at 15 knots. Again, conclusions drawn from feigned engagements were seen to be of doubtful value, the outcome of these particular trials being virtually ignored.[105] Such reactions were not necessarily a manifestation

101 John Buckley, *The RAF and Trade Defence, 1919–1945: Constant Endeavour* (Keele, 1995), p. 44.
102 Minute by the Director of the Naval Air Division, 1 June 1928, TNA, AIR 2/1101.
103 Letter from Air Officer Commanding Coastal Area to Director of Operations and Intelligence, 11 March 1929, TNA, AIR 2/1101.
104 Letter from the Admiralty Secretary to the Secretaries of the Cabinet Office and the Committee of Imperial Defence, 26 July 1939, TNA, AIR 2/1910.
105 O'Connell, *Sacred Vessels*, p. 278.

of recklessness or blind conservatism. No peacetime training could ever fully replicate the physical and psychological ordeal of real combat and, particularly in an era when technical innovations emerged every couple of years, history was an increasingly uncertain guide to the future. As one MP who had served in the Royal Navy was to remark in a debate about imperial defence held in 1934:

> [W]e in our lifetime have seen the most revolutionary changes in the technique of war which have ever taken place. The changes are so revolutionary and are coming so quickly that it is impossible to foresee what particular effect a new weapon, or the application of a weapon, may have on the conduct of a war. . . . All your plans . . . may be wrecked by some small event which could not possibly have been foreseen.[106]

For much of the interwar era, the IJN, like the Royal Navy, also regarded the torpedo as the most promising, cost-effective weapon for assailing ships. In the 1930s it was to invest heavily in developing new varieties, including aerial torpedoes, even though there was no scope whatsoever for using such armaments in the Sino-Japanese War, a conflict in which Japan's sea- and land-based aviators acquired useful experience in other types of bombing missions. Whilst bigger, more sophisticated weaponry called for commensurately more powerful planes to carry them, there were as many tactical and doctrinal obstacles to be surmounted as technical ones if aerial torpedoes were ever to fulfil their potential. It was feared that modern battleships especially could generate so much AA fire that engaging them would prove a costly if not futile affair; many planes would be lost and far fewer hits scored than mere exercises suggested. Executing efficacious raids under combat conditions would demand not only sacrifices but also meticulous choreography: aircraft would have to approach on at least two opposing axes so as to counter any evasive manoeuvres by the ship; the planes would have to squeeze into a comparatively small area of sky while avoiding collisions and flak; and torpedoes would have to be dropped from unprecedented altitudes and distances from their targets if the aircraft were to have any chance of evading destruction.

All of this would call for further technical innovations, not least torpedo designs with suitably robust casings and greater closing speeds. Variants that could be safely launched into shallow water would also be essential should it ever be necessary to assail vessels in key anchorages – among them, Hong Kong, Manila, Pearl Harbor, Singapore and Vladivostock – with aerial torpedoes. (The tail fins of those that were to be used in the raid on Pearl Harbor were fitted with a wooden assembly that broke the weapon's plunge towards the seabed, enabling it to level off quickly.) By 1939, the crews of Japanese naval bombers were practising unleashing torpedoes at altitudes of up to nearly seven hundred

---

106 *Commons Debates*, 21 March 1934, vol. 287, col. 1,288.

feet. Attacks *en masse* and in conjunction with land-based planes were also being rehearsed, as were nocturnal raids, guided by pathfinder units, and the employment of torpedoes in shallow waters.[107] High-level, horizontal bombardment was, by contrast, to be resorted to ever less after 1941, the IJN finding dive- and torpedo-bombing far more efficacious.

Competing commands and concerns

Military capabilities take years and decades to develop, but political intentions can change within just hours and days. In building the air force that she was barred from possessing by the Versailles Treaty that had ended the First World War, Germany, like Japan, was to emphasize the acquisition of tactical rather than strategic bombers. These were seen as having more utility in the new concept of *Blitzkrieg* – which closely integrated aerial and ground forces – and, plainly, were more cheaply and readily produced than bigger machines. These were important considerations, both economically and politically: built up as clandestinely as it was swiftly, the *Luftwaffe* was, once revealed to the world, intended to further overawe the guarantors of the Versailles accord, notably the French, whose large land forces were the principal restraint on German recidivism. The emphasis in Nazi Germany was to be on designing and producing a handful of sound, adaptable airframes, from which new variants might be devised comparatively quickly and cheaply, a prime example being the Junkers Ju.88, which, recast, was to fulfil an assortment of roles, including acting as a torpedo-bomber.

Overall, however, the Nazis were to neglect naval aviation as such. Although Hitler's *Kriegsmarine* – which was as much a violation of the Versailles settlement as the *Luftwaffe* – was configured for a *guerre de course*, few resources could be spared for the development of specialist air power that might support Germany's U-boats and surface raiders. As in the Great War, the priority was to try to meet the needs of, above all, the army. *Reichsmarschall* Hermann Göring, the head of the *Luftwaffe*, was to reject the calls of *Großadmiral* Erich Raeder, the head of the *Kriegsmarine*, for a discrete naval air arm, merely promising, in January 1939, that planes would be allocated for maritime reconnaissance purposes or in the event of a clash between battle-fleets. The numbers and types of aircraft involved, together with the training of their crews, ineluctably remained the responsibility of the *Luftwaffe*.[108]

As in Italy, this arrangement left the navy at the mercy of the air force so far as the development and introduction of aircraft were concerned. Rather than design bespoke planes, the Germans adapted commercial machines for use in maritime settings or relied on *Luftwaffe* units that were essentially configured to support operations by land. Both of the suitably large aircraft that they were to employ in, for instance, long-range maritime reconnaissance during

---

107 See Peattie, *Sunburst*, pp. 143–7.
108 Karl Doenitz, *Memoirs: Ten Years and Twenty Days* (London, 2000), pp. 132–3.

the Second World War – the Focke-Wulf Fw.200 *Kondor* and its successor, the Junkers Ju.290 – were fashioned from ageing airliner designs.[109] Whilst these – together with some medium bombers that were intermittently allocated to such missions – provided some scope for mounting torpedo attacks, for an air force that had to focus on supporting troops, tailor-made torpedo-bombers and their concomitant weapons were just too *sui generis* in nature. By the same token, Germany was not to have any functioning aircraft carriers at her disposal and did not establish a discrete fleet air arm. A carrier, the *Graf Zeppelin*, was launched in 1938 but, owing to the very lack of experience within the infant *Kriegsmarine* regarding aeronautics, she never became operational.[110]

Unlike Britain and Italy in the second half of the 1930s, Japan hardly needed fleets of large bombers with which to deter strategic aerial attacks and, should that fail, to repay the aggressors in their own coin. The establishment of an autonomous air force was resisted by her admirals and generals alike, each of the two existing services having its own peculiar and competing tactical concerns. The navy feared that an independent air force would be dominated by soldiers and would – as had occurred in Britain following the creation of the RAF – divert essential resources from the maritime environment, diluting tactical capabilities and complicating command arrangements and the formulation and execution of strategy. (Although, in 1937, the Royal Navy was to secure control over all ship-based planes, Coastal Command remained, somewhat incongruously, within the RAF's purview.) Above all, the IJN did not want to end up in a similar dilemma to that faced by its Italian counterpart, which had no organic aviation and was thus ineluctably subject to the whims of the RAI's leadership.

Indeed disputes – largely centred on control of what became the Fleet Air Arm (FAA) – were to dominate the relations between the Royal Navy and the RAF for much of the interwar era with the result that very little cooperation on the development of aircraft types or operational tactics was to occur. In 1922, Trenchard, head of the latter service, predictably stressed the indivisibility of air power: 'The science of the employment of aircraft is in the broad principles the same over the sea as over land', he insisted. Past experience had 'proved that those whose sole duty is to study aerial problems developed such employment faster and to a higher pitch than it could be developed by any individual service (such as the Navy in the war) which had many other grave problems to consider'.[111] The following year, Admiral Sir Charles Madden was to exclaim: 'The trend of policy is now to develop a First Class Air Force at the expense of the Navy. This may lead to London and our other big towns being safe from air attack, while the Second Class Navy is overpowered until the Empire is starved into submission'.[112]

109 See: Kenneth Poolman, *Focke-Wulf Condor: Scourge of the Atlantic* (London, 1978).
110 Roger Chesneau, *Aircraft Carriers of the World, 1914 to the Present: An Illustrated Encyclopedia* (London, 1992), pp. 76–7.
111 Quoted by Geoffrey Till, *Air Power and the Royal Navy*, p. 31.
112 'Remarks', Admiral Sir Charles Madden, 1923, TNA, ADM 116/3432.

Following the absorption of the RNAS into the autonomous RAF towards the end of the First World War, there were few within the Admiralty who had much knowledge or experience of aeronautics. A Naval Air Section was set up in July 1920, but, headed by a mere captain, it lacked its antecedent's influence. Unlike in the USA – where the Bureau of Aeronautics (BOA) was established in 1921 under Rear Admiral William Moffett – in Britain there was no unified authority to coordinate the assortment of agencies that concerned themselves with air power's various facets, not least the production of planes. Through the BOA, the US Navy was to maintain close links with the aviation industry, backing, for example, Pratt & Whitney's development of 'Wasp' and 'Hornet' radial engines in the late 1920s, whereas the Royal Navy's links with British manufacturers were channelled through and regulated by the Air Ministry.[113] Further, although the Admiralty had the responsibility for drawing up the specifications of aircraft and their operation, designing and building the planes was part of the Air Ministry's remit.

Shared control between the Admiralty and the Air Ministry was to exacerbate the situation in other respects, too.[114] The Trenchard–Keyes agreement of 1924 stipulated that, of the officers attached to the FAA, 70 per cent were to come from the Royal Navy and 30 per cent from the RAF. However, if only because it had a glut of young candidates at the time, the Admiralty quickly filled its quota. This prompted the Secretary of State for Air, Samuel Hoare, to grumble to the Prime Minister, Stanley Baldwin, in 1926 that: 'The excessive navalisation of the FAA makes it a continued focus of unrest. . . . [It is] swamped by naval officer pilots and observers. . . . [T]his festering sore in the relations of the Air Force and the Navy carries its infection elsewhere'.[115]

Notwithstanding this, although personnel who understood aeronautics were palpably indispensable to the evolution of the FAA, the Air Ministry was to obstruct the incorporation of RAF officers into the Naval Staff. Because of the system of joint superintendence, pilots serving in the FAA also remained as dependent upon the RAF for promotion as on the Admiralty. Whereas the Air Ministry was reluctant to appoint sailors to senior positions, the Admiralty declined to admit NCO pilots from the RAF to the FAA. Similarly, the Air Ministry refused to accept naval ratings for training as pilots. These selection processes helped dictate that, come the 1930s, almost all of the Royal Navy's serving fleet commanders – Chatfield, Cunningham, Fisher and Forbes – would have a background in gunnery, not aircraft. In fact it was not until 1960 that an aviator, Caspar John, was to be appointed First Sea Lord.

In America, by contrast, the challenge posed by Mitchell, together with the tribulations experienced by the Royal Navy, swiftly moulded and hardened the US Navy's policy regarding the control of maritime aviation. The remit of the

113 Hone et al., *Carrier Development*, p. 153.
114 Till, 'Airpower', p. 121.
115 Quoted in Till, *Air Power and the Royal Navy*, p. 42.

BOA encompassed advising on naval air operations, coordinating the design and manufacture of machines, and the selection and promotion of pilots. In 1936, its head, Rear Admiral Ernest King, was to stress that: '[I]t takes a great deal more than the ability to fly an airplane to meet the demands of naval aviation. An efficient pilot must have a thorough knowledge of the Navy, its doctrine, customs and needs'.[116] Whereas in 1926 the US Navy's hierarchy included a vice-admiral, three rear-admirals and two captains receiving flying pay, all of ten years later the Royal Navy was to have just one rear-admiral and a few commanders and junior captains who had risen through the ranks as aviators.[117]

Whereas Britain's Air Ministry was particularly loath to invest in aircraft best suited to maritime operations precisely because these might fall under naval jurisdiction, her Admiralty pardonably agitated for the restoration of unfettered control over what, from 1924, was known as the FAA. Certainly, the maritime environment did not attract much favourable attention from the Air Ministry. In 1924, the Coastal Area's disposable forces amounted to less than a dozen flying boats. Designed to carry neither bombs nor torpedoes, these, furthermore, possessed greater totemic significance than martial utility. Having supplanted the army in policing many parts of the empire, the RAF wanted to use these planes for demonstrative purposes, notably the majestic, flag-waving tours of the Far East that were carried out by *Southampton* aircraft in 1927 and 1928.[118] Nevertheless, such machines came to dominate thinking within the Coastal Area's hierarchy, not least because of financial stringencies. Whereas the prospect of another major war seemed very remote, Britain's economic problems were immediate and acute. Amidst the austerity of the late 1920s and early 1930s, there was insufficient funding for the creation of more flying boat units, let alone the development of expensive land-based alternatives, which would also need suitable aerodromes to accommodate them. As late as 1936 the Royal Navy was to continue to enjoy the largest slice of Britain's defence budget, whereas the RAF had the smallest.[119] Capable of operating from existing harbours, flying boats constituted a tried, tested and seemingly cheap platform for long-range reconnaissance.

Although the widespread expectation that Asdic would in any event reduce the requirement for shore-based planes to support the fleet at sea, the retention of ageing flying boat designs helped to suppress the advancement of land-based machines that had the necessary reach and firepower to contribute significantly to every aspect of maritime operations. The backbone of Coastal Command in the late 1930s was the *Anson*, which had worrying shortcomings in these respects, although other aircraft – among them *Whitley* and *Hudson* medium

---

116 Quoted in Ernest J. King and Walter M. Whitehall, *Fleet Admiral King: A Naval Record* (New York, NY, 1952), p. 250.
117 Till, *Air Power and the Royal Navy*, p. 45.
118 Buckley, *The RAF and Trade Defence*, pp. 27–30.
119 Malcolm Smith, *British Air Strategy Between the Wars* (Oxford, 1984), p. 336.

bombers and huge *Sunderland* flying boats – were becoming available as British rearmament programmes gathered pace. However, the organization's core mission had not been hammered out until as late as October 1935. This was to be the monitoring of hostile incursions, be they by aircraft, ships or submarines. Actively combatting these would most likely require the assistance of both RAF Bomber and Fighter Commands.[120]

Predictably, Coastal Command's inability to muster the conceptual and material wherewithal to destroy submarines in particular was to spawn consternation and exacerbate existing tensions between the services. A meeting presided over by the Fifth Sea Lord barely a year before Britain declared war on Germany concluded that the only solution to the problem was for the Royal Navy to take the matter in hand and assist the RAF in developing anti-submarine techniques.[121] This particular dilemma was to deepen within a few months of hostilities breaking out, for the swift deterioration in the wider geostrategic situation was to rip huge holes in Britain's distant blockade of the sea lanes and ports of mainland Europe. Certainly, countering U-boats with land-based aircraft was to turn all the more problematic. Their rapid occupation of France's western coastline was to give the Germans direct access to the Atlantic, while, from captured harbours in Norway, U-boats and surface raiders were to mount forays into both that ocean and the Arctic. This was to jeopardize Britain's communications with, among other countries, the USA and USSR, the only major states that might still actively aid her in fighting Italy and Germany. Long before this vicious circle arose, however, gaps in the coverage and capabilities that shore-based RAF aircraft might bring to bear were apparent enough. It was clear that, if planes were going to assist in denying the *Kriegsmarine* and the RMI control of the high seas, these would have to come primarily from bases afloat.

There were interwar squabbles about these, too. The Air Ministry took a dim view of aircraft carriers, dismissing them as vulnerable and arguing that they should only be used for the transportation of planes to shore installations overseas, from where the trade routes might then be adequately defended. The Royal Navy was thus depicted as having rather less importance than the RAF or even Britain's disposable land forces. In fact during the preparations for the London Naval Conference in December 1929, Lord Thompson, the Secretary of State for Air, whilst conceding that the navy did 'of course' need some aircraft, advised the prime minister, Ramsay Macdonald, that 'the Fleet Air Arm is an integral part of the RAF' and that planes allocated to it would reduce the number of those 'immediately available for Home Defence'. Responding, the First Lord of the Admiralty was adamant that: 'The FAA exists for the Navy. Its machines are designed to meet naval requirements and are paid for by naval vote. . . . It will be required to operate with the Fleet in any part of the world in which the

---

120 Minute, Director of Staff Duties to Chief of Air Staff, 4 October 1935, TNA, AIR 2/8875.
121 Report of meeting held by the Fifth Sea Lord, 13 September 1938, TNA, ADM 116/4038.

Fleet is needed. . . . [T]he Admiralty must be concerned with a broader aspect, namely Imperial Defence'.[122]

It was against this backcloth that, in September 1932, the Air Ministry seized the opportunity proffered by the Geneva Disarmament Conference to propose 'drastic' reductions and limitations regarding the aircraft carrying capacity of fleets. Although the Admiralty's reply set out minimalistic requirements of only 60,000 tons of carrier shipping and just 210 planes, it emphasized that aircraft remained essential for the navy's efficiency.[123] Although proposals were put forward that the UK and USA might reduce their carrier fleets from 135,000 to 110,000 tons each, the Royal Navy would not agree to start making the cuts before December 1936, when the Washington and London Treaties were due to expire. In any event its plans to create a balanced structure of up-to-date forces quickly encountered opposition from the Air Ministry. During January 1934 Sir Philip Sassoon, the Under-Secretary for Air, objected to the construction of a new carrier – the future *Ark Royal* – on the grounds that it would also necessitate some investment by the Air Ministry in shore facilities for the ship's squadrons and that expansion of the FAA was wrong in principle insofar that it diverted resources from the defence of the homeland. The Admiralty countered that, as the first of a new generation of bespoke carriers, *Ark Royal* was indispensable; none had been constructed since *Hermes* had been commissioned all of ten years earlier.[124]

Although, in 1937, Sir Thomas Inskip, Minister for Coordination of Defence, was finally to recommend that, much to the Air Ministry's chagrin, control of the FAA should be turned over to the Admiralty, it was not until the end of May 1939 that this measure was fully enacted. The concurrent formation of the Department of Air Materiel to oversee procurement also released the Royal Navy from the obligation to blindly accept Air Ministry advice about future FAA aircraft.[125] Plane production had reached a low ebb in the late 1920s and early 1930s, the strength of the FAA increasing by only 18 machines between 1929 and 1934. Just two firms – Fairey Aviation and Blackburn – dominated the manufacture of British naval aircraft, but even these had been awarded only tiny orders. There were qualitative shortcomings, too; in 1928, the Fairey IIIF was the only FAA aircraft that was both specifically designed for bombing missions and duly equipped with an up-to-date sight.[126] Such companies were in any case unaccustomed to, and ill-prepared for, the mass-production of planes: in the words of Correlli Barnett, Britain's entire aircraft-manufacturing sector before rearmament began in the mid-1930s was 'a cottage industry with obsolescent products; sleepy firms with factories little more than experimental aircraft shops employing hand-work methods, and centred on their design departments'.[127]

122 Quoted in Roskill, *Reluctant Rearmament*, p. 197.

123 Ibid., p. 137.

124 Roskill, *Reluctant Rearmament*, pp. 203–4.

125 Till, *Air Power and the Royal Navy*, p. 126.

126 Letter from Commanding Officer, HMS *Furious*, to Rear Admiral Commanding Battlecruiser squadron, 27 August 1928, TNA, AIR 2/1071.

127 Correlli Barnett, *The Audit of War* (London, 1986), p. 130.

The lack of technical expertise within the Admiralty also helped to preclude the production of carrier aircraft of the same quality as those of the USA and Japan, where the whole process was undertaken by the navies themselves. With so few new ones to hand, there had been a temptation to configure available airframes for a variety of roles. In 1936, the FAA contemplated a design for a three-seater floatplane that might serve as both a spotter and a fighter. The S.9/36, it was envisaged, would be launched from battleships and cruisers but would also be capable of operating from carriers. Unlike that of the two-seater fighter version, the cockpit would provide the observer with panoramic views.[128] Air Chief Marshal Edward Ellington, the Chief of the Air Staff, criticized these specifications as being symptomatic of 'the confusion of thought and changing policy of the Admiralty in the matter of aircraft design. . . . It is difficult enough to combine two or more functions with land-based aircraft where the limitations are only those connected with take-off and safe landing. . . . This so-called fighter will almost certainly be inferior in speed to the best land-based bomber . . . and it is likely, also, not to be able to bring to action the flying boat of the future'.[129] Criticism of multi-tasking also came from within the Admiralty. Admiral Backhouse, the Controller, observed in 1931 that: 'If we continue to provide in one aircraft for every possible duty we shall merely defeat our object of getting a type which is of utility at sea, having regard to what we want them to do and what they are likely to meet in the way of enemy aircraft'.[130]

In November 1937 the Admiralty asked the firms designing the Torpedo Spotter Reconnaissance Monoplane to consider whether producing a spotter-fighter variant based on the same basic airframe would be feasible.[131] Further evidence of the Admiralty's naivety in the sphere of aircraft design followed in the form of a suggestion that the prospective Dive-Bomber-Reconnaissance and Torpedo-Bomber-Reconnaissance machines might be fashioned from the self-same airframe.[132] Notwithstanding the rejection of this proposal, the Admiralty's Specification S.24/37, issued in September 1937, did call for a single-engined monoplane capable of undertaking not only reconnaissance and spotting duties but also torpedo strikes and both horizontal and dive-bombing attacks.[133] (The machine that emerged was the Fairey *Barracuda*. Fabricated entirely from metal and intended to supplant the *Albacore* and *Swordfish*, it proved too much of a

128 Air Staff Requirement for a Three-Seater Spotter-Fighter, Specification S.9/36, for Fleet Air Arm, 15 July 1936, TNA, AIR 2/1748.

129 Minute from the Chief of the Air Staff to the Secretary of State for Air, 16 November 1936, TNA, AIR 2/1748.

130 Quoted in Till, *Air Power and the Royal Navy*, p. 102.

131 Letter from Secretary of the Admiralty to the Secretary of the Air Ministry, 29 November 1937, TNA, AIR 2/1748.

132 Minute from the Deputy Director, Operational Requirements, to Deputy Chief of the Air Staff, 29 July 1936, TNA, AIR, 2/1749.

133 Requirements for TBR Monoplane for Fleet Air Arm from Secretary of Admiralty for Secretary of Air Ministry, 2 September 1937, TNA, AIR 2/2080.

jack of all trades and master of none, although several 'specialist' variants were turned out.)

Obtaining aircraft that were commissioned with discrete roles in mind also had its pitfalls, however. When casting around for a shore-based torpedo-bomber in May 1935, the RAF was to conclude that it would have to order a bespoke machine. Although FAA aircraft could of course operate from dry land if necessary, the very characteristics that made them suitable for carrier operations hamstrung them in other environments. In this particular case none of the FAA's existing planes were capable of accommodating the massive 1,900 lb torpedo that was being devised (by the Admiralty) for the RAF.[134] Meanwhile, the Air Ministry, still enamoured with high-level bombardment techniques, opposed the FAA's repeated requests for a dive-bomber. Exasperated, the Admiralty eventually circumvented normal procedure, ordering the Blackburn *Skua* – a fighter-cum-dive-bomber – straight off the drawing-board. Three months before this aircraft actually entered service in September 1938, the Admiralty's Air Materiel Department was to suggest cancelling the enterprise entirely, for the *Skua* was, by this juncture, feared to be outmoded.[135] So, too, was the bulk of the 232 machines that, together with six carriers, the FAA was to muster at the outbreak of war just twelve months later. By June 1945 there were over fifty carriers of various types and the FAA's front line comprised 1,336 planes divided between 73 squadrons. (These were numbered from 800 upwards to distinguish them from RAF units.)

Britain, a global as well as a European power, was torn between the numerous and interlocking interests and liabilities that made up her peculiar geostrategic situation. Her resources, intermittently overstretched in the past, were, by the 1930s, fast becoming woefully insufficient to maintain her standing in a world where the balance of power was shifting, not least because of technological innovation and economic and political upheaval. Whereas more authoritarian regimes might, if only for a time, ruthlessly channel resources into military might, gambling that territorial aggrandizement would soon redress any economic deficiencies, democratically elected politicians were hampered by, not only a shortage of hard cash, but also by the nature of the societies they were trying to govern and preserve. A combination of the Great Depression and the 'Ten Year Rule' had left Britain's armaments industries short of both the machinery and the skilled labour that were essential for any rapid, swift expansion of output. Bottlenecks and delays in the manufacturing of, among other essentials, aircraft and AA weaponry were common in the late 1930s, as was the danger of a balance-of-payments crisis and rampant inflation. Her creditworthiness had always been one of Britain's greatest strategic assets and when, in 1934, she began rearming in earnest, she financed many new programmes

---

134 Note from Deputy Director of Intelligence to Deputy Chief of the Air Staff, 13 May 1935, TNA, AIR 2/2715.

135 Till, *Air Power and the Royal Navy*, p. 101.

through borrowing.[136] The Treasury, led by Neville Chamberlain, who was to succeed Baldwin as prime minister in 1937, quickly took the lead in laying down priorities for defence. The security of the homeland and of the empire was emphasized at the expense of any commitment to a conflict waged on the European mainland; here, as had occurred in Napoleon's day, allies would have to undertake most of the fighting.

Although investment in each of Britain's three armed services now rose dramatically, by 1938 the army's share of the overall budget had duly become the smallest, whereas increased expenditure on air power – widely seen as an alternative to the repugnant trench warfare of 1914–18 – had given the RAF the largest slice of the spending cake.[137] Much of this money was invested in Britain's unique integrated air–defence network, which combined radars, command cells, shelters and searchlights with weapon systems, notably new, highly capable fighters like the *Spitfire* and *Hurricane*. As the threat of a sudden, knock-out blow diminished, so too did the importance of bombers for deterrence and, should that fail, the aerial counteroffensive that had been the cornerstone of RAF doctrine for so much of the interwar era. However, one major repercussion of this shift in policy and of the preference for 'limited liability' was that the RAF was to neglect the provision of close support for land forces, just as it had accorded comparatively little importance to maritime aviation. Although, once war began, the BEF was to be shipped across the Channel without incident, it and its continental partners were to be swiftly overwhelmed by *Blitzkrieg*. In September 1940, as the Battle of Britain was nearing its climax, Churchill – the erstwhile First Lord of the Admiralty who had succeeded Chamberlain at the government's head in May – was to tell the cabinet that: 'The Navy can lose us the war, but only the Air Force can win it. Therefore our supreme effort must be to gain overwhelming mastery of the air. . . . The Air Force . . . must . . . claim the first place over the Navy or the Army'.[138]

Poorly suited to a self-contained aerial campaign, the *Luftwaffe* was to be thwarted in the Battle of Britain by forces specifically configured for such an eventuality. This was a crucial but insufficient victory. Twenty years earlier, Admiral 'Jacky' Fisher had cautioned that it was not so much invasion that the British had to fear if their navy was beaten but starvation.[139] The battle for the Atlantic and other sea areas was thus to prove every bit as important as the Battle of Britain. In the meantime, the British Isles were to serve as a base for air, land

136 See: R.A.C. Parker, 'British Rearmament 1936–9: Treasury, Trade Unions and Skilled Labour', *English Historical Review*, 96 (1981): pp. 306–18; and G.C. Peden, 'Keynes, the Economics of Rearmament and Appeasement', in W. Mommsen and L. Kettenacker (eds), *The Fascist Challenge and the Policy of Appeasement* (London, 1983), pp. 142–56.

137 See: G.C. Peden, *British Rearmament and the Treasury, 1932–1939* (Edinburgh, 1979), p. 205; and R.P. Shay, *British Rearmament in the Thirties* (Princeton, NJ, 1977), p. 297.

138 War Cabinet: Munitions Situation, memorandum by Prime Minister, 3 September 1940, TNA, CAB 66/11, WP (40) 352.

139 Fisher, *Records*, p. 135.

and amphibious operations against opponents in mainland Europe. As RAF Bomber Command and the USAAF were the only means by which the enemy's heartlands could be struck directly for most of the remainder of the conflict, it ineluctably fell to such forces to hold the ring until such time that Allied armies could be got onto Italian and German soil. However, for those who had hoped that aviation would offer an alternative to gory, protracted surface operations, the Combined Bomber Offensive and the wider events of the Second World War were to prove a rude awakening: air power became not so much an alternative to attrition as another instrument of it. In the RAF alone, between 1939 and 1945 over ninety-nine thousand personnel were to be killed, wounded or captured, most of them in Bomber Command. This is half as many casualties again as were to be sustained by the Royal Navy.[140]

The impact of arms-limitation on the expansion of carrier fleets

For much of the interwar era, many nations looked to the control and limitation of armaments by international accord as the primary means of consolidating their security. As this approach faltered, however, ever more states turned to deterrence and counterforce policies to fill the resulting void. Besides the disarmament and arms-limitation clauses of the Versailles Treaty, and the (ultimately futile) endeavours of the Disarmament Commission of the League of Nations, the Five Power Washington Naval Treaty was to have significant ramifications for air power in the maritime environment.

The construction of bespoke aircraft carriers, such as Britain's *Hermes* and Japan's *Hosho*, was the exception rather than the rule in the decade that followed the First World War. This was largely due to the limits set on carrier tonnage by the international agreement that emerged from the Washington Conference of 1921–22, namely 135,000 tons for Britain and the USA, 81,000 tons for Japan, and 60,000 for France and Italy. The displacement of individual carriers was initially restricted to 27,000 tons, but later raised to 33,000. Predictably, these tonnage parameters became targets to which the respective nations aspired. As their navies sought, moreover, to squeeze as many capabilities out of their permitted assets as was possible, the efficiency of aircraft carriers as floating aerodromes became a prime consideration. After all, even in the early 1920s there were those who were persuaded that aviation was already emerging as the battleship's nemesis, rendering the tonnage ratios embodied in the Washington Treaty a doubtful guide to maritime strength. Although the strategic and doctrinal rationale for the possession of sea-based aircraft differed from one nation to another, the potency of any such planes remained the product of the complex interactions between them, their parent vessels and their wider environment.

---

140 'Casualties and Medical Statistics', in W. Franklin Mellor (ed.), *History of the Second World War: Medical Series* (London, 1972), pp. 829 ff.

Among the numerous factors that influenced the efficiency of carriers was their size, architecture and the number and characteristics of the aircraft they were intended to support.

The Washington accord stipulated that capital ships currently under construction should, unless granted a special exemption, be scrapped. Both the Japanese and the Americans converted unfinished hulls into aircraft carriers: the US Navy re-engineered the battlecruisers *Lexington* and *Saratoga* (each of which had a displacement of 36,000 tons), while the IJN did likewise with the *Akagi* and *Kaga*.[141] The American ships were fitted with a battery of eight 8" guns, the maximum permitted under the treaty, notwithstanding the protestations of Admiral Moffett, the BOA's head, who insisted that: 'The necessary defence of an airplane carrier against aircraft should be the aircraft carried on the carrier. It should therefore not be necessary to install anti-aircraft guns aboard an airplane carrier'.[142] The US Navy's General Board disagreed, however, arguing that the envisaged, primary role of carrier aircraft was not that of protecting their parent vessels; heavy AA guns were thus indispensable. Moffett also had reservations about the placing of so many of the navy's aviation eggs into just two baskets. Whilst acknowledging that, acting *en masse*, the aircraft from big carriers like the *Lexington* and *Saratoga* would muster a reassuringly large amount of firepower, he suggested that having a fleet of four carriers – each of 16,000 rather than 32,000 tons – would endow the navy with greater operational flexibility.[143] Certainly, *Lexington* and *Saratoga* did take up a great deal of the overall tonnage assigned under the Washington Treaty and, in the late 1920s, the US Navy duly turned to experimenting with smaller designs. *Ranger*, the first of these, entered service in 1934. At 14,000 tons she was capable of housing a fairly large air group, but her slow speed precluded her from functioning well as a Fleet Carrier. Displacing 20,000 tons and accommodating 80 aircraft, ships of the subsequent *Yorktown* class, proved to be a good compromise when they started to enter service in the mid-1930s.[144] Nevertheless, there remained a degree of uncertainty within the US Navy about the respective merits of large and small carriers.

For its part, the British Admiralty planned to convert HMS *Furious*, *Glorious* and *Courageous* and then build four new carriers of 17,000 tons each using tonnage margins liberated through the scrapping of *Argus*, *Hermes* and *Eagle*. Although it was intended that the new vessels be laid down in 1926, 1929, 1932 and 1935 respectively, owing to financial stringencies construction of the first did not commence until as late as 1934; *Argus*, *Hermes* and *Eagle* had to be

---

141 David C. Evans and Mark R. Peattie, *Kaigun: Strategy, Tactics and Technology in the Imperial Japanese Navy, 1887–1941* (Annapolis, MD, 1987), p. 249.

142 Quoted in Norman Polmar, *Aircraft Carriers: Volume One, 1909–1945*, p. 51.

143 Ibid., p. 51.

144 Phillips Payson O'Brien, 'Politics, Arms Control and US Naval Development in the Interwar Period', in Phillips Payson O'Brien (ed.), *Technology and Naval Combat in the Twentieth Century and Beyond* (London, 2001), pp. 158–9.

retained until 1939.[145] This protracted reliance on older ships made it difficult for the Royal Navy's carriers to operate as a coherent tactical unit.

At the London Naval Conference in 1930, the Americans suggested that the carrier tonnage for Britain and the USA permitted under the Washington accord should be reduced from 135,000 to 120,000 tons. The US Navy's General Board, supported by Moffett, also wanted the *Lexington* and *Saratoga* granted experimental status. This measure would, the Americans envisaged, release a mass of tonnage that they wished to use in building several smaller vessels akin to *Ranger*, thereby nearing their goal of establishing parity in numbers of carriers with the Royal Navy.[146] Owing to the 81,000-ton restriction imposed on them by the Washington Treaty, the Japanese also pursued the small carrier route in the early 1930s with the construction of the 12,500-ton *Ryujo*. In order to accommodate all of 48 aircraft, she had two hangar decks, but this architecture adversely affected her seaworthiness and extensive modifications had to be made to her structure following her launch in 1933.

In the late 1920s, the Royal Navy gave priority to the construction of cruisers over carriers and even failed to fulfil the quota of aircraft it was permitted. By 1930, it had only 150 planes, roughly half the number in service with the US Navy.[147] In 1929, the Naval Staff planned to increase the FAA's front-line strength from 141 machines to 251 over eight years. In the interim, the US Navy, by contrast, secured approval to raise its number of carrier aircraft to 750 by 1940. By 1938, it had 670 machines – almost triple the number in the FAA.[148] Indeed during the March of that year, Duff Cooper, the First Lord of the Admiralty, conceded that American naval aviation 'is supposed to be the best in the world, whereas ours is a source of grave anxiety'.[149]

The ratio of planes to ships also varied with differing national doctrines. The US Navy calculated the aircraft capacity of its carriers by the size of the flight-deck on which most of the planes would normally be stowed; the hangar was seen as being solely for maintenance work. This was a fundamentally different approach from that taken by the British and Japanese. They regarded the size of the hangar (or hangars) as the critical factor when deciding upon the number of aircraft to be allotted to the vessel, since they both favoured a policy of housing each and every plane below deck. In designing carriers, the IJN and the US Navy sought to maximize the offensive potential of their aircraft at the cost of furnishing the ships themselves with protective plating and other architectural safeguards. The Royal Navy's decision to, from 1936, incorporate armoured

---

145 Till, *Air Power and the Royal Navy*, p. 65.

146 Roskill, *Reluctant Rearmament*, pp. 46–7.

147 Jon T. Sumida, 'British Naval Procurement and Technological Change, 1919–1939', in Phillips Payson O'Brien (ed.), *Technology and Naval Combat in the Twentieth Century and Beyond* (London, 2001), p. 132.

148 Roskill, *Reluctant Rearmament*, p. 22.

149 Quoted in Roskill, *Reluctant Rearmament*, p. 366.

flight-decks into its next generation of carriers was, by contrast, predicated on an assumption that it would not always prove possible to repel enemy aerial attacks. It was therefore essential to help preserve the ship's own machines by encasing them in a protective box, namely the hangar. On the other hand, however, the addition of this extremely heavy sheathing had to be reconciled with the need to keep within total tonnage parameters. The room available for planes declined accordingly: whereas HMS *Ark Royal*, commissioned in 1938, could accommodate 60 machines and had a displacement of 22,000 tons, HMS *Illustrious* had space for only around half that number of aircraft, despite weighing in at 23,200 tons. With less aviation overall, the carrier was also unavoidably more reliant on its passive defences, should it have to protect itself from hostile aircraft.

One consequence of all of this was that the US Navy's carriers were effectively more capacious than similarly sized counterparts in other navies. However, lacking armoured decks, American and Japanese carriers were that much more vulnerable. This was particularly so of the latter, since their hangars were habitually used for, not just storage and servicing, but also refuelling and rearming planes.[150] This made for an exceptionally hazardous environment. The venue and other details of such processes could in any event prove of critical importance for both a carrier's survivability and efficiency, not least because they partly determined the speed with which aircraft could be launched, recovered and 'turned around' for another sortie. The number, skill, energy and stamina of technicians and the amount of time and space within which they had to perform their tasks; the quantity, speed and capacity of elevators for shuttling aircraft to and from the flight-deck; the exact shape and size of that platform and the maximum number of planes that could safely be 'spotted' – positioned for take-off – upon it; the distribution and quantity of refuelling hoses and ports; the location and size of magazines, munition lockers and the hoists that connected them; the stipulated procedures for the safe handling of munitions and combustibles; the load-bearing capacity of harnesses, jacks and carts; the quantities and dimensions of the trolleys that were essential for moving torpedoes, bombs and other heavy items to and from planes; the precise design of the differing brackets used in mounting the various types of ordnance on aircraft; the availability and layout of ventilation systems for expelling exhaust fumes and the (mostly highly inflammable if not toxic) vapours given off by fuels, solvents and lubricants: all of these formed but a fraction of the minutiae of operating planes from an aircraft carrier.[151]

As the size, weight and power of aircraft increased, changes had to be made to the decks from which they took off and on which they landed. So-called short flying from small platforms became increasingly impracticable and had died out

---

150 See: Parshall and Tully, *Shattered Sword*, pp. 119–20, 196–7, 244–8; Peattie, *Sunburst*, pp. 63–5.

151 An early British perspective on the need for certain procedures and material assets for effective carrier operations in war can be found in Document 70, 'Letter from Vice Admiral, Aircraft Carriers to Secretary of Admiralty: Lessons from wartime experience', 20 August 1940, in *FAA in Second World War: Volume One*, pp. 217–18.

by the 1930s. The rapid recovery of aircraft was facilitated by the development of mechanisms that brought them to a standstill more promptly and safely. Following tests conducted aboard the USS *Langley*, in 1926 the US Navy became the first to adopt a system that comprised a barrier and longitudinal arrester wires. Whereas it had hitherto taken at least 35 minutes to land ten aircraft, this could now be accomplished in less than half that time. The Royal Navy, too, experimented with fore-and-aft arrester wires in the 1920s, but abandoned all such mechanisms in 1927. It was not until 1932 that arrester wires, which ran athwart their flight-decks, were to be incorporated into its carriers.[152] The abandonment of arrester gear, although temporary, severely retarded the fulfilment of the force's potential: planes could not be safely left on the flight-deck; the number of machines that could be activated was constrained by the amount of time it took for them to land and be stowed away; and, when recovering planes, the carrier had to accelerate to maximum speed so as to generate sufficient wind resistance to enable the aircraft to touch down.

The potential of carrier aircraft was transformed in the mid-1930s with the introduction of monoplanes powered by large, radial engines. The dramatic improvement in performance can be illustrated by a simple comparison of two American strike aircraft, the Curtiss F8C-4 of 1929 and the Curtiss SBC-4 of 1937. The F8C-4, powered by a 450 horsepower engine, had, when carrying two 100 lb bombs, a range of 400 miles at a maximum speed of 145 mph. The SBC-4, powered by a 950 horsepower engine, had, when carrying one 500 lb bomb, a range of 590 miles at a speed of 237 mph.[153] When allowance was made for the time required to launch a large number of aircraft and form them up, the average reach of American carrier strikes was, by 1941, some two hundred miles. The Japanese equivalent of the SBC-4 was the Aichi D3A1, which, when laden with one 551 lb bomb under the fuselage and two 132 lb bombs under the wings, had a range of 795 miles at a maximum speed of 240 mph.[154] Although Britain's FAA had little if anything to compare with these machines at the start of the Second World War in Europe, by the conflict's conclusion the service had hundreds of very capable planes, many of them procured from the USA.

It is worthwhile recalling that the Washington Conference, which began in November 1921, was spawned by, above all, the naval arms-race between Japan, the USA and Britain that had gained momentum in the Pacific in the aftermath of the First World War. Certainly, the initial concern was the balance of nautical strength between this triad of powers. Held at the Americans' behest, the negotiations – which ultimately attracted delegates from Britain, Belgium, China, France, Italy, Japan, the Netherlands, Portugal and the USA – ineluctably broadened in scope, however; for maritime rivalries were as much a symptom as

---

152 For a comparative discussion of landing procedures and mechanisms, see: Peattie, *Sunburst*, pp. 65–72.

153 Thomas C. Hone and Trent Hone, *Battle Line: The United States Navy, 1919–1939* (Annapolis, MD, 2006), p. 98.

154 Peattie, *Sunburst*, p. 284.

a cause of wider tensions in the Far East especially. As well as establishing a ratio of British, American and Japanese capital ships of 5:5:3 and one of 1.75:1.75 for the French and Italians, the conference secured a pledge from Britain and the USA to the Japanese that they would not strengthen their respective naval bases in Hong Kong and the Philippines. Furthermore, the formal alliance that had existed between London and Tokyo since 1902 was discontinued and, besides the naval conventions, new treaties were introduced: Japan undertook to restore Kia-ochow to China and, together with all the other conference participants, agreed to respect her independence and maintain the 'open door' policy; while Britain, the USA, France and Japan all guaranteed one another's territories in the Pacific.

The fledgling USSR – the successor to tsarist Russia and as much an Orien-tal as a European polity – did not participate in the conference and was not a signatory to any of its accords. Moreover, whilst the Washington Treaty and the arms-limitation regime it engendered were to have a substantial impact on the shape and size of naval forces for ten years and more, they did not affect armies and only touched on maritime aviation indirectly. So much of the latter was in any case shore- rather than ship-based and officially deemed to be part of an autonomous air force or even the responsibility of soldiers, not sailors. Indeed air power's 'indivisibility' was one of the factors that bedevilled any attempt to regulate its ownership and employment through multilateral agreements. Although the Washington Conference did establish a panel of jurists to examine the concept of universal rules for aerial warfare, their recommendations were not incorporated into international law, not least because the authors' definitions of several fundamental precepts were perforce so woolly as to be self-defeating. If the moderation of submarine attacks through covenants was an intractable business, then trying to control war in and from the skies was all but impossible.

The fleet ratios that were set down in the Washington Treaty and were entrenched by the London Naval Conferences of 1930 and 1936 proved increas-ingly intolerable to the IJN, which, by the time of the Versailles settlement, had become the world's third largest navy. After Japan's government accepted the outcome of the first London conference, a rift opened up between her cabinet and the Admiralty. Having failed to prevent the accord being ratified, the latter pressed, successfully, for Japan to rescind the treaty. Due notice was given in December 1934 and, by 1937, the Japanese had shaken off the arms-control regime that had done so much to mould the IJN.

Similarly, the British, faced with an alienated Japan, turned to bilateral measures in a bid to shore up their nautical superiority in Europe. The Anglo-German Naval Agreement of 1935 effectively endorsed Berlin's repudiation of the Ver-sailles settlement and helped to undermine the so-called Stresa Front – the reaf-firmation of the Locarno Treaties by France, Italy and Britain – without gaining anything tangible in return. When Mussolini attacked Ethiopia in October 1935, the British and French – obliged to uphold collective security by enforcing the League of Nations' punitive sanctions against Italy – only succeeded in antago-nizing Mussolini. By 1937, he had openly sided with Hitler's Germany, while the USA, the instigator of the Washington Treaty, was seeking safety in splendid isolation. Although she was to soften this stance over the next four years, it was not until the attack by Japanese carrier aircraft on Pearl Harbor in December 1941 that she became fully embroiled in the Second World War, by which juncture so many of the globe's sea lanes were under threat from Axis forces.

Visions of aircraft carrier operations

All of the countries that developed carrier arms did so with a view to them fulfilling a clutch of broad functions. These were: the provision of reconnaissance and aerial escorts for friendly warships and merchant vessels; the 'fixing' of hostile fleets in order that surface units might more easily ensnare and destroy them; attacks against both targets on dry land and at sea; the transportation of planes to other, distant aerodromes, be they ashore or afloat; and the mounting of counterforce operations to thwart any bid by an adversary to make use of the skies. In those cases where an opponent possessed carriers of his own, the last of these tasks might well include the disabling of any such vessels. However, not all putative foes had carriers in their inventories, any more than they had remote airfields that could only be reinforced and replenished through maritime conduits. Whilst Britain, Japan and the USA had some difficulties in common, each of these powers had its own peculiar concerns and priorities, too. Above all, in trying to determine the size and composition of their carrier fleets and the numbers and characteristics of the aircraft they should accommodate, the differing navies had to try to foresee the principal tactical challenges their respective forces would encounter in any war.

Until equipped with radar, carriers, much like airstrips ashore, had to rely upon barrages from AA guns and on essentially random patrols by fighters to keep enemy planes at a safe distance. Throughout the 1920s the US Navy attached carriers to fleets essentially as a defensive measure; their aircraft were to help safeguard the ships, not least the carriers themselves. This practice was turned on its head by Rear Admiral Joseph Reeves during an exercise off Panama in January 1929. Reeves had taken over as Commander, Aircraft Squadrons, Battle Fleet, a few years earlier, operating from the small carrier *Langley*. Now in command of the *Saratoga* – a much more capable ship, with all of 110 aircraft and 100 pilots – he deviated from the scheduled plan of events and set out for the Panama Canal with a single escort vessel. Approaching at high speed and under cover of night, he cut through the cordon of forces – among which was the carrier *Lexington* – protecting the waterway and unleashed a strike with 70 planes. The attack achieved complete surprise and both the canal and the adjacent airstrips were adjudged to have been extensively hit. Admiral Pratt, the fleet commander, praised Reeves's plan as 'the most brilliantly conceived and most effectively executed naval operation in our history'.[155]

155 Quoted in C. Melhorn, *Two-Block Fox: The Rise of the Aircraft Carrier, 1911–1929* (Annapolis, MD, 1974), p. 114.

Although, precisely because it was insufficiently realistic, Pratt was unpersuaded that pertinent lessons could be drawn from the exercise itself, it did highlight the tantalizing potential of large carriers as autonomous strike platforms; possessing great speed and firepower, they only needed to be used with audacity and imagination. However, the drill also underscored the vulnerability of such vessels, it being deemed that *Saratoga* would have been fatally damaged in each of three separate incidents. Whilst there were those within nautical aviation who perceived an independent role for carriers, this view was not widespread within the navy as a whole. There were also disputes over whether mastery of the skies was best secured through offensive action against an opponent's carriers, or whether a fleet should employ its planes reactively to preserve its control of the surrounding airspace.[156]

By the late 1930s the US Navy had concluded it would be impossible to protect its carriers from aerial attacks. Control of the skies would have to be secured through pre-emptive raids on hostile carriers. A carrier group was duly allotted four squadrons of planes, half of which comprised strike aircraft, fighters and scouts making up the remainder. It was envisaged that, once the reconnaissance squadron had located any enemy carrier, the torpedo- and dive-bombers would assail it in concert while a fighter escort staved off any interceptors.[157]

Prior to the introduction of radar, carriers had to depend upon a familiar mixture of passive and active defences to protect themselves from aerial attack. Concealment and pre-emption had to be emphasized, otherwise, as was noted in the wake of exercises conducted in 1934: 'Under actual war conditions, it is quite possible that all of the carriers engaged . . . [will be] lost or put completely out of action. With opposing air forces of equal efficiency this is by no means an impossible result of the opening movements of a naval campaign'.[158] As late as July 1939 Admiral Ghormley, the head of the War Plans Division, gave the US Navy's General Board a similarly pessimistic assessment, stressing that 'the vulnerability of our carriers constitutes the Achilles Heel of our Fleet'.[159] It was not until mid-1941 that trials with radar were to suggest that this innovation might transform the effectiveness of American carriers. The question as to whether such ships should operate in discrete carrier divisions or as part of a larger taskforce also lingered. After exercises that were held in 1939, an advocate of the latter solution, Vice Admiral Ernie King, Commander, Aircraft, Battle Force, deduced that as many as nine carriers might function alongside one another.[160] However, it was not until 1939 – when *Yorktown* and *Enterprise* were commissioned, joining the *Lexington*, *Saratoga* and *Ranger* – that the US Navy had sufficient carriers for its quotidian commitments; even in peacetime, ships

156 Hone et al., *Carrier Development*, pp. 48–50.
157 Hone and Hone, *Battle Line*, p. 102.
158 Quoted in Hone et al., *Carrier Development*, p. 54.
159 Quoted in ibid., p. 66.
160 Hone and Hone, *Battle Line*, p. 103.

could seldom be spared for impromptu concentrations. Moreover, smaller and slower than the others, *Ranger* would have struggled to keep up with a flotilla of such vessels.

A momentous development regarding Britain's carriers was the appointment of Reginald Henderson as Rear Admiral, Aircraft Carriers, in June 1931. (A similar post had existed between 1917 and 1918, the occupant serving as the principal adviser on aviation matters and having responsibility for the operations of the Atlantic Fleet's carriers and aircraft.) Henderson had commanded HMS *Furious* between 1926 and 1928 and, as far as the Royal Navy was concerned, pioneered the agglomeration of carriers as a strike force, bringing together *Glorious*, *Courageous* and *Furious* for exercises held in 1933.[161] Later, as Controller, Henderson was also responsible for the design of the *Illustrious* class of armoured carriers.[162] For the British at this juncture, the main preoccupations for organic aviation in a fleet engagement were to be reconnaissance followed by attacks. The latter were essentially intended to slow down hostile battleships sufficiently for them to be overtaken and destroyed. The escape of the bulk of the *Hochseeflotte* at Jutland continued to haunt the Royal Navy's planners, while many of the capital ships that had since been added to, above all, the IJN were known to be significantly faster than their British counterparts.[163]

Besides planes that were attached to its principal flotillas, the Royal Navy also needed ship-based aircraft for the day-to-day defence of Britain's sprawling trade routes. In 1936 it was gauged that five carriers would be required for such commitments, although, it was thought, they could safely be significantly smaller and less formidable than their brethren in the battle-fleets. Some forethought was given to converting liners into auxiliary carriers in the event of hostilities – an expedient that, without all sorts of preliminary work, could only have proved far too laborious and expensive to implement in a suitably timely fashion. Unsurprisingly, nothing came of this idea.[164] On the other hand, the notion of furnishing sufficient numbers of small but bespoke vessels to safeguard commerce looked increasingly like a false economy as well, for the price of constructing diminutive carriers could scarcely prove proportionate to their capabilities. Indeed it was quickly recognized that, if only because of the paucity of large ones, any small carriers would have to be capable of undertaking more onerous tasks, too. As it was considered 'impracticable' to build cheap, small, unarmoured carriers possessing the requisite capacity and speed,[165] it was decided that four sizeable ships with 100 aircraft between them should be allocated for this task instead. Deployed to focal zones within the trade web, these would be capable

161 Till, *Air Power and the Royal Navy*, p. 164.
162 Ibid., pp. 75–7.
163 Hone et al., *Carrier Development*, p. 108.
164 Requirements for new naval construction, 1939–1941, Part II: Requirements, Outbreak of War to Dunkirk, TNA, CAB 102/536.
165 Notes of meeting held by Assistant Chief of the Naval Staff, 28 April 1936, TNA, ADM 1/11971.

of operating with a battle-fleet if necessary.[166] This approach, it was pointed out, would 'obviate the need for special ships for Trade Routes work and lead to increased flexibility in regard to [the overall] carrier force'.[167] Henceforth, it was noted: 'The use of the terms "Fleet Carrier" and "Trade Protection Carrier" should be discontinued, since the latter type should be capable of work both with the Fleet and on the Trade Routes'.[168]

The Japanese, by contrast, gave almost no thought to the defence of commerce in the event of hostilities with other maritime nations. This might appear peculiarly incongruous, for their quest for raw materials formed the single most important factor in accounting for the expansionist foreign policy that ineluctably brought them into conflict with other major powers. Japan was hard hit by the economic upheaval that followed the First World War. With few indigenous resources and a soaring population, she was dangerously reliant on precarious foreign trade. Rocked by crises in banking and agriculture, her economy was increasingly subjected to centralized delineation and control at a juncture when the armed services were acquiring more influence in domestic politics. Even before President Roosevelt's 'New Deal' for America of 1933 and John Maynard Keynes's publication of *The General Theory of Employment, Interest and Money* in 1936, Japan's government was trying to haul her out of economic lethargy with increased spending, more than a little of it on armaments. Certainly, one way or another, the economy, increasingly militarized and planned, shook off the Great Depression's worst effects faster than those of both Britain and the USA, further fuelling Japan's need for imported resources and the feeling within her people that, like other major powers, she, too, should have extensive colonial possessions.

Meanwhile, the army and navy argued over their respective budgetary allocations and over where, exactly, fresh territorial acquisitions should take place: whereas the IJN favoured a policy of holding fast in the north, the army wanted to continue making encroachments here and was sceptical about the navy's desire to push southwards, if only because this would further antagonize the USA and would in any event necessitate significantly more spending on the fleet and its supporting aircraft. Any additional expansion northwards was, on the other hand, simultaneously becoming more hazardous, not least because of the ambiguous attitude of the USSR. Since Tokyo had joined Berlin in forming the Anti-Comintern Pact in 1936, the sudden announcement of the Nazi–Soviet accord of 1939 was to irritate the Japanese as much as it startled the French and British. Although the IJN especially resisted any formal alliance with Germany and Italy, fearing that it could only provoke active collaboration between the Royal Navy and the US Navy, by September 1940 the success of Hitler's *Blitzkrieg* had lured Tokyo into forming the Tripartite Pact with Rome

---

166 Requirements for new naval construction, 1939–1941, Part II: Requirements, Outbreak of War to Dunkirk, TNA, CAB 102/536.

167 Notes of meeting held by Assistant Chief of the Naval Staff, 28 April 1936, TNA, ADM 1/11971.

168 Minute by Assistant Chief of the Naval Staff, 11 May 1936, TNA, ADM 1/11971.

and Berlin – the 'Axis'. Nevertheless, instead of joining the Germans and Italians in attacking the USSR in 1941, the Japanese strove to consolidate their grip on the Pacific's western rim; they occupied the whole of French Indo-China and finalized their preparations for the conquest of Malaya, Burma and the sprawling Netherlands East Indies, which included Celebes (Sulawesi), Dutch Borneo and New Guinea, Java, Moluccas and western Timor.

The push southwards was to commence with the crippling of the US fleet at Pearl Harbor and attacks on the Philippines. For all its tactical success, which was appreciable, the air raid on Pearl Harbor was a strategic disaster insofar that it ignited a conflict Japan had little prospect of concluding on favourable terms, if only because of her economic weaknesses. Admiral Yamamoto himself had warned Prince Konoe Fumimaro, then the prime minister, that: 'If we are told to fight, regardless of consequences, we can run wild for six months or a year, but after that I have utterly no confidence. I hope you will try to avoid war . . .'.[169] Indeed Konoe – reluctant to resort to such desperate measures but having failed to reach a peaceful compromise with either Japan's own armed forces or the USA's diplomats – was to resign in October 1941, just weeks before his successor sanctioned the attack on Pearl Harbor. As in Imperial Germany in 1914, however, there were those in Japan's labyrinthine political system who were persuaded that fighting was the only way she might remain a major actor in an era when the balance of power was undergoing a transformation, particularly across the Eurocentric world. Among its other consequences, Hitler's *Blitzkrieg* in the spring of 1940 was to wrest effective control from the Dutch and French over their respective possessions in the Orient, while Britain was to be plunged into a fight for her very survival. By February 1942 the sort of 'great disaster' foreseen by John Colomb as early as 1880 had befallen her: Its lines of communication severed by sea and land, Singapore, the Royal Navy's cardinal base in the Far East, had surrendered – not to sailors but soldiers.

Although, on paper, Japan's new conquests secured her an abundance of the natural resources she craved, her attempts to exploit these assets were severely constrained by deficiencies in her merchant fleet and in the provisions for its protection from active and passive threats. There was not even a plan for the introduction of rudimentary precautions, such as the formation of convoys, and a chronic shortage of commercial tonnage – most notably of oil tankers, the cargoes of which were indispensable in an age of mechanized warfare – was aggravated by very heavy losses to mines and submarines especially. Whilst the IJN was to be granted control over all ship-building programmes in February 1943, Japan never possessed adequate capacity to construct enough replacement freighters sufficiently quickly. Whereas the *Kriegsmarine* was to fail in its bid to starve Britain into submission through the protracted Battle of the

---

169 Quoted by Roger Pineau in 'Admiral Isoroku Yamamoto', Sir Michael Carver (ed.), *The War Lords* (London, 1976), pp. 396–7.

Atlantic, the lengthy *guerre de course* in the Pacific was to unfold very much in the Allies' favour.

But the IJN had always anticipated a short bout of hostilities that would culminate in a decisive naval engagement somewhere on the maritime glacis surrounding the Japanese home islands. The disposition of precious aviation assets had been shaped accordingly, as had procurement policies. Land-based planes had long been envisaged as the cornerstones of metropolitan Japan's defensive perimeter and, by 1938, such aircraft outnumbered those aboard her carriers by 563 to 332.[170] The focus of these ships, moreover, was resolute, high-intensity combat against hostile flotillas, not patient, unglamorous counter-blockade missions. Accordingly, these vessels were normally separated into divisions of two or three ships and attached to different fleets. In April 1941, however, no fewer than seven carriers with 474 planes were to be assembled to form the First Air Fleet, a force of unprecedented size and potential.[171]

Massed carrier formations were to be employed against Pearl Harbor in December 1941 and in the Battle of Midway six months later. This approach deviated somewhat from interwar doctrine, insofar that this had envisaged that carriers would, for their own security, be dispersed across several axes of operation. Their respective aircraft would then agglomerate in mid-air, seeking to mount a pre-emptive attack *en masse*. Hostile carriers were to be the premier target. The air groups on those of the Japanese duly included a large proportion of strike planes, especially torpedo-bombers.[172] Just as these – deployed either in line abreast or ahead – would approach the target on both sides so as to counter any attempts at jinking and to divide the defenders' attention and fire, so too would dive-bombers. As the cruisers HMS *Cornwall* and *Dorsetshire* and the carrier HMS *Hermes* were to discover in April 1942, such tactics could prove lethal.[173] However, as the conflict dragged on, the IJN was destined to struggle to produce sufficient carriers to replace its losses and meet its swelling commitments, let alone furnish new ships with aviators and technicians who were as proficient as those who had struck the opening blows of the war in the Pacific.

170 Peattie, *Sunburst*, p. 29.
171 Evans and Peattie, *Kaigun*, p. 349.
172 Ibid., pp. 333–4.
173 For details of the fate of *Cornwall* and *Dorsetshire*, see: Report by the Director of Naval Construction, 13 November 1942, TNA, ADM 267/84. For details of the loss of *Hermes*, see: Report from Senior Surviving Officer, HMAS *Vampire*, to Commander-in-Chief, Eastern Fleet, TNA, ADM 267/106.

3 Air power in maritime environments

Some key episodes, 1939–1945

Invasions and evacuations, 1939–1941: Scandinavia, France, Greece and Crete

The Nazi-Soviet Pact of August 1939 had serious ramifications for, among others, the states that encompassed the Baltic Sea. In the wake of his agreement with Hitler, Stalin was anxious to bolster the USSR's defences against any invader approaching from the west. He duly sought to persuade the Finns to relinquish strips of territory on both the Karelian isthmus and the Arctic so as to deepen the strategic glacis in front of Leningrad (St Petersburg) and Murmansk, respectively. He also hoped to acquire several islands within the Gulf of Finland and to lease the port of Hanko (Hangö) at its mouth as a naval base. From these outposts, Soviet warships and submarines would, he anticipated, be able to dominate the narrow sea lanes to, not only Leningrad, but also to Finland's capital, Helsinki, and the entire coastline of Estonia. Nevertheless, the rationale for any such alteration to the USSR's boundaries to the north of the Gulf of Finland also dictated that corresponding changes be made along the Baltic's southern shores as well. Indeed by the end of September 1939 Estonia – an independent republic since the fall of tsarist Russia – had been compelled to conclude a Treaty of Friendship and Cooperation with the USSR, whereby Soviet troops were to be allowed onto her soil. This accord was destined to be invoked in June 1940 and, before the summer was out, the indigenous government had been subverted and ousted, Estonia formally becoming part of the USSR. Essentially the same pattern of events occurred concurrently within Lithuania and Latvia – Estonia's neighbours to the south – with the result that these countries, too, were herded into the Soviet fold.

Meanwhile, Stalin, having failed to get the Finns to accede to his demands through negotiations, had resorted to force, unleashing the so-called 'Winter War' at the end of November 1939. By February 1940 France and Britain were pondering the possibility of intervening in the Baltic, ostensibly in aid of Finland, a fellow member of the League of Nations. However, the harsh fact was that they were unable and unwilling to give the Finns much practical assistance. They would have had to channel most of any aid overland through neutral Norway and Sweden – a contentious step for all involved – and were in any event pardonably more anxious about the wider strategic situation developing across the Nordic area, not least the trade in iron ore. Much of Germany's demand for this essential commodity was satisfied by Swedish exports via Luleå at the Gulf of Bothnia's northern tip. Whenever that harbour was ice-bound – which was normally the case for six to seven months of the year – cargoes were redirected through the Norwegian port of Narvik. The Allies, though predictably eager to curtail these deliveries, lacked nearby bases from which they might physically interdict the shipments across the Baltic, much of which was in any event too overshadowed for comfort by Berlin's martial might.[1] A patchwork of protagonists, the press-ganged and vulnerable non-belligerents, the whole region was by this juncture a diplomatic minefield, too – a dubious distinction it was to retain for much of the rest of the Second World War. Only Sweden was destined to avoid hostilities, occupation, or both.

A large country with a small population, feeble armed forces and a dangerous reliance on bulky imports from overseas, notably fossil fuels, Sweden had not participated in an armed conflict since Napoleon's day and was forgivably eager to avoid being embroiled in the fighting erupting all around her. Nevertheless, her own security was inextricably bound up with events in the neighbouring states of Finland, Norway and Denmark. Although these polities, together with her own geographic remoteness from the war's epicentre, wrapped Sweden in something of a *cordon sanitaire*, she was left isolated and vulnerable as surrounding countries were engulfed by what inexorably escalated into a global conflict. Essentially unenforceable, her neutrality had, perforce, to be somewhat pliable from time to time, particularly up to 1943, when Germany's ascendancy peaked; Stockholm was, for instance, intermittently pressured into granting transit rights to Axis forces and materiel traversing the Nordic region. Sweden was also a leading source of strategically important commodities and artefacts, notably ores and high-quality ball bearings. Since these metallic components formed the very foundation of so much mechanical engineering, the belligerents were predictably anxious to secure their fair share – and more, if possible – of Sweden's output. However, neither the Allies nor their enemies stood to gain much from provoking the Swedes to the point where they either ceased trading or took up the sword against one party or both. Acknowledging, if not always respecting, Sweden's professed non-alignment constituted the least disagreeable course open to her own citizens and the belligerents alike.[2]

Within a few months of Britain and France commencing hostilities against Germany, Norway and Denmark were, essentially because of their geostrategic location, to find themselves in far worse political straits than Sweden was ever to experience in the whole of the Second World War. No sooner had Stalin successfully concluded the Peace of Moscow, whereby Finland ceded a tenth of her

1 See: Thomas Munch-Petersen, *The Strategy of Phoney War: Britain, Sweden and the Iron Ore Question, 1939–1940* (Stockholm, 1981).

2 See: Wilhelm Mauritz Carlgren, *Swedish Foreign Policy during the Second World War* (London, 1977).

territory to the USSR, than fighting broke out at the western end of the Baltic. Early on 9 April 1940 the Germans invaded neutral Denmark, the gateway to that sea and a stepping-stone to Norway, the seizure of which was Hitler's real priority and which commenced on the selfsame day. Denmark's small and rather decrepit armed forces could offer little or no resistance: while the tiny air force was neutralized through pre-emptive aerial attacks on its bases, a German troopship steamed, unhindered, into Copenhagen. By 06:00, the city had fallen. Reluctantly, the Danish government acquiesced in their country's occupation, in return for Hitler pledging not to interfere in her internal affairs. This undertaking was to be steadily pared down until, in August 1943, the Germans seized absolute control. In the interim, the British took the precaution of occupying both Iceland – an independent polity that retained the Danish monarch as its head of state – and the Danish colony of the Faroe Islands. Jurisdiction over Greenland, another colony, was passed to the USA through a bilateral accord in April 1941 and, later that same year, American troops replaced the British garrison in Iceland.[3]

The occupation of Denmark was merely the opening phase of Operation '*Weserübung*', an enterprise conducted by land, sea and air that was intended to bring the whole of Scandinavia's western rim within Hitler's grasp. The Norwegian coast between the Skagerrak and Nordkapp alone extends for over a thousand miles as the crow flies and is bisected by the Arctic Circle. Its potential significance had first caught the imagination of German strategists shortly after the Second World War had begun: if this region could be conquered, Britain's flank would be turned, her merchantmen and warships would be completely banished from the Baltic approaches and her links to the USSR via the Arctic put in jeopardy. From pens erected along Norway's shores, even small U-boats would be able to menace shipping in the North and Norwegian Seas, while bigger ones might more easily slip past British aerial and nautical patrols and break out into the Atlantic.[4] Already – on 16 March 1940 – several Ju.88s had ventured across the North Sea from Germany to bomb both the Home Fleet in its main anchorage at Scapa Flow and the nearby FAA aerodrome. The cruiser *Suffolk* was slightly damaged in this attack, which Winston Churchill – who, once again, was serving as First Lord of the Admiralty at the time – dismissed as a 'petty and ill-directed raid'.[5] But from bases in Norway, bombers would be more readily able to reach targets in many parts of Scotland and on the north-eastern fringes of England as well as supporting the *Kriegsmarine* across a large tract of the waters bounded by the UK, the Faroes and Iceland. Equally, *Luftwaffe* fighters could furnish a protective umbrella for German land and naval forces, not least surface raiders that, in between harassing convoys in the Atlantic, might take shelter in

---

3 For a study of Denmark's dilemma during the Second World War, see: R. Petrow, *The Invasion and Occupation of Denmark and Norway, April 1940–May 1945* (London, 1974).

4 See: Barnett, *Engage the Enemy*, pp. 102–3.

5 See: *Commons Debates*, 20 March 1940, vol. 358, cols 1,973–4.

Norway's many capacious fjords. Conversely, strands of Germany's own supply web, notably the sea lanes to and from the Norwegian port of Narvik, would be rendered that much more secure.

Much like that of the Swedes, the Norwegians' government was persuaded that their country's geographic remoteness on Europe's periphery would, together with her protestations of non-alignment, keep the belligerent powers at bay. Measures for the active defence of Norway had long been neglected and, although some new equipment – notably aircraft to replace the hotchpotch of rickety biplanes in service with her undersized fleet and air force – was on order, in 1940 her armed services suffered from a range of qualitative and quantitative deficiencies. Still, Oslo reasoned, others could be expected to shield Norway, if only indirectly. The British Royal Navy's apparent domination of the North Sea would surely deter Berlin from even contemplating any substantial military venture in this quarter. Providing Franco-British infringements of her sovereignty did not antagonize the Germans too much, Norway was certain to be left in peace – an attitude that seemed to be vindicated by the *Altmark* Incident of February 1940.[6]

All of this was cruelly deceptive, however, if only because neutrality rests more on the perceptions of those who behold it than of those who profess it. For so much of the 1920s and 1930s, numerous states had turned international law – not least arms-control accords – into the bedrock of their security policies. This approach had been undermined by a gradual and calculated resort to *force majeure* that was bound to have repercussions for the world as a whole and for certain actors in particular. Yet, if only because they had failed to develop any viable alternative during the interim, several countries persisted in entrusting their safety to little more than jurisprudence. Whereas this left most non-belligerents merely talking of their legal rights, it fell to the British and French especially to actively safeguard, not only their own liberty, but also that of much of humankind. The situation vis-à-vis Scandinavia in general and Norway in particular posed a legal, moral and practical quandary with which Paris and London grappled for some months before concluding that any direct military intervention here would, however well-intentioned, most likely prove counterproductive.[7] Germany, by contrast, had no such scruples. Far from regarding the neutral Nordic countries as sacrosanct, she clearly saw parts of this region (among others) as being of such overriding geostrategic importance that they had to be brought under her immediate control through either coercive diplomacy or outright conquest. Sure enough, just hours after the Royal Navy started

6 On 16 February 1940 a British destroyer entered Norwegian waters in order to waylay and board an auxiliary tanker of the *Kriegsmarine*, the *Altmark*. Held captive on this vessel were 303 Allied seamen, sailors from merchantmen sunk by the surface raider *Graf Spee*. Whereas their detention by the Germans in neutral territory violated international accords, the interception of the *Altmark* was equally controversial from a legal perspective.

7 See Barnett, *Engage the Enemy*, pp. 99–102.

laying minefields along the Norwegian coast in a bid to drive passing German freighters into international waters, Hitler's legions suddenly swept northwards, invading both Denmark and Norway on 9 April. The so-called 'Phoney War' that had set in since the partition of Poland by the USSR and Germany at the end of September 1939 was evidently at an end.

What ensued was to constitute an authentic test of many of the interwar theories about the respective merits of the battleship and the bomber, insofar that the Norwegian campaign of spring 1940 was to prove the first sustained naval operation to be deeply affected by air power if not decided by it. Under real combat conditions, the Germans and the British alike had to strive to utilize their own aviation efficaciously while evading the attentions of their opponents'.

From the outset, the German forces achieved a high degree of both strategic and tactical surprise. The preliminary dispositions of the *Kriegsmarine* for the seaborne invasion of Norwegian soil were initially misinterpreted by the British, who concluded that what they were witnessing was the start of a grandiose foray into the Atlantic. In fact the Germans were poised to seize several of Norway's principal coastal settlements in one fell swoop: soldiers were to be simultaneously put ashore from warships at key points, ranging from Narvik in the north to Oslo and Kristiansand on the Skagerrak. At the same time, airborne units were to overrun the country's two major airports – those at Stavanger and Oslo – through which supplies and reinforcements might then be poured. Similarly, convoys of merchant ships were to disembark more personnel and materiel just as soon as the vanguards of the invasion force had secured the necessary harbours.

Most of this bold undertaking ran like clockwork. Norway's skeletal, ageing navy and sparsely-equipped coastal batteries were generally no match for the *Kriegsmarine*, although shore-based guns and torpedoes did manage to sink the heavy cruiser *Blücher* as she tried to thrust up the Oslofjord to overawe the capital. The pocket battleship *Lützow* (formerly the *Deutschland*) was also damaged by shells.[8] Within a few hours the invaders had, as planned, consolidated their various footholds and their spearheads were fanning out, disrupting the Norwegians' efforts to muster and concentrate what very few reserves were to hand.

Whereas resistance in Denmark had all but ended within a couple of hours of the start of Operation '*Weserübung*', the subjugation of vast, mountainous Norway was never likely to be completed overnight, particularly if there was any prospect of intervention by British and French forces. Rejecting Berlin's calls to surrender, Norway's king and government fled northwards from Oslo, readily accepting offers of assistance from Paris and London. However, the Allies were poorly placed to respond quickly and effectively to a crisis of such magnitude. A small expeditionary corps was available and was to start disembarking near

8 See: Thomas Kingston Derry, *The Campaign in Norway: History of the Second World War: United Kingdom Military Series: Official Campaign History* (London, 1952), p. 36.

Trondheim in central Norway and near Narvik by the middle of the month. In the meantime, air and naval forces – notably parts of the Home Fleet at Scapa Flow and Rosyth and planes based along the eastern edge of the British Isles – would have to take the lead in trying to hold the ring.

Some small but potentially significant clashes had already occurred. On 7 April RAF *Blenheim* bombers executed a high-level raid on a cluster of German vessels that had been spotted north of the Skagerrak. Predictably, no hits were scored.[9] Early the next day, a Polish émigré submarine, the *Orzal*, torpedoed an enemy transporter off Kristiansand.[10] Then, just hours later and west of Trondheim, a destroyer, HMS *Glowworm*, which had been detached to retrieve a sailor who had been lost overboard in the dreadful weather, came across some hostile warships, including the heavy cruiser *Admiral Hipper*. Despite gallant resistance, they overwhelmed the British vessel, which finally foundered after ramming the *Hipper*.[11] But whether these developments presaged a sortie into northern waters by sizeable units of the *Kriegsmarine* or an assault on central Norway remained unclear, partly because the Allies had yet to develop fully the electronic eavesdropping capabilities that they were to enjoy later in the war and which were to have such a seismic impact on its course, particularly in maritime environments.

During the First World War the Allies had gained an invaluable peephole into German naval plans from cipher manuals that were seized from the *Magdeburg*, a cruiser that ran aground off the Estonian coast and was captured by the Russians. However, the mechanization of encryption over the following years had, by 1940, made code-breaking much more a matter for mathematicians and engineers than the wordsmiths of yesteryear. Gradually, brilliant cryptographers and banks of increasingly sophisticated, (re)programmable computers at the British Code and Cipher School at Bletchley Park were to unravel the contents of ever more messages that were enciphered by machines, notably Germany's 'Enigma'. Exchanged via the airwaves between parts of the enemy's political hierarchies, bureaucracies and armed services, these transmissions – which were believed to be incomprehensible to any foe trying to listen in and which, by the same token, could be relied upon as factually impeccable – were harvested by very sensitive radio receivers and then decrypted, translated and passed on to relevant analytical agencies, among them those of the Royal Navy and RAF Coastal Command. Together with data from High Frequency Direction Finding (HF/DF) – networks of monitoring stations that, through triangulation, established the rough coordinates of the sources of Axis radio emissions – this 'Ultra' intelligence, as it came to be known, furnished the Allies with invaluable insights into the distribution and allotted missions of even the stealthiest of elements within the enemy's forces, such as submarines.

9　Till, *Air Power and the Royal Navy*, p. 13; Derry, *Norway*, p. 29.
10　Barnett, *Engage the Enemy*, p. 110.
11　Ibid., p. 110.

The most fallible link in the Nazis' command, control and communication network was not so much the enciphering devices themselves, which were extremely advanced, as the people who operated them. A part of the security regime protecting 'Enigma' was daily alterations to the machines' settings, with the result that such 'keys' were constantly being introduced, added to or discontinued. Furthermore, each of the German armed services utilized differing variants of 'Enigma', the navy's being the most complex. 'Enigma' encoding was awesomely convoluted, since a keyboard console employing three (of five) alphabetical rotors and a plugboard with ten cable sets could generate 158,962,555,217,826,360,000 permutations. Encrypting attachments – *Geheimschreiber* – were also affixed to the German high command's teleprinter networks so as to cloak their functional (Baudot) code in dauntingly labyrinthine ciphers.

Developing the wherewithal to unravel such transmissions at all, let alone do so with anywhere near sufficient timeliness, was in its infancy in spring 1940, as was the process of assimilating radar suites and other electronic aids into warships. (The proliferation of, for instance, miniature HF/DF sets – commonly referred to as 'huff-duff' – proved impracticable until 1942, and even then demand for such instrumentation exceeded supply.) Belligerents were in any case mindful that their radio transmissions were being monitored by friends and foes alike and duly did their best to avoid disclosing information to their enemies, not least through the use of encryption. This dictated that any intelligence obtained through the cracking of such codes had to be exploited with scrupulous care; for, should an adversary so much as suspect that his supposedly secret messages were being read, alterations to their encipherment were bound to follow.

Obtaining and preserving an advantage in this respect was thus an endless task that called for a blend of great skill, discrimination, heroism and sacrifice on the Allies' part. In the course of 1941 and 1942 code manuals, setting schedules and 'Enigma' network components were to be plucked from three *Vorpostenboote* – the armed trawlers *Krebs*, *Lauenburg* and *München*, all of which were stationed in northern waters – and snatched from within two (sinking) submarines, the *U-110* and *U-559*. The cipher customarily generated by the (three-rotor) 'Enigma' console was to be broken fairly quickly by Bletchley Park and the *Geheimschreiber* (Lorenz) code was also to be penetrated before 1942 was out. All of this was accomplished without the leadership of the Axis powers truly realizing what had occurred. In fact so well disguised was the real motivation behind some of the pillaging especially that the smokescreen surrounding it was to linger for decades after the war's end. When, early in 1942, the addition of a fourth rotor to the 'Enigma' machines utilized by U-boats in the Atlantic was to temporarily blind Allied cryptanalysts, part of the response was the mounting of Operation 'Jubilee' six months later. This elaborate, combined-arms raid on the French port of Dieppe was long regarded as having been a test bed for the amphibious landings that were later staged in Normandy and elsewhere – events that this ill-fated expedition indubitably helped shape. But this rationale had always been more apparent than authentic. The assault on Dieppe was actually Britain's most spectacular attempt at smash-and-grab espionage: at the incursion's heart was a quest to spirit away documents and devices that were assumed to be in the German naval headquarters located by the harbour.[12]

In the days before 'Ultra' intelligence became available, finding hostile vessels amidst the vastness of the seas and trying to establish their intentions were problems that had to be solved mainly with the old methodologies of visual and photographic reconnaissance, deduction and intuition. By the start of the second week of April 1940, in a bid to intercept groups of German vessels that were thought to be progressing up the coast, numerous British ships were converging on central Norway especially. Bringing up the rear was the solitary aircraft carrier immediately available, HMS *Furious*. She had had to hurry across from the Clyde, where she had been refitting, and had been obliged to sail without her fighter squadron, which was stationed at a distant aerodrome when the crisis erupted; the only planes aboard her were 16 *Swordfish*.[13]

Sir Charles Forbes, the Commander-in-Chief, Home Fleet, had sensed as early as 8 April that the Germans were bent on invading Norway, but he still had little information as to the strength and movements of the participating forces, partly because the stormy, cloudy weather was making reconnoitring such a lengthy, convoluted littoral that much more difficult and partly because he had few maritime scouting planes at his disposal. *Furious* came up too late to conduct anything like a comprehensive search and so initial sweeps had to be performed by a *Walrus*, which was launched from the battleship *Rodney*, and a couple of flying boats dispatched by RAF Coastal Command. By the time Forbes's main body had gathered to the west of Trondheim, the opening phase of '*Weserübung*' had been completed, yet the British had still to verify the size and pinpoint the location of some of the flotillas that had been involved in the operation. The Germans had succeeded in achieving the very high degree of surprise upon which the feasibility of their plans had always hinged. In such a dynamic, complex situation, many of the intelligence reports available to Forbes and to his colleagues in the military and political hierarchy in London were too ephemeral to serve as sure guides as to what was actually happening along Norway's rambling shoreline.[14] There was, moreover, a nagging concern that the Germans were seeking to infiltrate the Atlantic with one or more of their capital ships, a fear that was heightened when HMS *Renown* glimpsed, pursued and fired on *Scharnhorst* and *Gneisenau* as they scurried past Narvik in the early hours of 9 April.[15] Turning to the northwest before disappearing into the sleet, these

---

12 See: David R. O'Keefe, *Dieppe Decoded: The Remarkable True Story Behind the Greatest Raid of World War II* (Toronto, 2014).

13 Document 39, 'Report of Proceedings from Commanding Officer, HMS *Furious*: . . . Operations off Norway, 8–28 April 1940', in *FAA in Second World War: Volume One*, pp. 96–7.

14 Derry, *Norway*, pp. 30–31.

15 *Renown* opened fire at all of 18,600 yards – some ten and a half miles. *Gneisenau* was mistakenly identified as the *Admiral Hipper*. See: 'Action between HMS *Renown* & Enemy Warships presumed to be *Scharnhorst* and *Hipper*, 9 April 1940': reports and enclosures, April–May 1940, TNA, ADM 199/474.

powerful battlecruisers, having succeeded in diverting Franco-British attention away from the incipient landing at Narvik, later doubled back, dashing behind Forbes's fleet to regain the sanctuary of Germany. Likewise, the *Hipper*, having concluded her work at Trondheim by dusk on 9 April, was to lose no time in steaming homewards.

Essential though the contribution of the *Kriegsmarine* to '*Weserübung*' was, Germany's warships (which did not include functioning aircraft carriers) could not themselves project power into the interior of Norway. Having seen the *Wehrmacht* safely ashore, those that still had sufficient fuel headed for home, leaving the battle that would ultimately seal the country's fate to others. Indeed '*Weserübung*' was one of only a handful of major undertakings during the Second World War in which all three of the German armed services could and did collaborate. At the heart of the doctrine of *Blitzkrieg* ('lightning war') was a partnership between soldiers – particularly very mobile, mechanized and motorized units – and airmen. In fact the *Luftwaffe* was designed to, above all, support ground operations, not least through the curtailment of an opponent's ability to make use of the medium of the skies. This was most easily and readily achieved by pre-emptive strikes, whereby any hostile air force was obliterated while it was still on terra firma. The neutralization of aviation can, however, also be accomplished through means short of destroying planes, for any active air force is, in the final analysis, the product of radicular installations and processes on the Earth's surface. Disrupting these can leave it grounded – the very antithesis of air power.

The successful exacerbation of this 'impermanence' – which is one of air power's inherent characteristics – often proved the fulcrum of *Blitzkrieg*: The *Luftwaffe*, with surface units following on close behind, would take the lead in trying to create inroads into an adversary's position. Any breakthroughs that were made would be consolidated as quickly as possible by the troops on the ground, who, besides being protected by an umbrella of friendly planes, would bolster their side's control of the surrounding skies with large numbers of AA weapons, belts of which would help foil any counterstroke by hostile bombers. Behind such shields and those formed by its own fighter screens, the *Luftwaffe* could then safely establish new airstrips closer to the front, reducing the time it needed to react to developments in any ongoing fighting.

This joint approach served the Germans very well in the offensives they conducted on the European mainland during the first two years of the Second World War. Certainly, from their onslaught against Poland in September 1939 until their *Blitzkrieg* ran into the sands (and snow) of the Russian steppes towards the end of 1942, they pulled off a sequence of stupendous triumphs, largely through the successful suppression of their adversaries' aviation on key sectors of the battlefront. If – as Douhet for one maintained – command of the air allowed wars to be won more quickly, by the same token loss of such control could hasten defeat. But on the whole this was only so if any gains achieved by aviation could be consolidated and tapped. Tellingly, when the *Luftwaffe* was obliged to wage a self-contained aerial campaign – that of the Battle of Britain in the summer of 1940 – German doctrine broke down in the face of hostile air and sea power especially; for, as long as it remained beyond the Channel, the *Wehrmacht* could not support the *Luftwaffe* and exploit any advantage it secured in its struggle against an integrated air-defence network, the principal weapon of which was a customized fighter force. Rapid, simultaneous thrusts both overland and through the skies into the hinterland of an opponent's lines were the essence of *Blitzkrieg*, since such blows swiftly undermined the material and psychological foundations of any resistance. As far as the conduct of aerial warfare was concerned, local aerodromes were the most obvious of linchpins in any defence and were, furthermore, vulnerable to a variety of threats, ranging from remote bombardment to seizure by ground forces or parachutists.

The proximity, quality and security of bases are important factors in determining the sustainability of aerial operations. The possession – or lack – of adequately convenient, sophisticated and safe aerodromes was indubitably a major consideration in the Norwegian campaign, where the Germans were especially well placed to pit the strengths of their own air power against the weaknesses of their adversaries'. Indeed the official British history of the expedition suggests that this issue was of critical significance in accounting for the Allies' wider failure, particularly in southern and central Norway.[16] From the very outset, parts of the latter region especially lay within striking distance of *Luftwaffe* units stationed in northern reaches of the *Reich*. Within a few hours of launching 'Weserübung' airborne troops had, moreover, seized control of the sizeable airport at Aalborg in northern Jutland and those beyond the Skagerrak in Oslo and Stavanger. Reinforcements, most notably combat aircraft, together with their supporting paraphernalia, were rushed into these bridgeheads by a fleet of some five hundred transport planes. The reach of the *Luftwaffe* was thereby extended into central Norway and across an ever wider swathe of the North Sea.

Among the first Allied vessels to be detected was a force of cruisers and destroyers. They had been heading for Bergen, hoping to pounce on a group of enemy ships that the RAF had spotted there but had proved unable to incommode. However, recalled from this increasingly doubtful venture, they were falling back on the main body of the Home Fleet when, on the afternoon of 9 April, they came under bombardment from numerous aircraft. In all, there were 41 Heinkel He.111s and 47 Ju.88s. The former dropped their payloads from high altitude, while most of the latter proved more adventurous, swooping over their prey in shallow dives.[17] The clash turned into a protracted one, with the ships battling against both their assailants and a very heavy swell that, especially in the case of the smaller vessels, sent waves surging over the forward AA batteries and enveloped the gun directors in spray. Focussing fire on the planes as the ships pitched and rolled proved extraordinarily difficult and, although the gunners expended much of the available ammunition, few hits were scored.

16 Derry, *Norway*, p. 82.
17 Derry, *Norway*, pp. 33–4; Till, *Air Power and the Royal Navy*, p. 15.

Exasperated, the commander of one of the destroyers, the *Gurkha*, eventually swung her out of formation, reduced speed and headed downwind in a bid to get into a better position to shoot at the convoy's fleeting tormentors. Yet, once out on a limb and beyond the canopy of fire being thrown up by the flotilla as a whole, *Gurkha* was an obvious and comparatively vulnerable target for any approaching German bomber. Her hull damaged, she started to flood and, after a struggle that dragged on for over five hours, was abandoned. HMS *Gurkha* thus acquired the dubious distinction of being the first warship of any appreciable size to be sunk by aircraft since the beginning of the war.[18]

Dogged by the hostile planes, the remaining vessels ploughed onwards, eventually joining up with the Home Fleet's main body. The German pilots now found themselves up against a bewilderingly big array of potential targets. Although *Furious* had no fighters with which to intercept the attackers, collectively the numerous ships could not only generate a substantial volume of AA rounds but also create several overlapping arcs of fire. Moreover, whilst the heaving sea continued to complicate matters for many of the gunners, the battleships especially were both that much more stable as firing platforms and impervious to any ordnance that came their way. As the last wave of aircraft turned for home, one plane managed to strike the flagship, HMS *Rodney*, with a single 1,100 lb bomb. This did not so much explode as disintegrate on her dense armour, injuring 18 men but doing negligible harm to the ship's structure.[19]

A few of the other vessels also sustained some slight damage in this running battle, which had dragged on for several hours and drawn many of the German planes over three hundred miles from their bases. The attacking machines had been unmolested by interceptors throughout, while the state of the sea alone had had a very debilitating impact on the effectiveness of what AA fire they had encountered. The *Gurkha* was, nonetheless, the only ship that had succumbed to the efforts of nearly ninety modern combat aircraft, at least four of which had been downed. This outcome was difficult to reconcile with the predictions made by Mitchell some twenty years earlier. (Even *Gurkha* took nearly six hours to sink after being crippled, during which time her crew were safely evacuated to the cruiser *Aurora*.[20]) Still, Forbes was pardonably concerned that quite so many enemy planes were on the prowl. The RAF and FAA might in principle be able to project some air power from aerodromes along the north-eastern fringe of the British Isles, but the distances involved could only attenuate any such efforts to a point where they threatened to be all but worthless. Crucially, without constant fighter cover, the Home Fleet would be unable to ward off the attentions of enemy reconnaissance and strike aircraft. These might prove increasingly lethal, if only because so many of the ships were already running low on AA

18 Ibid.
19 Derry, *Norway*, p. 34; Till, *Air Power and the Royal Navy*, p. 15.
20 'Loss of HMS *Gurkha* by enemy air attack. . . . Rescue of Survivors by HMS *Aurora*', 9 April 1940, TNA, ADM 199/474.

ammunition. In any case the epicentre of any fighting the Royal Navy might helpfully participate in was surely shifting further northwards. If there were any German vessels still in the ports of south-western Norway, land-based aviation would have to go after them; *Furious* and her few *Swordfish* were needed elsewhere.

Indeed that same evening, 24 RAF *Wellington* and *Hampden* bombers executed an ineffectual raid on Bergen.[21] The submarine HMS *Truant* was far more successful, fatally torpedoing the light cruiser *Karlsruhe* as she left Kristiansand, her part in 'Weserübung' satisfactorily completed.[22] Around the same time, a nautical battle also loomed on the northern flank of the Germans' operations. Here, five British destroyers were nearing Narvik, hoping to intercept a larger flotilla of similar vessels that were evidently intending to disembark troops there. Although much too late to prevent this landing, the British squadron, blanketed by heavy snow squalls, advanced up the Ofotfjord towards Narvik the next morning, catching their adversaries unawares and inflicting heavy losses on the destroyers and merchantmen in the anchorage before being driven off.[23] 10 April also saw 16 *Skua* aircraft make a round trip of some six hundred miles from their base on the Orkneys to sink the *Königsberg* – the sister ship of the *Karlsruhe* – at her moorings in Bergen.[24] The dive-bombers surprised the defenders, encountering little flak and no enemy fighters. Three hits were scored and the light cruiser foundered rapidly.

In the interim, the core of the Home Fleet had moved further to the north-west. Here, Forbes hoped, it could avoid observation and fire from the skies while mounting offensive operations against the Germans in Narvik and central Norway. At dawn on 11 April *Furious* started launching her *Swordfish* to strike at enemy cruisers that had been reported by the RAF to be in Trondheimsfjord. Certainly, the *Hipper*, for one, had been active in the area, sinking HMS *Glowworm* just three days earlier. The planes found the anchorage to be empty of capital ships, however. What were thought to be two destroyers were assailed, one with torpedoes and the second with bombs, but both attacks went awry: the torpedoes grounded and detonated in shallow water, while the bombs failed to find their mark, not least because hills lining the fjord compelled the *Swordfish* to bank away prematurely as their target raced for shelter.[25] British endeavours further south proved more productive. Combing the Skagerrak, the submarine HMS *Spearfish* did find a target that was as viable as it was valuable: the *Lützow* was making her way home from 'Weserübung'. So badly damaged was the pocket

21 Derry, *Norway*, pp. 43–6; Barnett, *Engage the Enemy*, p. 15.

22 Derry, *Norway*, p. 34.

23 Barnett, *Engage the Enemy*, pp. 114–16.

24 See: Documents 38 and 38a 'Sinking of *Königsberg* at Bergen, 10 April 1940', in *FAA in Second World War: Volume One*, pp. 93–5.

25 See: Document 39, 'Report of Proceedings from Commanding Officer, HMS *Furious*. . . . Operations off Norway, 8–28 April 1940', ibid., p. 97.

battleship – a torpedo almost tore off her stern – by her invisible assailant that she had to be towed back to Kiel for repairs.[26]

In the light of what had transpired in the raid on Trondheim, the airmen aboard *Furious* were now directed to employ bombs rather than torpedoes against their next target, the German destroyers known to be stranded in Narvik. This mission, executed on 12 April, called for a flight amounting to some three hundred miles over the Norwegian Sea. Much of the journey was undertaken in abysmal weather, with a combination of oppressively low cloud and thick snow flurries slashing visibility to as little as 250 yards in places.[27] Narvik lies, furthermore, around forty miles inland on a dendritic network of waterways set amidst mountainous terrain. This compounded the navigational difficulties faced by the aircraft. If only because no preliminary reconnaissance had been possible, from the outset the crews of the *Swordfish* were unaware of the exact locations of their objectives within the rambling fjord. The only maps available to them of this particular area – and, indeed, of Norway in general – were photographic copies of maritime charts. Useful though these might prove in fathoming the depths of adjacent stretches of sea, they seldom displayed the contours of land masses. As such, they were of limited utility to aviators, particularly those seeking a safe route across an unfamiliar landscape of peaks, ridges and glaciated valleys in machines that had open cockpits. In any event, as the carrier's captain, Thomas Troubridge, subsequently stressed, the plodding *Swordfish* was 'a most unsuitable machine for low flying attacks unless the element of surprise is present'.[28]

In view of all of this, it is unsurprising that the raid proved almost counterproductive. The aircraft that made it to Narvik at all were funnelled up the Ofotfjord, where they were quickly spotted and fired on. They found few viable targets and hit few of these, sinking nothing more than a couple of Norwegian steamboats. Three machines were lost, although their crews were recovered.[29]

One of air power's main attractions at this time was its ability to strike at shipping that was ensconced in harbour – an undertaking that, with the perfection of the mine and torpedo, had generally become too risky for large surface vessels. The FAA having failed at Narvik, nine destroyers and the mighty *Warspite* were now detached from the fleet to annihilate the hostile vessels that remained there. However, the British squadron was preceded up the Ofotfjord by the battleship's seaplane, a *Swordfish*, which, flying at just 1,000 feet, pinpointed the enemy's positions and relayed this intelligence to the flotilla as it followed on behind. Any chance the Germans might have had of surprising their opponents was ruined by this classical use of a spotter aircraft in support of naval gunnery. All eight of the German destroyers left in the locality were wrecked, as was the

26 See: Barnett, *Engage the Enemy*, p. 117.
27 Till, *Air Power and the Royal Navy*, p. 17.
28 See: Document 39, 'Report of Proceedings from Commanding Officer, HMS *Furious . . . .* Operations off Norway, 8–28 April 1940', in *FAA in Second World War: Volume One*, p. 107.
29 Ibid., p. 98; Till, *Air Power and the Royal Navy*, p. 17.

U-64, which was caught on the surface and assailed by the seaplane. As much by luck as design, one 100 lb bomb disappeared into the submarine's belly through an open hatch, ripping her apart from within. By contrast, attacks by some of the *Swordfish* from *Furious* against the pirouetting destroyers inflicted more psychological than physical harm. Indeed two of these planes were downed by AA fire. An attempt by some He. 111s that had rushed up from Trondheim to molest the British ships was equally ineffectual, not least because low cloud bedevilled their customary technique of high-altitude bombing.[30]

The triumph of the *Warspite* left some two thousand German troops marooned and blockaded in Narvik. However, help was approaching from the south, radiating from the bridgehead at Trondheim. This encompassed the nearby airfield at Vaernes, which was to be fully adapted as a forward base for the *Luftwaffe* by the middle of April. Large numbers of Ju.88s and some fighters were already operating out of Sola – the aerodrome at Stavanger – and more *Luftwaffe* units were arriving in these enclaves with every passing day. At this point in the year, such northerly latitudes fall under the sun's rays for a growing proportion of the time, allowing aircraft to operate almost round the clock. When the Allied expeditionary corps began landing in central Norway – at Namsos between Trondheim and Narvik, and at Molde and Åndalsnes, southwest of Trondheim – troops, ships and harbours alike came under almost incessant raids from planes.

The three small ports were all but razed, but it was the Allies' shipping that, as the logistical conduits for their land forces, attracted most attention. However, though frequently harassed, vessels were seldom hit, not least because the local topography complicated matters for incoming aircraft. Nobody had waged an aerial battle in an environment comparable to much of Norway since the fighting in and around the Alps during the First World War. Just as ships had little scope for either concealment or evasive manoeuvres in the constricted waterways, so too were low-flying planes impeded by the peaks and ridges around them. These eminences also truncated the vision of both people and radars. Their approach concealed from the electronic and human eyes on the British ships, aircraft might suddenly loom over a crest and espy a potential target in the valley below, but they and its defenders alike would be left with little time to bring their weaponry to bear. By contrast, attack runs executed lengthways along the fjords usually held fewer surprises for all concerned.[31]

Nevertheless, whatever the tactical complexities and outcomes of these clashes, it was the Germans who squeezed the most advantage from them in the wider operational context: the mere threat of aerial raids disrupted the Allied navies'

---

30 Ibid., pp. 17–19; and Document 39, 'Report of Proceedings from Commanding Officer, HMS *Furious*. . . . Operations off Norway, 8–28 April 1940', in *FAA in Second World War: Volume One*, p. 99. Also see: 'Report from Vice Admiral Commanding Battle Cruiser Squadron to Commander-in-Chief, Home Fleet, 25 April 1940', in 'Report of attack by British naval forces on Narvik, 13 April 1940', TNA, ADM 199/473.

31 Till, *Air Power and the Royal Navy*, p. 19.

endeavours to support the army they had put ashore, which was where the campaign was ultimately going to be decided. A dearth of AA artillery and inadequate, intermittent fighter cover scuppered attempts to cling on to any gains made by the Franco-British troops, particularly around Trondheim, the key to central Norway. This in turn prevented the capture of better ports and airfields that might have enabled the agglomeration of superior forces both on the ground and in the skies. The resulting inability to take the initiative and then consistently pursue any objective left the Allies with little option other than to withdraw.[32]

The Allied forces in central and southern Norway soon found themselves in the rapidly closing jaws of German pincer movements. Whereas their adversaries enjoyed the backing of substantial *Luftwaffe* units stationed nearby, friendly air power had to be projected either from carriers operating at a safe distance off the coast, or from aerodromes far away in the British Isles. Support for Allied surface units ineluctably proved patchy and sporadic. For instance, on the night of 22–23 April RAF *Whitley* bombers ventured all of four hundred miles across the North Sea in search of the enemy airfield at Vaernes, only to fail to locate their objective.[33] Even if they had had more luck in this regard, it is improbable that such spoiling attacks would have had much lasting effect. As the Germans' own experience was to confirm during the Battle of Britain later that year, disabling a sizeable aerodrome through high-level bombing alone was difficult and, even if accomplished, would most likely prove ephemeral.

From the outset of 'Weserübung', the Germans had, through their seizure of Norway's few large airports, denied their opponents much chance of getting any unshakeable grip on the neighbouring skies. The Allies struggled to redress the balance. A cruiser, the *Suffolk*, was detached to shell the *Luftwaffe* in its base at Sola, Stavanger. She steamed into range under cover of darkness and commenced firing her 8" guns at dawn on 17 April, the fall of shot being monitored by her *Walrus* seaplane. But meteorological conditions suddenly tossed a spanner into the works: atmospherics caused intermittent breakdowns in the link between the spotting plane and the ship, leading to several salvoes missing the aerodrome entirely. Worse still, what was supposed to be a cursory hit-and-run raid was turned into a dangerously protracted foray. *Suffolk* was originally instructed to bombard Sola and then make her escape westwards out to sea where she would rendezvous with RAF *Blenheim* fighters dispatched from the Orkneys. These, it was reasoned, would ward off any pursuing German bombers. However, the Admiralty in London succumbed to the temptation to try to micromanage matters through long-distance communications: the *Suffolk* was directed to head northwards to intercept some enemy destroyers. Moving parallel to the coast instead of away from it, she remained in reach of hostile planes for far longer

---

32 See: Derry, *Norway*, pp. 78–80, 82.
33 Till, *Air Power and the Royal Navy*, p. 19.

than was strictly necessary, while the fighters that were supposed to protect her were left unsure of her whereabouts. *Suffolk* and her four escorting destroyers came under attack over thirty times during the next seven hours and expended virtually all of their AA ammunition. One 1,000 lb bomb sliced into the cruiser and detonated under a turret, inflicting tremendous damage. Partially flooded and with her speed slashed to 18 knots, she limped away to Scapa Flow, lucky to have survived.[34]

In an effort to contest their opponents' growing command of the skies, the Allies now shipped 18 RAF *Gladiator* fighters into the theatre of operations. Brought in aboard the carrier *Glorious*, these antiquated machines were sufficiently nimble to be able to alight on and take off from the only inland airstrip available, a small runway extemporized from a frozen lake near Åndalsnes. Predictably, however, no sooner was it detected than this makeshift base became a prime target for *Luftwaffe* counterforce operations and was thereby rendered untenable within a couple of days of being established.[35] Notwithstanding this, the Allies did have something of a fighter contingent to hand before the month was out, for the Royal Navy had brought up not only *Glorious* but also *Ark Royal*. Planes from these carriers struck at the aerodrome in the swelling German pocket around Trondheim as well as providing some close support for the soldiers who were struggling to contain the enemy on the ground. Indeed for all their slowness, *Swordfish* twice pulled off the remarkable feat of raiding the formidable *Luftwaffe* base at Vaernes without sustaining a single casualty.[36] The FAA, however, stockpiled little ordnance that was ideal for destroying anything other than maritime targets, while their nautical charts were of only marginal assistance in a battle that swayed across a complex, alien landscape.[37] Further, communication links with friendly troops were primitive in comparison to those between *Luftwaffe* and *Wehrmacht* units; closely coordinating the efforts of the soldiers fighting on the front line with those of the Allied aviators proved difficult if not impossible. Unlike the FAA – or even the RAF at this stage in the war – the Germans possessed the requisite doctrine as well as the hardware for such joint undertakings on terra firma. Indeed the last days of April were to see them intensifying the pressure on the Franco-British and Norwegian forces with one of the most formidable components of *Blitzkrieg*, the Ju.87 *Stuka*.

This flying artillery enabled the invaders to dominate great chunks of territory that their ground forces then overran, often with virtual impunity. Manoeuvred or simply blasted out of any positions they sought to hold, the Allied troops were

34 Derry, *Norway*, pp. 74–5; Till, *Air Power and the Royal Navy*, pp. 20–21.

35 David Brown, *Carrier Operations in World War II: Volume One: The Royal Navy* (London, 1974), pp. 16–17; Derry, *Norway*, pp. 115–18.

36 Document 51a, 'Report from Vice Admiral, Aircraft Carriers, HMS *Ark Royal*. . . . Operations off Norway, 23 April – 3 May 1940', in *FAA in Second World War: Volume One*, p. 141.

37 Document 39, 'Report of Proceedings from Commanding Officer, HMS *Furious*. . . . Operations off Norway, 8–28 April 1940', ibid., p. 107.

either encircled and eliminated or compelled to fall back towards their toeholds along the coast. The Royal Navy's carriers – the only aerodromes in Allied hands available within the combat theatre – were, for their own safety, likewise obliged to pull further away from the epicentre of the fighting. This increased the amount of time (and fuel) planes expended in simply travelling to and from the front line, progressively diluting the amount of air cover that could be maintained above it. Vice Admiral Lionel Wells, in charge of *Glorious* and *Ark Royal*, had to forsake his attempts to support the troops around Namsos and Åndalsnes because of the threat posed to the carriers by enemy bombers. Both ships also needed to take on new aircraft.[38] Meanwhile, *Furious* – which had no machines aboard that could be employed as fighters, was running low on oil and which had been sheltering in a fjord far to the north near Tromso – had received orders on 23 April to leave Norwegian waters altogether. A few days earlier, she had been espied by a lone German plane flying at high altitude and had come under attack, two large bombs landing close by. These damaged the carrier's turbines, cutting her maximum speed to 20 knots. After a series of 'exasperating' delays, she was able to refuel and escape her predicament, setting out for the Clyde on 25 April.[39]

By this juncture, the main Franco-British army was on the verge of relinquishing central Norway entirely, although another contingent of troops had been assembled and was now poised for an assault on Narvik. On 29 April Allied shipping, acting as far as possible under the cover of darkness and mist, commenced the embarkation of the soldiers in the southernmost bridgeheads. Those retreating via Namsos followed close behind, the last of the units here withdrawing on 3 May. The *Luftwaffe*, frequently in the form of *Stukas*, disrupted operations at every opportunity, pursuing the convoys far out to sea and sinking a French and a British destroyer. As they worked their way further northwards in order to cover the retreat, *Glorious* and *Ark Royal* also received some unwelcome visitors. These were, however, kept at bay with a combination of AA fire and interceptors.[40] Whereas, on 1 May, both these carriers had to withdraw to Scapa Flow for replenishment, *Ark Royal* had rejoined the core of the Home Fleet before the week was out.[41] Meanwhile, German troops, backed by aircraft based at Trondheim, continued up the coast, seeking to relieve their beleaguered colleagues at Narvik.

Throughout these events, the FAA struggled valiantly to protect the fleet, not least the carriers – the only forward aerodromes available for the time being – and to assist the Allied troops that remained on Norwegian soil. It was a rather unequal contest, partly because the British planes were grossly outnumbered – the *Luftwaffe* had around six hundred combat machines to hand from the start of

38 See Till, *Air Power and the Royal Navy*, p. 22; Brown, *Carrier Operations*, p. 17.

39 Document 39, 'Report of Proceedings from Commanding Officer, HMS *Furious*. . . . Operations off Norway, 8–28 April 1940', in *FAA in Second World War: Volume One*, pp. 102, 105–6.

40 See: Document 51a, 'Report from Vice Admiral, Aircraft Carriers, HMS *Ark Royal*. . . . Operations off Norway, 23 April – 3 May 1940', ibid., pp. 155–9.

41 Ibid., p. 161; and Brown, *Carrier Operations*, p. 17.

May[42] – and partly because the FAA's aircraft were on the whole technologically inferior to those they were called upon to counter. The Germans marching to the relief of Narvik's garrison soon progressed beyond the range of the *Stukas*, but He.111s and Ju.88s were capable of reaching far into the Arctic Circle. There were also some Messerschmitt Bf.110s – twin-engined '*Zerstörer*' that had a lengthy reach and packed a hefty punch – on the prowl and a few fearsome Me.109s, although the latter lacked the combat radius of the former. HMS *Furious* had already been impelled to retreat to a respectful distance from the enemy because of her utter lack of fighter cover and many of the machines on the carriers closer to hand were asked to take on roles for which they were not especially suited, notably the interception of enemy aircraft. In this regard, Vice Admiral Wells – the Vice Admiral, Aircraft Carriers – stressed in a report to Forbes that the *Skua* was primarily designed as a dive-bomber, not a fighter like the RAF's *Hurricanes* and *Spitfires*. In fact the *Skuas* were slower 'on the climb, level and dive' than the German bombers they were called upon to confront.[43] The Ju.88s and He.111s especially all too often identified and attacked targets from high altitudes, where they were both on the edge of the operating ceiling of the *Skuas* and very difficult to espy from the surface. Frequently, they released their payloads and made off before their adversaries were even alerted to their presence.[44] 'Our FAA aircraft are hopelessly outclassed by everything that flies . . . and the sooner we get some efficient aircraft the better', Forbes subsequently fumed to the Admiralty.[45] 'Our fleet fighter aircraft are outclassed in speed and manoeuvrability', he emphasized in another missive, insisting that 'it is only the courage and determination of our pilots and crews that have prevented the enemy from inflicting far more serious damage'.[46]

This last remark highlights an important consideration within aerial combat in particular and warfare in general. In the final analysis, a lot of technology is only as good as the people using it. Aircraft cockpits are technological interfaces between humans and their surrounding environments. The product of appreciable training and preparation in peacetime and during the 'Phoney War', many of the FAA's aircrews were accomplished aviators, even at this early stage of the conflict.[47] They went on to prove themselves courageous combatants, too. To some extent they were able to offset their machines' limitations with skill and bravery, never hesitating to engage superior numbers of hostile planes. Indeed, as Vice Admiral Wells later remarked, it was believed that 'on many occasions the

---

42 Barnett, *Engage the Enemy*, p. 122.
43 Document 51a, 'Report from Vice Admiral, Aircraft Carriers, HMS *Ark Royal*. . . . Operations off Norway, 23 April – 3 May 1940', in *FAA in Second World War: Volume One*, p. 141.
44 See: Till, *Air Power and the Royal Navy*, pp. 23–4.
45 Document 51, 'Letter from Commander-in-Chief, Home Fleet, to Secretary of Admiralty', 15 June 1940, in *FAA in Second World War: Volume One*, p. 140.
46 Document 50, 'Letter from Commander-in-Chief, Home Fleet, to Secretary of Admiralty', 11 June 1940, ibid., p. 121.
47 See, for instance, Brown, *Carrier Operations*, p. 15.

enemy were under the impression that they were being attacked by *Spitfires* – this speaks for itself'.[48] Although the mere threat of interception could often prove sufficient to break up formations of incoming bombers, driving them off course and rendering the execution of a synchronized attack impracticable, then as now the infliction of physical rather than psychological damage on opponents demanded the concentration of adequate firepower in time and space.

Ever since the 'Fokker Scourge' of 1915–16, thoroughbred fighter planes had been evolving as single-seater machines with forward-firing armaments. The aircraft the FAA relied upon for dogfighting bucked this trend, essentially because they were hybrids developed in compliance with distinctive doctrinal reasoning. Primarily configured as a component of battle-fleets that were intended to counter hostile flotillas and individual vessels on the high seas, British ship-based aviation was never really expected to have to take on large numbers of enemy aircraft, particularly bespoke fighters emanating from terra firma. Consequently, the interceptors the Royal Navy's carriers accommodated were principally designed, not so much with speed and agility in mind, as endurance, sturdiness and firepower.

At the advent of the Second World War, the sole single-seater machine available was the *Sea Gladiator*. This was the last and best biplane fighter to be introduced and the first with a fully enclosed cockpit, albeit one that, in common with those of other FAA planes, did not have a bullet-resistant canopy. A marinized version of the RAF's Gloster *Gladiator*, which entered service in 1937, its endurance averaged just two to three hours. This was a serious handicap in carrier operations, if only because relatively large amounts of time could be wasted in launching and recovering aircraft, with the ship being obliged to head into the wind while doing either. Although it was a customized interceptor, the *Sea Gladiator* was teetering on the brink of obsolescence when hostilities dawned. Indeed it was already being supplanted by a pair of monoplanes, the Blackburn B.25 *Roc* and the Fairey *Fulmar*, the former of which was really an experimental type while the latter was the descendant of a light bomber, the Fairey *Battle*. To ease the burdens imposed by quotidian, transmarine patrols on their occupants, both the *Roc* and the *Fulmar* were conceived and fabricated as two-seaters, with the second crewman acting as navigator, radio-operator and observer. If necessary, he could serve as a gunner, too, manipulating a coaxial machine gun mounted in the cockpit's stern. (The fuselage of the *Roc* – a spin-off of the *Skua*, which was designed as a dive-bomber – was so elaborate in this respect as to incorporate a rotatable turret, an innovation pioneered by the British.[49]) The *Fulmar*, being derived from a small bomber, was a relatively roomy machine that,

48   Document 51a, 'Report from Vice Admiral, Aircraft Carriers, HMS *Ark Royal*. . . . Operations off Norway, 23 April – 3 May 1940', in *FAA in Second World War: Volume One*, p. 141.

49   The *Roc*, as 'a free gun fighter, with an excellent multi gun turret' was regarded as superior to fixed-gun fighters such as the *Skua*, which, unless endowed with far superior aerodynamic characteristics than opponents, were 'confined in attack to a limited arc of approach'. See: Document 32, 'Letter from Commander-in-Chief, Home Fleet, to Secretary of Admiralty: Suitability of *Roc* and *Skua* Aircraft', ibid., pp. 81–2.

it was tempting to believe, could be weighed down with two occupants without suffering much loss of performance.[50]

By the end of 1940 the FAA was also substituting the *Martlet* – an adaptation of the Grumman *Wildcat* – for its *Sea Gladiators* especially and, in the course of the next two years, added the *Sea Hurricane* and, later, the *Seafire* (the nautical variants of the *Hurricane* and *Spitfire* respectively) to its inventory. Nevertheless, good fighters though they were, *Sea Hurricanes* and *Seafires* were not bespoke naval aircraft but stop-gap adaptations of machines that were optimized for service from aerodromes on terra firma rather than from carrier flight-decks; they ineluctably brought handling and operating complexities with them, as well as being comparatively lacking in endurance. Demand for a two-seater aircraft that could act as an interceptor as well as fulfilling other roles persisted, the *Fulmar* being superseded by the Fairey *Firefly* when that came on stream during 1943.[51]

By that time the ungainly *Roc* had long since been withdrawn from front-line duties and most of the FAA's fighter planes were bespoke single-seater machines with forward-firing armaments – an arrangement that fused the roles of pilot and gunner. Whereas this configuration was generally more efficient in a dogfight, in order to bring the fighter's weaponry to bear at all it was necessary to align the plane itself with the target and approach to within a suitably short distance. This, perforce, put as much of a premium on the speed and other aerodynamic characteristics of interceptors as it did on the amount of firepower at their disposal. (In fact during the opening weeks of the Norwegian campaign the RAF intermittently sought to supplement the FAA's machines with squadrons of Bristol *Blenheim* bombers that had been recast in the role of long-range fighters.) Regardless of the type of technology concerned, however, the human component remained of critical importance in aerial warfare. Not least because the manoeuvrability of a plane is partly a function of the skill of the person behind the controls, the quality of aircrew was ineluctably a key factor in the overall effectiveness of fighter units in particular.

The mounting demands that were to be thrust upon, above all, Japan's naval pilots and the fighter *Staffeln* of the *Luftwaffe* during the course of the Second World War were bound to have dire consequences for the ability of the Axis powers to retain at least some control over the skies. This was so if only because both Berlin and Tokyo had gambled that their respective military doctrines offered viable alternatives to the sort of protracted struggle witnessed between 1914 and 1918. For all their initial triumphs, *Blitzkrieg* and Japanese expansionism went insufficiently far to clinch a climactic victory. The conflict's other principal protagonists – the British commonwealth, the USSR, the USA and, to a lesser extent, China – were able and willing to persevere with it, largely because the roots of so much of their might lay safely beyond their enemies'

---

50 Minute by the Director of the Naval Air Division, 4 January 1939, TNA, ADM 1/9720.
51 For an insight into the FAA's views regarding the respective merits of two-seater and single-seater fighters, see Document 54, 'Memorandum from Fifth Sea Lord to Admiralty Board', 21 June 1940, in *FAA in Second World War: Volume One*, pp. 167–9.

reach, protected by a daunting combination of geography and active defences, not least sea and air power. As time passed and ever more of the Allies' resources were mobilized, the Axis struggled to contain its enemies, let alone defeat them, be it on the land, the sea or in the air, but attrition among relatively tiny bodies of highly trained, specialist personnel, such as aircraft technicians and pilots, had particularly pronounced ramifications for the sustainability of the contest as a whole. Casualties incurred in almost relentless combat against ever increasing numbers of opponents left neither Japan nor Germany with any choice but to replace seasoned, accomplished aviators with novices who seldom survived long enough to develop any latent skills they might have possessed. Such ten-derfoots were pardonably incapable of getting the best out of the technology at their disposal, which, in the case of the IJN, came to include new, sophisticated planes such as the Kawanishi N1K2-J *Shiden* and the Mitsubishi J2M *Raiden*.[52] Indeed so desperate did the Japanese become that, towards the end of 1944, they resorted to giving new pilots the most elementary training and forming them into *Kamikaze* and *Tai-atari* squadrons, the sole function of which was to ram enemy ships and bombers, respectively.[53]

Some twelve months prior to this, the *Luftwaffe* was to find itself the target of a campaign that was specifically devised as a means of annihilating German air power.[54] In trying to counter the Combined Bombing Offensive in Europe, interceptors were repeatedly sucked into engaging big formations of USAAF bombers that had equally large (and very capable) fighter escorts. A crippling rate of attrition proved inescapable. During November and December 1943, casualties among the pilots of the *Luftwaffe* amounted to 21 and 23 per cent, respectively. By the spring, half of its remaining fighters and a quarter of its pilots were being lost each month.[55] The resulting decline in the calibre of the aviators struggling to defend the *Reich* was the single most important factor in account-ing for their demise.[56] Certainly, countermeasures that included quantitative and qualitative improvements to their equipment failed to stop the rot. Besides switching more manufacturing capacity to the production of fighters in order to meet the growing threat, during 1943 and 1944 Germany also introduced new machines that were technically outstanding, notably the formidable Focke-Wulf 190.D9 and the world's first jet fighter, the Messerschmitt Me.262A.

The USAAF's counterforce campaign was to encompass the systematic bom-bardment of the material foundations of German air power. These included aircraft-assembly lines as well as facilities for the production, storage and distri-bution of aviation fuel and lubricants. Some airfields were bombed, too, together

52 See: Peattie, *Sunburst*, p. 93.

53 See: Albert Axell and Heideaki Kase, *Kamikaze: Japan's Suicide Gods* (New York, NY, 2002).

54 See Biddle, *Rhetoric and Reality*, pp. 215–16, 226–7, 232; Haywood S. Hansell, *The Air Plan That Defeated Hitler* (Atlanta, 1972), pp. 206–9.

55 Williamson Murray, *Strategy for Defeat: The Luftwaffe, 1933–1945* (Maxwell, AL, 1983), pp. 211–15.

56 Hansell, *The Air Plan*, pp. 208–9.

with their concomitant command and control complexes, munition dumps and maintenance workshops. The material effects of these attacks tended to be ephemeral, however. The bigger and more labyrinthine a target, the harder disabling it with conventional explosives is likely to prove. For instance, by using angled, if relatively short, trajectories, planes might still be able to take off from a runway that is quite badly pitted with bomb craters or liberally sown with mines. With sufficient time, labour and other resources, physical damage can in any event be made good. The upkeep of morale – the single most important factor in warfare – and of the corporeal and mental stamina of combatants and those who sustain them are often more challenging tasks.

Most of the British aviators disputing control of Norway in the spring of 1940 were operating from carriers. These were that much more vulnerable and inflexible than the airfields being used by the *Luftwaffe*. Stationed on terra firma, the Germans were less restricted in terms of the numbers and sizes of aircraft they could employ, which, in turn, affected the quantities and types of ordnance they could bring to bear. Whereas two strikes by *Swordfish* and *Skuas* against Vaernes and the seaplanes based in Trondheim's harbour were deemed by the FAA to be 'most effective', such raids could at best inflict sufficient physical and psychological damage to slow down the tempo of the enemy's sorties.[57] A ship, by contrast, although harder to find than fixed bases, was, if discovered, exposed to potentially deadly threats: a solitary, well-placed bomb might render the flight-deck of even the most robust of carriers inutile, even if it did not destroy the vessel outright. This imbalance of risks could only favour the *Luftwaffe*: it was imperative that the FAA safeguard its floating aerodromes against potential threats, the size and nature of which was all but impossible to gauge, while simultaneously trying to maintain some capacity for offensive action. Yet, if only because of the need to top up the carriers with fuel, munitions and other essentials, there was normally only one actually cruising off the Norwegian coast at any given time. What is more, of this ship's available aircraft, some had to be kept back to help protect it in particular and the Home Fleet in general against opponents whose own machines were far more numerous and, in many cases, more versatile.

The aerial battles that erupted over the Norwegian seaboard in April and May 1940 were unprecedented and, broadly speaking, unforeseen. Not only were friend and foe alike uncertain at this juncture as to how potent modern air power might actually prove in combat, but also the forces of both sides had been moulded with rather different missions in mind. Whereas the *Luftwaffe* had but a handful of personnel who had much understanding of nautical matters per se and was primarily designed for operations against opponents other than ships, much the reverse was true of the Royal Navy: its hierarchy contained few officers who had a background as aviators, while British carriers and their planes had been developed primarily as aids to surface forces confronting, not land-based

---

57 See: Document 51a, 'Report from Vice Admiral, Aircraft Carriers, HMS *Ark Royal. . . .* Operations off Norway, 23 April – 3 May 1940', in *FAA in Second World War:Volume One*, pp. 141, 146 and 153.

adversaries, but flotillas on the high seas. Here, the quality and quantity of the aviation housed aboard the vessels of putative foes had dictated the requirement for countervailing units, 'since the functions of reconnaissance and shadowing, air striking, and air fighting, absorb the majority of the aircraft with the fleet. In this respect air inferiority cannot be counterbalanced by any superiority in cruiser strength, nor can the advantage obtained from air spotting and . . . observation be denied except by the possession of similar facilities'.[58] In such settings, if only because neither Germany nor Italy disposed of carriers of their own, any enemy aircraft that were encountered would, it was assumed, be likely to be few in number and limited in their capabilities.

By January 1940, however, this view was undergoing refinement, largely because of the maturation of aviation in general and of the *Luftwaffe* in particular. A committee chaired by Vice Admiral Guy Royle, the Fifth Sea Lord, and charged with formulating policies on fighter procurement concluded that, notwithstanding the lack of serviceable carriers within the *Kriegsmarine*, the FAA's existing interceptors would be likely to have to contend with some comparatively advanced aircraft over areas of the North Sea for sure. Besides having a lengthy reach, planes such as the Ju.88 and Me.110 were perceived to be substantially faster than the *Fulmar*, the FAA's latest accession, production of which had in any event fallen behind schedule. It seemed reasonable to suppose that, by the time any freshly conceived FAA fighter entered service in, perhaps, two or three years, enemy shore-based bombers might be able to reach speeds of 300 mph, while fighters might attain 400 mph or more. To deal with these threats and to protect its bases around the globe – a task that the RAF might prove too slow, reluctant or unable to perform – it was suggested that the Royal Navy should acquire a batch of *Hurricanes* or *Spitfires*, ideally with folding wings so that they could be transported aboard carriers, if not operated from them. Yet that any such variants could be delineated and manufactured in under nine months seemed implausible.[59] After all, the RAF's competing demand for *Spitfires* especially was already intense and marinizing an existing design was not merely a matter of bolting on a tail-hook. 'In the light of experience to date, of the characteristics of the present war, and of possible future developments', Captain Charles Daniel, the Director of Plans, also reappraised both the demand for and distribution of carriers for the first time since 1938. The strength of the *Kriegsmarine* in 1940, Daniel opined, made a major fleet action in the waters to the east and north of the British Isles an 'improbability', permitting a reduction in the number of carriers retained with the Home Fleet. More extensive use of land-based planes for maritime patrols would also alleviate the burden on the Royal Navy's cruiser squadrons, which were in need of time for 'training, repair and

58 Document 28, 'Memorandum by Director of Plans: Requirements for Aircraft Carriers', 24 January 1940, ibid., p. 75.
59 Document 27, 'Notes by Fifth Sea Lord of Meeting held on 4 January 1940', 22 January 1940, ibid., pp. 72–5.

rest'; machines with better endurance would, he pointed out, shortly start sup-planting those aircraft currently to hand – notably the Avro *Anson*, which could barely reach across the North Sea – and this could only lead to an improvement in the overall situation.[60]

The absence from Daniel's musings of any reference to Germany's burgeon-ing air power and any implications this might have for Britain's maritime grip on the North Sea was conspicuous. Certainly, in the Norwegian campaign, the Royal Navy found itself confronted with a set of circumstances that differed markedly from those that had long been used as the basis for peacetime plan-ning. Unlike many of the Admiralty's strategists far away in London, Forbes and his colleagues in the fleet off Norway quickly became only too mindful of the sheer size and sophistication of the threat that could suddenly materialize in the skies above them and the Allied troops ashore. If the balance of forces in this regard was to be redressed, land-based planes, notably bespoke interceptors, would have to supplement if not replace the FAA's ill-suited machines. This, however, would necessitate the establishment of airfields on terra firma some-where in the theatre and the importation of RAF fighters (and their logistical entourage) from the UK.

But where could airstrips with the requisite proximity and security be created and maintained? By this stage of the proceedings, the Allies were focussing their attention on the capture and retention of Narvik, all of central and southern Norway having fallen to the invaders. This town also had relatively developed harbour facilities that, moreover, played a pivotal role in the export of iron ore and other valuable commodities from Sweden – trade that the French and British were anxious to deny the Germans if not commandeer for themselves. A couple of sites for airfields were identified, notably Bardufoss, some fifty miles north of Narvik.[61] An anchorage for *Walrus* flying boats was also set up alongside the Allies' main depot at Harstad on the Lofoten Islands, just to the west.[62] HMS *Glorious* and *Furious* were duly assigned to ferrying in two RAF squadrons – one equipped with 16 *Hurricanes*, the other with 18 *Gladiators* – from Britain. Of older design, these were the only carriers with flight-deck lifts that were spa-cious enough to accommodate fixed-wing machines with the dimensions of a *Hurricane*.[63]

Transporting these fighters across the North Sea and installing them at their respective airstrips took until the last week of May, whereupon much of the responsibility for safeguarding the skies around Narvik passed from the FAA to

60 Document 28, 'Memorandum by Director of Plans: Requirements for Aircraft Carriers', 24 Janu-ary 1940, ibid., pp. 75–8.

61 See, for instance, Brown, *Carrier Operations*, p. 17.

62 A diary of the activities and experiences of this force can be found in 'Fleet Air Arm Force Bishop: Letter of Proceedings', 17 June 1940, TNA, ADM 199/480.

63 See: Clause 12, Document 27, 'Notes by Fifth Sea Lord of Meeting held on 4 January 1940', 22 January 1940, in *FAA in Second World War: Volume One*, pp. 74–5.

the RAF. The carriers were duly withdrawn to Scapa Flow for recuperation and replenishment.[64] In the interim, planes from the *Ark Royal* had had to bear the brunt of the aerial battle alone. For her own safety, the ship had had to be kept well to the north and all of a hundred miles from the coast. The sheer remoteness of their base from the epicentre of the battle on the ground ineluctably meant that her aircraft could only provide support for the Allied troops sporadically, particularly at times when engagements were raging on scattered sectors of the front. It was noted that, whereas requests for fighter assistance came from far and wide, patrols could not be simultaneously maintained over the entire theatre 'unless the carrier operated from a position too close inshore to be acceptable'.[65] This dilemma spawned others. To name but one, lengthy stretches of beach were left unobserved, let alone protected, for much of the time, with the result that on at least one occasion the Germans were able to get behind entrenched forces by disembarking parties of soldiers from seaplanes and coastal vessels.[66] The elements, too, helped cover these turning movements, just as they hampered the efforts of the FAA to protect the Allied armies from aerial bombardment. Vice Admiral Wells attested that: 'Generally the weather was unfavourable for the operations of fighter aircraft, a great deal of fog and low cloud being encountered. Often the clouds were low at sea and over . . . the coast, while conditions further inland . . . were [more] favourable. . . . [Enemy] bombers often appeared to be making use of a route along the Swedish border clear of the low coastal cloud, and thus they were able to reach their objectives while our fighters could not penetrate from seaward'.[67]

These adverse circumstances deeply affected the frequency, variety and outcomes of missions mounted by the FAA during this episode of the campaign, the hostile environment often posing more problems for the British aircrews than their opponents did. The cardinal restraints on the *Luftwaffe* – which largely confined itself to aiding the *Wehrmacht* through raids on the Allied troops encircling Narvik and the convoys that supplied them – were the amounts of ordnance and fuel its numerous planes could carry. Meteorological conditions, not least the almost perpetual daylight experienced in such northern climes at this time of the year, favoured its operations more than those of the British aviators. Simply regaining the *Ark Royal*, particularly in gales or periods of poor visibility, could prove hard enough. Returning from lengthy, hazardous sorties to far-flung objectives, aircraft and their occupants were frequently at the end of their tethers and were obliged to make forced landings on a pitching, rolling

64 Document 50a, 'Report from Vice Admiral, Aircraft Carriers, to Commander-in-Chief, Home Fleet: Operations of HMS *Ark Royal* off Narvik, 4–24 May 1940', ibid., p. 122.

65 Ibid., pp. 137–9.

66 Till, *Air Power and the Royal Navy*, p. 23.

67 Document 50a, 'Report from Vice Admiral, Aircraft Carriers, to Commander-in-Chief, Home Fleet: Operations of HMS *Ark Royal* off Narvik, 4–24 May 1940', in *FAA in Second World War: Volume One*, p. 121.

flight–deck lashed by spray. Several machines were lost or damaged in mishaps of various kinds. Neither was a carrier that, for days on end, had to linger within striking distance of enemy submarines and land-based bombers the best of places for men to rest and relax in between bouts of combat.[68]

Indeed occasional attempts by the *Luftwaffe* to shadow or assail *Ark Royal* and her escorts gave rise to a novel dilemma that she and the *Glorious* had first encountered during the evacuation of central Norway at the beginning of May. Few in the British Admiralty had envisaged carriers having to choreograph aerial operations of the size, complexity and variety witnessed along the Norwegian littoral. Assimilating and processing the information generated by protracted dogfights especially could tax personnel and equipment alike. The commander of one carrier noted that the volume and intricacy of information submitted by pilots and observers on returning from sorties could prove overwhelming. 'The evidence they produce of the same event is amazingly varied and requires the analysing powers of a King's Counsel', he observed. Small committees should be established on carriers to sift the reports of returning aircrews just as soon as they landed, he recommended. A coherent account of what had transpired might then be compiled in a suitably timely manner. Such essential work was, he insisted, 'quite beyond the Captain or his flying staff who are unceasingly engaged in handling the ship and the aircraft'.[69]

Similarly, the growing reliance on technological interfaces between people and their wider surroundings as a mechanism for gathering and disseminating information could prove a mixed blessing. Although radars mounted aboard some of the larger ships helped detect the presence of planes, the difficulty experienced in distinguishing friend from foe in an aerial melee could increase in proportion to the number of 'signatures' received. RAF fighters geared to the policing of the UK's airspace were equipped with radio emitters – 'pip-squeak' – in order that their location could be monitored within the country's integrated air-defence system. Carriers and their escorts could not as yet replicate such an operating framework, partly because FAA planes had no automatic means of identifying themselves either to one another or to the vessels with which they were supposed to act in concert. Not only did this compound the complexities inherent in vectoring interceptors against putative opponents but also it added to the chances of fratricide. While covering the withdrawals from Åndelsnes, Molde and Namsos, it was noted that *Glorious* had considerable difficulty in directing her *Gladiators* towards enemy aircraft. Although radio links with the interceptors worked satisfactorily enough, messages were so numerous as to be dangerously bewildering: many of the targets pointed out to the fighters were 'undoubtedly the *Gladiator* patrols themselves. Section Commanders of . . .

---

68 See: Document 51a, 'Report from Vice Admiral, Aircraft Carriers, HMS *Ark Royal*. . . . Operations off Norway, 23 April – 3 May 1940', ibid., p. 141.

69 Document 39, 'Report of Proceedings from Commanding Officer, HMS *Furious*. . . . Operations off Norway, 8–28 April 1940', ibid., p. 107.

[those] patrols also reported being repeatedly fired on by . . . [their] own ships'.[70] Captain Guy D'Oyly-Hughes, the commander of HMS *Glorious*, subsequently pleaded that there had been 'such a constant stream of bearings and distances' from the radar plotters aboard the flotilla surrounding the carrier that it was extremely hard to decide what pieces of information to pass on to the *Gladiators* and what to give to the lookouts and AA gunners aboard the surface vessels.[71]

There were implications here, too, for the number, calibre, distribution and control of any AA weaponry affixed to such vessels. In seeking to hit ephemeral, aerial targets from platforms that were themselves not motionless, gunners aboard ships had to resolve some perplexing problems, not least some convoluted geometric equations. Faced with rapidly moving adversaries, in order to have much prospect of proving comprehensive any static air-defences had to be wieldy and potent enough to engage targets at long, medium and close range as well as at differing bearings and angles of elevation. This, however, was easier said than done, particularly before AA warheads with dependable proximity fuses – rather than timer or impact variants – were perfected. So-called barrage fire actually amounted to the timely generation of envelopes of mid-air explosions that, it was anticipated, attacking planes would traverse. Yet the surrounding sky was bound to include zones through which approaching planes might travel too low and fast for heavy guns to be trained upon them but which also lay beyond the range of armaments that, though more manoeuvrable, lacked reach and clout. The number of targets that could be engaged at any distance was, furthermore, constrained as much by the number of fire-control directors as it was by the quantity of armaments that was nominally available.

Such considerations had always dictated that, ideally, a blend of weaponry with contrasting performance characteristics be employed to generate concentric circles of fire. For much of the interwar period, however, the countering of aircraft that were flying straight and level and at either relatively high or low altitudes had been accentuated. Together with the substantial increase in the average speed of combat planes, the proliferation of dive-bombers during the late 1930s complicated matters enormously. Whereas torpedo attacks had to be delivered from within a few hundred feet of the surface and horizontal bombing was normally done from several thousand feet above the target, dive-bombers manoeuvred far more in three rather than two dimensions, plunging towards their prey at extraordinary velocities. Since even tachymetric director systems that could track individual planes with considerable precision required several seconds at least to compute a fire-control solution for a selected target, even the best of fixed AA defences could take too long to react to such attacks in particular.

The vagaries of trying to hit moving targets with projectiles released from moving platforms were such that a split second or a fraction of a degree made

---

70   Document 51a, 'Report from Vice Admiral, Aircraft Carriers, HMS *Ark Royal*. . . . Operations off Norway, 23 April – 3 May 1940', ibid., p. 159.

71   Ibid.

all the difference between success and failure. For both sides, engagement windows were seldom open for long. When granted adequate warning, vessels often resorted to radical manoeuvre in a bid to dodge torpedoes and bombs. Most warships and a lot of merchantmen had a sufficient turn of speed to simply outrun many designs of torpedo and even relatively sluggish ones might sidestep approaching ordnance. But the act of jinking complicated affairs for both attacking aircraft and any active AA defences aboard the target vessel. Abrupt turns could also incur the risk of collisions with any craft sailing alongside. Judgements had to be made as to whether evasion would prove a better safeguard than gunnery and whether vessels would derive more protection from moving in formation than independently. Optimally, any manoeuvres would be executed in such a way as to bring the maximum amount of AA fire to bear. Indeed, particularly in the Second World War's early years, many ships were neither willing nor able to rely on firepower alone, if only because of shortages of customized armaments and their requisite paraphernalia, including sophisticated targeting sensors and trustworthy fuses.

For much of the conflict the IJN, for instance, was to continue to pin its hopes principally on the passive defences of individual ships.[72] Since AA gunnery was regarded as too prone to failure to be anything but a weapon of last resort, protective formations of the kind habitually used by British battle-fleets were not even incorporated into Japanese doctrine until the middle of 1943. Rather, proactive defence was emphasized, whereby fighter aircraft were expected to take the lead in thwarting aerial attacks on friendly flotillas.[73] Under this system, outlying destroyer pickets monitored the surrounding skies and helped vector the patrolling interceptors against any foes that materialized. Space was thus traded for time, allowing the core of the fleet to ready itself for action, should that prove necessary.[74]

Whilst this approach was plainly more flexible than any reliance on passive and reactive defences alone, it did hinge on the provision of fighter cordons that were dense enough to keep any residual threat down to manageable proportions. This called for not only the assignment of sufficient carriers to a given fleet but also the circumspect deployment and meticulous control of however many planes were to hand. Yet knowing in advance of actual events just how many carriers would prove enough was no more possible than identifying exactly how many fighters would have to be concentrated in time and space in order to win a dogfight. Just as warfare per se – as Carl von Clausewitz, the doyen of military philosophers, stressed in *Vom Kriege* – is more a matter of chance and probabilities than certainties, aerial battles are peculiarly dynamic affairs in which a solitary plane, particularly if left unopposed, might do tremendous harm.

The goal of the concept of air-defence is the limitation of any damage. Although the system favoured by the IJN for so doing was to fall short of

---

72 Parshall and Tully, *Shattered Sword*, p. 144.
73 See: Peattie, *Sunburst*, pp. 155–7.
74 Parshall and Tully, *Shattered Sword*, p. 138.

expectations on numerous occasions as the war unfolded, the earliest instance of catastrophic failure occurred barely six months after Pearl Harbor at the height of the Battle of Midway in June 1942. The debacle concerned stemmed essentially from an ephemeral imbalance in the force-to-space ratio of the fighter patrols cocooning the Japanese fleet: in countering threats that not only followed one another very quickly but also developed at widely differing altitudes, too many of the aircraft devoted to guarding the carriers from which they operated either exhausted their ammunition or were drawn dangerously far from their charges.

The sequence of attacks that culminated in the wrecking of three of those carriers in the space of a few minutes commenced at about 09:15 on 4 June. The first assailants were numerous torpedo planes, nearly all of which were eventually destroyed. Several managed to push through the aerial curtain, however, dragging the Japanese interceptors down to the surface and obliging the straggling line of four carriers at the heart of their fleet to take evasive action. Before these ships could turn back into the prevailing wind and launch additional fighters, American dive-bombers hove into view amidst the broken cloud high overhead. With next to no interceptors left at this level, the only armaments the Japanese could train on this new threat were the relatively few AA weapons aboard their vessels. The majority of these were mounted on the carriers themselves and, their crews evidently caught unawares, many were slow to begin shooting. Targeting dive-bombers was in any case notoriously difficult at the best of times. Certainly, the volume and accuracy of the AA fire proved insufficient to parry the *Dauntlesses*, most of which descended on the pirouetting *Kaga* and *Sōryū* while three swerved towards *Akagi*. Although several bombs went astray, *Sōryū* was struck thrice and *Akagi* once for sure, while *Kaga* sustained at least four direct hits.[75]

The compromise between proactive and passive defence now showed itself to have been too lopsided. Whereas British armoured carriers might have withstood punishment on a similar scale, the structural weaknesses of the Japanese vessels helped finish what the Americans' bombs had started; fires – many caused by burning petrol (which was very hard to extinguish with water alone) – and secondary explosions snuffed out any attempts to save the stricken ships. Dispersion as a means of protection had likewise been neglected. In assembling four carriers at one point in space rather than agglomerating aircraft flying from widely scattered ships, the Japanese had gambled that their entire strike force would not be vulnerable to a knockout blow.[76]

On the other hand, clustering surface units together like this should in principle have enhanced their ability to actively resist attacks with overlapping arcs of AA fire.[77] At Midway, however, the risks that came with concentration outweighed the benefits, for the AA weaponry aboard the Japanese fleet as a

75 See: ibid., pp. 205–43.

76 See: Peattie, *Sunburst*, pp. 158–9; 148–52.

77 For some discussion of the IJN's doctrine and capabilities regarding ship-borne AA fire, see ibid., pp. 157–8.

whole was insufficient in either quantity or quality to make a significant difference. Only four battleships and heavy cruisers, eleven destroyers and a light cruiser surrounded *Kaga, Sōryū, Akagi* and the fourth carrier, *Hiryū*. So sparse and maldistributed was their collective AA armament that the carriers' guns alone accounted for nearly sixty per cent of the aggregate 'throw weight' – the amount of shot that could be discharged in a given period – with any one of the carriers generating twice the volume of fire produced by the weapons of the light cruiser *Nagara* and all the destroyers put together.[78] One particularly thorough analysis of the battle suggests that more American planes fell victim to accidents at Midway than were downed by ships' guns. Of the 146 US aircraft lost on 4 June, only two were definitely claimed by fire from surface forces.[79]

This performance contrasts unfavourably with either that of the American warships involved in the battle or the AA batteries protecting Midway's shore installations.[80] Impressive though the IJN's fighter planes were, the Japanese had entered the Second World War trailing the USA and even the hard-pressed British when it came to increasing the quantity, variety and quality of AA defences available to maritime forces. Surveys by both the US and Royal Navies of what had occurred at Midway not only stressed the importance of adequate fighter cover for shipping and friendly strike aircraft but also recommended providing carriers with larger escort screens to help keep hostile planes at bay.[81]

By this juncture the potency as well as the number of AA defence systems incorporated into many British and American vessels had also been growing incrementally for some years. Well before the 1930s were out it was being acknowledged that, if the volume and acuity of defensive shot were going to remain sufficient to threaten aircraft that were getting ever faster and sturdier, advances in AA technology and the art of gunnery would have to keep abreast of some of those in aviation, notably the supplanting of mechanical devices and human muscle and faculties by electrical engineering: automated and semi-automatic mountings had to supersede those controlled by hand; single barrel guns had to give way to more complex variants (which devoured that many more shells); electromagnetic sensors had to be exploited in not only surveillance instruments but also projectile fuses; target-acquisition and fire-control had to be centralized; and there had to be refinements to the ergonomics of operating AA armaments, if only because of the growing mechanization of reloading procedures and the expansion and acceleration of ammunition consumption.[82]

But where were all these guns and their concomitant infrastructure to be accommodated? The components of any air-defence network can be as mutually

---

78 Parshall and Tully, *Shattered Sword*, p. 144.

79 Ibid., p. 145. Also see ibid., p. 178.

80 See: ibid., pp. 202–4; 295.

81 See: Director of Admiralty Air Cooperation Division [Captain J.P. Wright], 'Tactical Lessons from the Battle of Midway', 24 August 1942, TNA, ADM 199/1302; and Ministry of Defence (Navy), *War With Japan, Volume II: Defensive Phase*, pp. 162–3. Also see: Peattie, *Sunburst*, pp. 157–8.

82 See: Andrew Gordon, *British Seapower and Procurement Between the Wars* (Basingstoke, 1988), pp. 196–7.

competitive as they are reinforcing. The Royal Navy as a whole, including its carrier arm, was essentially configured for countering hostile shipping on the high
seas, including the protection of trade routes from surface raiders and submarines.
Was it to be assumed that, henceforth, much more capacity for dealing with
enemy aircraft would be needed? Should destroyers, for instance, be designed to
help ward off U-boats and capital ships, or long-range, land-based reconnaissance
planes and bombers? Should some – or all – of their customary armament of
torpedoes and gun-turrets give way to AA artillery and racks for depth-charges?
Or should the traditional naval gun (and its projectiles) be modified so as to cope
with targets other than submarines caught on the surface and with other ships?

Part of – by way of an illustration – the Royal Navy's solution to this conundrum was to create a new breed of warship, one that specialized in air-defence.
The prototypes of these were rejigged veterans of the First World War, the light
cruisers *Curlew* and *Coventry*, which were transformed during the late 1930s
into AA platforms through the substitution of a mixture of quick-firing, small
bore cannon and 'pom-pom' guns for their original complements of heavy, anti-
shipping weaponry.[83] These and a second tranche of re-engineered hulls were
gradually supplemented with new, bespoke AA cruisers, namely the 'Dido' class,
which started entering service in 1940. It was originally envisaged that these
would have as many as ten 5.25" dual-purpose guns mounted in turrets in order
that they might retain some capacity for action against surface targets as well as
against aircraft. However, at this juncture competing demands – not least from
Britain's army – for common or garden medium guns were proving insatiable.
Devising and manufacturing a new, more versatile weapon and tailoring it to
nautical use was ineluctably a slow, relatively costly process. If only because fully
dual-purpose versions were often in short supply, the Royal Navy was obliged to
choose which elements of the fleet should receive them as a matter of priority
and which, for the time being at least, should run the risk of persevering with
armaments that were fundamentally monofunctional.

In addition to having distinctive complements of weaponry, the Royal Navy's
AA cruisers were also among the very first vessels to pioneer the exploitation of
radar for reconnaissance and targeting at sea. The *Curaçoa* and the *Carlisle*, for
instance, both of which participated in the Norwegian campaign, were respectively equipped with Type 279 and 280 early-warning suites. Located very close
to the surface, these primitive electronic eyes had a surveillance envelope that was
shaped by the Earth's curvature and their sensitivity was heavily dependent upon
atmospheric conditions and on the attributes of any object that fell under their
gaze. Still, in cool and temperate climes they could usually detect aircraft flying
at 10,000 feet at around sixty miles distant; in hotter, more humid conditions,
they were likely to prove less efficacious.[84] When harnessed to AA batteries and

---

83 For details of these ships see: Brown, *Nelson to Vanguard*, pp. 155–6.
84 For an analysis of the application of radar to the maritime environment, see: Derek Howse, *Radar at
Sea: The Royal Navy in World War 2* (London, 1993).

fast monoplane interceptors, radar transformed the capacity of fleets to protect themselves from aerial attacks. Nevertheless, it should be remembered that, firstly, electronic sensors did not in themselves have the ability to destroy things and, secondly, that such instruments were mere adjuncts to the organic receptors of the human brain, the entity that actually interpreted any visual information and thereby strove to segregate reality from mirages. The explication of the relationship between blips and squiggles on a radar monitor and the physical environment these phenomena were believed to reflect could easily give rise to illusions. 'Clutter' could dupe combatants, sometimes spectacularly so, one of the most notorious instances of which being the so-called 'Battle of the Pips' of 26 July 1943, whereby ships and planes of the US Navy detected and fired on what was thought to be a Japanese flotilla moving under the cover of dark-ness. In fact the dots on the Americans' radar screens were not enemy vessels but return echoes from distant mountain peaks. On other occasions, however, the Allies were to turn such misperceptions to their advantage, most notably on D-Day in 1944, one of their principal stratagems for deluding their adversaries as to the true site of the impending landings being the deliberate generation of misleading radar signatures. Through the laying of meticulously devised pat-terns of reflective foil above the English Channel, aircraft conjured up a virtual invasion fleet on the monitors of the Germans' *Seetakt* coastal radars around Boulogne and Le Havre.[85]

Britain's AA cruisers played a significant role in the expedition to Norway. Besides helping to shield transport vessels and the naval task-forces from enemy planes, they frequently loitered close to shore, extending their protective umbrel-las landwards over some of the Allied troop contingents, many of whom were badly lacking in AA weaponry of their own. In this regard these cruisers were particularly active in trying to thwart the raids by the *Luftwaffe* on the footholds of Åndalsnes, Molde and Namsos, HMS *Carlisle* serving as rearguard to the evacuation convoys when the last of these ports was grudgingly relinquished at the beginning of May.[86] Likewise, later that month the *Curlew* ventured into the confined waters of the Ofotfjord, seeking to protect the shipping and soldiers converging on Narvik. Here she foundered after being bombed by one of the numerous Ju.88s that were overflying the area. Narvik fell to Franco-Norwegian troops two days later, the German garrison fleeing towards the Swedish frontier.

As a counterpoise to hostile air power – exposure to which was expected to be all but minimal, certainly as far as the open seas were concerned – the generation of British AA cruisers developed between 1935 and 1939 was something of a contingency. Useful though their combination of sensors and firepower might prove in constricted waters especially, AA cruisers were ines-capably somewhat inflexible in comparison to many other escorts, not least because any defence they offered was essentially reactive, rather than proactive.

85 See: Gates, *Sky Wars*, p. 72.

86 Till, *Air Power and the Royal Navy*, p. 22.

They were also vulnerable to attack from other surface vessels equipped with torpedoes and bigger guns. On the other hand, if deployed as part of a balanced force they might liberate accompanying platforms from the stressful distractions of active air-defence.

Indeed first the desire and then the need to ring new and older ships alike with more AA weaponry had a significant effect on the architecture of fleets as a whole as well as on that of many individual vessels. Alterations and innovations varied with national priorities, particularly once the 'Phoney War' turned into a 'total' conflict that was as universal as it was intense. Whereas the British started stripping torpedo tubes, turrets and even some boilers from older destroyers to create specialist mercantile escorts that were primarily geared to anti-submarine and anti-aircraft operations, the US Navy made dual-purpose guns the standard armament of its destroyer designs as well as introducing a class of AA cruisers, the first of which, the USS *Atlanta*, entered service in 1941.

Such vessels can be pictured as a comparatively cheap alternative to sea-based air power, notably fighters. Indeed among the issues at the heart of the doctrinal debates about the respective merits of bombers and battleships was the question of whether planes were generally better instruments for the delivery of direct and oblique fire than artillery pieces. As the Second World War dawned, the head of America's BOA, Rear Admiral William Moffett, decried the adding of heavy AA guns to aircraft carriers, insisting that these ships' own aviation should be relied upon to shield them against assaults from above the waves.[87] The Royal Navy's latest carriers were, by contrast, primarily designed less with aerial than sea battles in mind – a presumption that was embodied in the number and characteristics of the machines these vessels bore as well as in the robustness of the carriers' passive defences. Stout hangars and armoured flight-decks left that much less room for aircraft, with the result that British carriers housed substantially fewer planes than their American and Japanese counterparts. The interwar arms-control regime and financial stringencies had, moreover, led to comparatively few carriers being constructed overall, while some of those that were started were never fully finished. None of the first-class maritime powers entered the conflict with more than seven sizeable vessels of this type, while neither the Germans nor the Italians had a single serviceable carrier between them. Although France's ageing *Béarn* – a vessel that was originally laid down as a battleship in 1914, partially finished and then slowly re-engineered as a carrier during the 1920s – was not absolutely inutile, commissioned as an experimental design she had long been scheduled for replacement by a new class of ship that had never actually materialized, not least because of inter-service rivalries.[88] As it was, the modernization of the French Navy in the late-1930s devoured around a quarter of all military expenditure, while proposals for the creation of a discrete

87 See: Polmar, *Aircraft Carriers: Volume One, 1909–1945*, p. 51.
88 Chesneau, *Aircraft Carriers of the World*, pp. 63–6.

corps of maritime planes, however modest, exacerbated ongoing disputes about the role and control of aviation per se; the all-important army was anxious that the air force – which had become an autonomous organization like the RAF in 1934 – concentrate on procuring machines and weaponry optimized for operations over dry land. Briefly pressed into front-line service in 1939, *Béarn* was to join the search for the *Graf Spee* but, if at sea at all, was normally used as a ferry for aircraft rather than as a platform for combat aviation. In fact at the moment when the French government was to capitulate in June 1940, she was on her way home from the USA with a cargo of American-manufactured planes. She was to spend much of the rest of the war laid up in Martinique.[89]

Any concentration of strength is by definition relativistic. At the tactical level, air-defence umbrellas – be they composed of airborne interceptors, fire from AA weaponry mounted on ships, or a mixture of the two – often resembled a poorly-fitting garment in that they were irksomely flabby in some places and uncomfortably tight in others. But experience in the Norwegian campaign suggested that, if deployed in a compact, protective formation, a balanced flotilla could hold its own against even quite large numbers of modern aircraft, primarily because it could exploit the synergy between its contrasting components. Blueprints for the architecture of fleets as a whole were, on the other hand, essentially strategic designs, being rooted in assumptions about the contemporary, geopolitical landscape. During the 1930s these, together with a harsh economic climate and arms-control accords, had moulded development and procurement policies that ineluctably took years to yield any tangible results. In the interim, incremental alterations to the geostrategic environment had invalidated some of the presuppositions about both the need for and the nature of martial might. In the Europe of the Locarno Treaties, there had been no hostile air force within striking distance of Britain. Nor had there been a navy centred on European waters that could pose a serious challenge to her own. Whereas the outbreak of war in 1939 was as much a symptom as a consequence of just how much matters had changed in the space of little more than three years, nobody could have foreseen the utter transformation of the map of the continent that was to occur during May and June 1940.

France in particular had hoped that the Norwegian campaign would divert German resources and attention away from her and the neighbouring states of Belgium, the Netherlands and Luxembourg, all of which had proclaimed themselves to be neutral when the Second World War began.[90] However, on 10 May 1940 – the day that Neville Chamberlain stepped down as Britain's prime minister and was replaced by Churchill[91] – scores of German armoured and

89 See: Jürgen Rohwer, *Chronology of the War at Sea, 1939–1945* (Annapolis, MD, 2005), p. 6; Max Hastings, *All Hell Let Loose: The World at War, 1939–1945* (London, 2011), p. 74.

90 See: Derry, *Norway*, p. 81.

91 For an analysis of this succession, which was not a foregone conclusion, see: Roy Jenkins, *Churchill* (London, 2002), pp. 570–610.

infantry divisions, backed by hundreds of aircraft, swept westwards across the frontiers of these countries. Tiny Luxembourg, which did not even have a standing army, was gobbled up almost instantaneously, much as Denmark had been just a month earlier. The Dutch – who, like the Swedes, had not participated in a war since Napoleon's time – lacked recent operational experience as well as up-to-date combat planes and land forces, particularly tanks, field artillery and AA weaponry. Their security rested primarily on passive defences and the goodwill of others: not only had Hitler issued assurances that the Netherlands' sovereignty would be respected but also the Dutch had long since been persuaded that their homeland's numerous waterways would severely hamper any invader, if only for a while. In the interim, help from outside – notably from the French, whose colossal army of 94 divisions was the principal counterbalance to German military might on the continent – would surely be forthcoming.

But the French did not intervene, if only because the German *Blitzkrieg* lived up to its name: it overcame the passive and active defences of the Netherlands in next to no time. Airborne troops seized key bridges, allowing armoured spearheads to penetrate the country's heart. On 14 May, having seen Rotterdam pounded mercilessly by the *Luftwaffe*, the Dutch government capitulated. Similarly, Eben Emael – the ostensibly formidable fortress that was the linchpin of Belgium's man-made defences – was overrun by airborne units in the opening hours of the offensive. Elsewhere, Belgium's small armed services offered spirited resistance but, virtually devoid of tanks, AA weaponry and modern combat planes, could scarcely dream of repelling the invaders. Although thousands of French troops and much of the BEF were promptly dispatched to contain the Germans in this quarter, this response played right into the hands of their enemies, who, bent on avoiding any repeat of the positional warfare seen in Flanders between 1914 and 1918, were planning to execute a massive turning movement far to the south. As 33 of the best Franco-British divisions shifted eastwards to occupy prepared lines in western Belgium, 44 German divisions swung sickle-like through Luxembourg and the Ardennes – a hilly, wooded region thought to be impassable to major forces – towards the unfortified French frontier around Sedan. Here, the principal obstacle they faced was a natural one, the River Meuse, over which, supported by some fifteen hundred aircraft, including many *Stukas*, they swiftly blasted their way before debouching northwards to cut off the Allied armies in Flanders. Seeking to evade the closing trap, these in turn hastened westward, abandoning Belgium to her fate.

On the 28 May – the day that Narvik was liberated by Franco-Norwegian soldiers – Belgium's king, Léopold III, hoping to obtain a similar settlement to that granted by Hitler to the Danes, surrendered both himself and his realm, notwithstanding the advice of many of his compatriots. Foremost among these were cabinet ministers and generals who had urged him to join Queen Wilhelmina of the Netherlands in continuing to inspire and orchestrate resistance from beyond the English Channel. Irrespective of Léopold's views, Belgian politicians did eventually follow the Poles and Dutch in setting up a government-in-exile in London. These bodies recruited and organized armed forces that, equipped by the British, were, together with individual citizens from various neutral and occupied countries, integrated into the Allies' order-of-battle. Among them were several Dutch warships that had been based in the Netherlands East Indies and numerous volunteers – many of whom served as airmen – from Czechoslovakia, Eire and the USA. There were significant numbers of merchant seamen as well, not least Belgian, Dutch, Polish and Norwegian crews who had spirited their vessels away, in some cases from under the Germans' noses.

Norway's cabinet and monarch were also destined to take refuge in London. By mid-May it was painfully clear that the French and British would have to disentangle themselves from Scandinavia and concentrate their efforts on the all-important western front, where the situation was getting worse with every passing hour. Strategically speaking, the Allies had been wrong-footed within days of the Germans commencing their offensive here and, wherever the hostile armies slammed into one another, the Franco-British and Belgian forces struggled to cope with the interplay of mechanized ground units and air power that was the essence of *Blitzkrieg*. If the damage sustained by the FAA's carrier-borne aviation during five weeks of operations in Norwegian skies was appreciable, that inflicted on the RAF in France in as many days was catastrophic. What happened at Sedan during the afternoon and evening of 14 May alone was crippling enough. Here, in one of the pivotal engagements of the Franco-Prussian War, sword-wielding French cavalry had, with reckless but useless courage, repeatedly charged into hails of German bullets in a desperate bid to contain the invaders' advance. Even King William of Prussia – soon to be proclaimed the first emperor of a united, victorious Germany – was moved to remark on the sheer bravery of the enemy horsemen.[92] Now, seventy years later, the descendants of those who had witnessed that fateful battle watched in bemused admiration as squadrons of bombers from the Advanced Air Striking Force (AASF) – the aviation component of the BEF – fought their way through screens of Me.109s and storms of AA fire to assail the Germans' pontoons and bridgehead on the Meuse.

Writers such as Italy's Gabriele d'Annunzio had eulogized early aviators as modern knights mounted on technological steeds. Indeed when the RAF – the world's first autonomous air force – was established in 1918 its trappings and equipment were a striking blend of cutting-edge machinery and venerable chivalric icons: modern, metallic-blue uniforms were adorned with accoutrements long associated with cavalry units especially.[93] Many British planes dating from the mid-1930s were hybrids, too, the products of fitful procurement programmes that were moulded by competing political, diplomatic and financial pressures as well as doctrinal ones. The machines that attacked the Sedan bridgehead on the afternoon of 14 May were predominantly Fairey *Battles*. These metallic, light bombers were not so much thoroughbred chargers as laggardly cross-breeds,

---

92 See: Michael Howard, *The Franco-Prussian War* (London, 1981), pp. 215–16.
93 See: Gates, *Sky Wars*, pp. 12–20.

being equipped with a solitary if good engine – a Rolls Royce 'Merlin', the same power plant that was installed in the *Spitfire* – but weighed down with a crew of three and up to 1,500 lbs of ordnance. Although it looked like an elongated fighter – it was in fact among the antecedents of the FAA's *Fulmar* – the *Battle* lacked the pace and agility of the German interceptors that opposed it.[94] Eight *Blenheims* were also committed to the raid. Of these, five were destroyed, while no fewer than 35 of the 63 *Battles* were shot down. This remains the highest rate of loss ever incurred by the RAF in any operation of comparable size.[95] Nevertheless, 28 more *Blenheims* renewed the assault in the evening, a quarter of them failing to return. When added to the casualties sustained among the French *Bréguets*, *Amiots* and *Léots* that had tried to strike targets around Sedan earlier that same day, by sunset the Allies had lost approximately ninety aircraft on this sector of the front alone. (The Germans believed the figure to be 112.) What was certain, however, was that: the pontoons had sustained negligible damage; the *Panzer* groups were continuing to roll onto French soil; the Allies' tactical bomber forces were all but ruined; and the commander of the flak batteries guarding the bridges, Colonel von Hippel, was awarded the *Ritterkreuz* for his outstanding services that day.[96]

Remarkable though the casualties at Sedan were, this clash encapsulated a wider failure by the Allies' aviation to contend with the close liaison between the *Luftwaffe* and German surface units. For the AASF alone the air-land battle that was *Blitzkrieg* had a truly ruinous cost. Expressed in relation to the total number of sorties flown, the rate of loss among its planes had been 40 per cent on 10 May and 100 and 62 per cent on the following two days respectively. By nightfall on 12 May the AASF's bomber fleet had shrunk from 135 serviceable machines to 72, prompting debilitating warnings from London that the corps needed to conserve its resources for what was unhelpfully described as the battle's 'really critical phase'; the next 24 hours were duly devoted to recuperation and refitting.[97] However, as the French premier, Paul Reynaud, realized as soon as fighting erupted there, the enemy's unexpected thrust through Sedan was potentially lethal. Henceforth, the wider battle would not be a linear affair like those of the last war. There were, he confessed to a shocked and horrified Churchill, insufficient strategic reserves with which to seal off the incursion.[98] Once across the Meuse, the invaders would be able to complete the demolition of France's defences from all sides, not least by compromising the security and operability of pivotal aerodromes. The *Panzer* divisions could only be halted if

94 See: Daniel M. March, *British Warplanes of World War II: Combat Aircraft of the Royal Air Force and the Fleet Air Arm, 1939–1945* (London, 1998), p. 105.

95 Denis Richards, *Royal Air Force 1939–1945: Volume 1: The Fight at Odds, 1939–1941* (London, 1953), p. 120.

96 Alistair Horne, *To Lose a Battle: France 1940* (London, 1979), pp. 376–9.

97 Ibid., p. 328.

98 Ibid., pp. 431–3, 444–7.

they were denied the backing of the *Stukas*, which the available Allied fighters were already struggling to restrain. If there was going to be much chance of the Germans' blow being parried, the RAF, Reynaud insisted, would have to divert ten more *Hurricane* squadrons from protecting the British Isles to the aerial battle beyond the Channel.

Anxious though he was to help, this was a risk that Churchill concluded he could not run, particularly after the head of RAF Fighter Command, Sir Hugh Dowding, showed him a projection of just how long the *Hurricanes* (and pilots) concerned would last if the current rate of attrition in France persisted. Dowding had also written to the Air Ministry's hierarchy, reminding them that their last estimate as to the number of fighters necessary to protect the UK was 52 squadrons, whereas the total available had now dwindled to the equivalent of 36.[99] In any case it was not apparent to which aerodromes any reinforcements should be sent, for the German counterforce operations were causing mayhem. At dawn on 15 May the French had no more than 237 fighters, 38 bombers and 38 night fighters with which to cover Paris and the entire north-eastern front. By midday half of the interceptors had been disabled by German bombing, while growing disruption within the communication webs was making scrambling the remainder ever harder. Similarly, the AASF found many of its forward airstrips so endangered by either the *Luftwaffe* or *Panzer* units, or both, that they had to be evacuated. An intricate undertaking made more so by ongoing enemy raids and limited transportation, this effectively put much of the AASF out of action at a critical juncture.[100] On 19 May the increasingly desperate French pitched 20 American-made naval aircraft into the fray. These *Vindicator* V.156F dive-bombers – recently purchased for future use aboard *Béarn* – were stationed around Boulogne, barely 40 miles from Dover. In their first combat sortie half of them were lost and the rest badly shot up to no avail; the enemy pincers continued to close on the Allied armies in Flanders. That same day, the decision to withdraw the AASF all the way to southern England was taken.[101] By dusk on 20 May one of the Germans' armoured spearheads had reached the Channel just south of Boulogne, having advanced all of sixty miles and more since the dawn. Within a week that port and Calais had both fallen, leaving the BEF's commander, Lord Gort, little choice other than to order a retreat on Dunkirk.[102]

By the time it regained England, of the 261 *Hurricanes* that had been allocated to the AASF only 66 remained. Just 75 had been destroyed in the fighting, the balance of 120 being made up of damaged, pilotless or otherwise immobilized machines that had had to be left behind on the continent. In ten short days of operations in the skies of northern France alone, the RAF

99 Ibid., pp. 422, 450; Jenkins, *Churchill*, pp. 594–7.
100 Horne, *France 1940*, pp. 420, 422, 531–2.
101 Ibid., pp. 531–3.
102 Ibid., pp. 550, 585–6, 596.

had lost 195 *Hurricanes* – around a quarter of its entire inventory of modern fighters.[103] Losses among its personnel were also appreciable. Besides the pilots and technicians killed or captured prior to the AASF's planes being extricated, some of the ground crews especially ended up amidst the numerous fugitives who sought to escape their pursuers via harbours in western France. There were more than a few of these men among, for instance, the thousands of soldiers and others crammed onto the commandeered liner *Lancastria* when, on 17 June, three days after Paris was occupied by the Germans, she was sunk by Ju.88s off St Nazaire. It is thought that roughly three thousand people perished in this incident – more than were lost aboard *Titanic* and *Lusitania* put together.[104]

There was no comparable, single loss among the vessels – which ranged from warships and sizeable merchantmen to little pleasure craft, Belgian fishing-smacks and Dutch *schuyts* – that, between 27 May and 3 June, rescued over a third of a million Allied soldiers from Dunkirk and the nearby beaches, although six British and two French destroyers were sunk in raids by enemy planes. The troops, furthermore, had to abandon every scrap of heavy equipment. The 'Miracle of Dunkirk' proved feasible essentially because of sea power and the determined resistance put up by the Allied armies, not only on the port's perimeter but also in remoter enclaves along and close to the Germans' axes of advance. However, benign weather and the RAF also played crucial roles, too, with all but a handful of the available interceptor squadrons taking a turn at covering the embarkations. In all, from comparatively secure bases beyond the Channel, fighters executed 2,739 sorties to keep the *Stukas* and Ju.88s at a distance, while bombers retarded the progress of enemy ground units, whose incessant demands over the preceding weeks had in any case brought the *Luftwaffe* close to exhaustion.[105] Notwithstanding this, that organization's head, *Reichsmarschall* Hermann Göring – aided, if only indirectly, by the field army's pre-eminent commander, the venerable Gerd von Rundstedt, who had his own, competing list of strategic priorities – succeeded in persuading Hitler that the *Luftwaffe* should be left to finish off the BEF and its entourage around Dunkirk while the *Panzer* divisions regrouped.[106] This decision coincided, moreover, with a change in what was widely being referred to as 'Göring's weather' – the atmospheric conditions that, since the start of the invasion of France, had so favoured aerial operations. Except on 27 May and 1 June clear skies gave way to bouts of fog and heavy cloud, the poor visibility over Dunkirk being exacerbated by palls of smoke. This further diminished the frequency and effectiveness of the German bombing raids.

103 Ibid., p. 532.
104 Roskill, *The War at Sea 1939–1945, Volume One: The Defensive*, p. 235.
105 See: Horne, *France 1940*, pp. 613–14, 617–20, 623.
106 For a discussion of the political and military factors that spawned this grossly controversial decision, see: Horne, *France 1940*, pp. 598–603; and Jenkins, *Churchill*, p. 598.

Above all, what transpired here underscored some of the handicaps inherent in pitting air power alone against a more rounded force. German aviation's limitations in this respect were to be highlighted even more vividly a few weeks after the BEF's extrication from Dunkirk. Mastery of the skies is desirable primarily because it can bestow some control over what occurs on the surface. This was certainly true of the Battle of Britain, a protracted, self-contained aerial campaign that was supposed to pave the way for the *Wehrmacht* to invade England. Yet the *Luftwaffe* was to fail in its mission not least because RAF Fighter Command was part of an integrated defensive system and because any gains Germany's pilots made could not be consolidated and exploited immediately by ground units. With the *Kriegsmarine* – which had been significantly enfeebled by the Norwegian campaign – equally unable to help bridge the Channel in the way that it had spanned the Skagerrak,[107] Hitler's mighty armies were left straining at the leash on the continent, unable to translate their tactical triumphs over the BEF into conclusive, strategic victory.

But success, though relative, is still success. Whereas in 1918, after four years of German offensives, the French had remained unconquered, in 1940 they were vanquished by a *Blitzkrieg* that lasted just six weeks. Britain was left isolated in Europe, with meagre prospects of forging a new coalition or of maintaining a toehold on the continental mainland, including her hard-won enclave around Narvik. Despite the competing demands imposed by the evacuation at Dunkirk, this, too, was relinquished during the first week of June, the Royal Navy managing to extricate the Allied troops and their equipment without significant interference from the *Luftwaffe*. At the same juncture, the strategic situation in the Mediterranean basin also deteriorated sharply. The British had long been haunted by the fear that the loss of just a couple of their capital ships off Norway would tempt Mussolini into actively joining with the Germans.[108] In the event, seeing the implosion of French martial might, he declared war on the Allies on 10 June. That same day France's political leadership fled from Paris. By 22 June it had been compelled to accept a humiliating armistice, the terms of which included the indefinite occupation of the entire western coast by German forces and the establishment alongside this '*zone occupée*' of a '*zone libre*' with its capital at Vichy. On paper the government set up here was autonomous, subject to the accord's wider provisions and, within the *zone occupée*, the German authorities in Paris. What this puppet regime might now do with France's outposts in Syria, Lebanon and North Africa and with her substantial navy were questions that pardonably troubled Churchill among others.

As a result of Hitler's conquests and diplomacy between April and June 1940, by the end of spring the whole of Europe's western littoral from the Norwegian-Finnish frontier at Petsamo on the Arctic to the Bay of Biscay had fallen under

107 In August 1940 the *Kriegsmarine* could muster just six cruisers and ten destroyers. See: Duncan Redford, *A History of the Royal Navy: World War II* (London, 2014), pp. 97–8.
108 Derry, *Norway*, pp. 75–6.

Axis control. Henceforth, from heavily defended bases in Norway and western France, U-boats and surface raiders would enjoy direct and relatively easy access to the Atlantic. Likewise, aircraft – among them the long-range Focke-Wulf *Kondor* – stationed along the seaboard would be able to harass vessels more or less at will. Suddenly (and contrary to interwar expectations), a widespread and significant aerial threat to Britain's warships and merchant convoys was in the offing. This could only have tremendous repercussions for the future configuration of escort vessels and carriers especially, not least the size and sophistication of their proactive and reactive air-defences. Making appropriate adjustments took time and resources, however, both of which were in short supply. During 1941 planes alone were to sink over a million tons of British shipping. This was nearly twice the tonnage destroyed in the previous year and rather more than the UK's construction-yards could make good.[109]

Britain's sea-based aviation was in any case sorely stretched in the course of the geostrategic revolution that swept across Europe during the spring and summer of 1940. By the time of her capitulation, France was to end up with rather more combat planes to hand than personnel who knew how to fly them. As the Battle of France neared its climax and that for Britain loomed, an even more convoluted dilemma arose north of the Channel. At the end of May, Albert V. Alexander, the First Lord of the Admiralty, approached Lord Beaverbrook, the Minister of Aircraft Production, pleading for more emphasis to be placed on the manufacture of Fairey *Fulmars*, the availability of which was already lagging six months behind the original forecast. 'I fully appreciate the urgent need for accelerating the production of RAF fighters and bombers', the First Lord emphasized. 'I learn, however, that Fleet Air Arm aircraft do not appear at all in the priority list . . . [circulated] to the Industry'. It was 'most important', he argued, that the Royal Navy should have good, fast carrier-borne fighters for its protection and 'for driving off shadowing aircraft and engaging shore based bombers, which . . . [*Skuas*] cannot do'. *Fulmars* might, he added, 'play a considerable part in the defence of these shores'.[110] Beaverbrook replied, however, that, as matters stood, the goals set were failing to meet 'the hourly need of the RAF in battle'. Consequently it might prove necessary to divert even more resources into trying to satisfy its requirements. This could only occur at the expense of other key programmes.[111]

As well as replacement machines, the RAF also desperately sought infusions of trained (and preferably experienced) pilots to offset those lost by Fighter Command in France and the Low Countries especially. Lest the Germans manage to get an invasion force ashore, there was a reluctance to transfer men from the

---

109 Edward P. von der Porten, *The German Navy in World War II* (London, 1972), pp. 174–8.
110 Document 46, 'Letter from First Lord of Admiralty to Minister of Aircraft Production', 26 May 1940, in *FAA in Second World War: Volume One*, p. 117.
111 Document 47, 'Letter from Minister of Aircraft Production to First Lord of Admiralty', 27 May 1940, ibid., p. 118.

Fairey *Battles* and other tactical bombers that had in any event been dreadfully mauled at Sedan in particular. Besides numerous *émigré* volunteers, the RAF received goodly loans of pilots from the FAA, prompting Alexander to warn Churchill of the impact such transfers were likely to have on the Royal Navy's capacity for aerial operations. Writing on 7 June he pointed out that calculations about personnel requirements were being predicated on 'the assumption that we shall suffer a casualty rate of 2½% whereas during the last two months it has been 10%, even though we were lucky in getting pilots back after forced landings in Norway'. If he approved further assignments to the RAF without there also being a significant reduction in the FAA's rate of attrition, it might, he cautioned, 'be necessary to pay off the *Hermes*, and perhaps the *Eagle*, in order to commission the new aircraft carriers we hope to get this year. . . . [W]e have just been able to provide squadrons for the *Illustrious* without paying off older Carriers, but if the casualty rate went up, even this would have to be reviewed'.[112] Barely had the ink of this missive dried than grim tidings arrived from the North Sea: a carrier had been sunk with virtually all its complement and some RAF personnel and planes to boot.

How had this disaster come about? Parts of the explanation remain elusive. The Germans, hoping to intercept the Allied convoys departing from around Narvik, had sent out *Scharnhorst* and *Gneisenau* again. Aerial reconnaissance had failed to detect them and, on 8 June, these swift, powerful battlecruisers had happened to come across the *Glorious*, one of their priority targets.[113] Steaming leisurely homewards, the carrier was evidently oblivious to the impending danger and, for reasons that remain unclear, was oddly unprepared for action. Certainly, she had not launched any aircraft for several hours, possibly because she was crowded with weary hitch-hikers; despite lacking tail-hooks, eight of the RAF *Hurricanes* that had first travelled to Norway aboard her had managed to alight on her flight-deck, as had ten of the *Gladiators* that had been ferried in during May on *Furious*.[114] *Glorious* had, moreover, a skeletal escort of just two destroyers, *Acasta* and *Ardent*; spiralling new commitments at Dunkirk and further afield had, together with combat damage, been draining vessels away from the Home Fleet for some time, spreading the remainder more thinly. In any event the *Kriegsmarine* had never seriously attempted to jeopardize the Allies' lines of communication throughout the Norwegian campaign. Until the abrupt

---

112 Document 49, 'Letter from First Lord of Admiralty to Prime Minister', 7 June 1940, ibid., p. 120. The subsequent movement of former FAA pilots to RAF Coastal Command spawned a similar warning within the Royal Navy's hierarchy. See: Document 115, 'Minute from the Fifth Sea Lord to the First Sea Lord: Withdrawal of Pilots from Coastal Command', 17 April 1941, ibid., pp. 391–2.

113 See the (translated) extract from the relevant report by the German admiral concerned: Document 58, 'Sinking of HMS *Glorious*, 8 June 1940', 1 July 1940, ibid., p. 179.

114 See: Document 55, 'Letter from Commander-in-Chief, Rosyth, to Secretary of Admiralty: Loss of HMS *Glorious*, 8 June 1940', 24 June 1940, ibid., p. 169; Document 55a, 'Report from Board of Enquiry to Commander-in-Chief, Rosyth: Loss of HMS *Glorious*, 8 June 1940', 22 June 1940, ibid., pp. 170–71; Barnett, *Engage the Enemy*, pp. 136–8; Till, *Air Power and the Royal Navy*, pp. 25–6.

transformation of the geostrategic situation at the start of June 1940, it was duly believed that much of the North Sea could be transited in comparative safety.[115]

Indeed it is not inconceivable that the carrier's captain thought the only likely threat to her was a receding one, namely the *Luftwaffe* aircraft she was leaving far behind. But the 11" guns of the *Scharnhorst* and *Gneisenau* were considerably more accurate than high-altitude bombers had normally proved. Approaching from her windward (western) side so that *Glorious* could not launch aircraft without turning towards her tormentors, the battlecruisers opened fire from almost sixteen miles away. In the interim, *Acasta*, having established that the oncoming vessels were hostile, raced towards them to mount a torpedo attack, generating a smokescreen to mask the carrier as she went. *Ardent*, too, tried to shroud *Glorious* in smoke, while both destroyers blazed away with their heaviest guns, although they could not hope to match their adversaries' reach and punch. Despite her escort's endeavours *Glorious* was soon listing and ablaze, as was *Acasta*, which foundered within the hour, followed by *Ardent*, which had remained with the sinking carrier. Some fifteen hundred men were lost aboard the three ships.

This was about a third of all the casualties sustained by the British in the expedition to Norway. *Glorious* was, furthermore, the second carrier to be lost in just nine months of hostilities, *Courageous* having been torpedoed by a U-boat as early as 17 September 1939.[116] Moreover, the Norwegian campaign's tail harboured further stings for the FAA especially. *Scharnhorst*, damaged by a torpedo from *Acasta*, withdrew to Trondheim where, on 13 June, she, together with *Gneisenau* and *Hipper*, was assailed by 15 *Skuas* launched from *Ark Royal*. A flight of RAF *Blenheims* were supposed to provide fighter cover during the raid, while *Beaufort* bombers executed a diversionary attack on the *Luftwaffe* base at Vaernes. However, synchronizing the actions of these different groups of aircraft proved impracticable: the premature bombing of Vaernes might well have put all the Germans in the locality on their guard, whereas the *Blenheims* failed to arrive from beyond the North Sea in time to rendezvous with the *Skuas*. The skies over Trondheim were, moreover, clear, exposing the dive-bombers to enemy observers as they ventured all of fifty miles inland in search of their quarry. Indubitably, the port's defenders were not taken by surprise. If only because of 'exceedingly fierce' fire from the numerous flak batteries ringing the anchorage and from AA weaponry aboard the ships themselves, little ordnance was released on target. The 500 lb armour-piercing bombs were in any case comparatively puny – one bounced off the *Scharnhorst* without detonating. Of the *Skuas* that emerged

---

115 Nevertheless, Barnett is scathing about the Home Fleet's dispositions and manoeuvres. See: Barnett, *Engage the Enemy*, pp. 133, 135.

116 It was believed that inadequate and inappropriate damage-control measures might have contributed to the foundering of *Courageous*. See: Document 18, 'Letter from Commander-in-Chief, Western Approaches, to Secretary of Admiralty: Loss of HMS *Courageous*, 17 September 1939', 15 October 1939, in *FAA in Second World War: Volume One*, pp. 42–5; and 'Report from Board of Enquiry . . . [regarding loss of *Courageous*]', 4 October 1939, ibid., pp. 45–9.

from their swoop, those that tried to regain altitude were waylaid by prowling Me.109s. The remainder, hugging the mist along the valley floor, managed to slip away. Only seven regained *Ark Royal*.[117]

It is very questionable whether, as has been suggested, not using torpedoes in this attack was 'a blunder' perpetrated by benighted admirals who were reluctant or unable to grasp the full potential of strike aircraft in maritime environments.[118] HMS *Furious* had, after all, deployed *Swordfish* against shipping in Trondheimsfjord two months earlier but with disappointing results – their torpedoes had grounded in the shallows. In fact this very mishap had spawned the decision to rely on free-fall bombs in the raid at Narvik that followed just a few hours later. That, too, had ended in planes being lost for negligible gains. On the other hand, after the sinking of the *Königsberg*, *Skuas* stationed on the Orkneys had made further forays against Bergen in the course of May; occasionally accompanied by RAF *Blenheims*, they repeatedly caught the German garrison unawares, bombing an assortment of targets in and around the harbour including several fuel tanks.[119] Destroying big, heavily armed warships at Trondheim was always going to be a more demanding undertaking, however. Even once *Scharnhorst* and *Gneisenau* had left the relative safety of the fjord for Germany and were that much more exposed, they proved daunting targets for aircraft. Whereas on 20 June *Gneisenau* was torpedoed and damaged by the submarine HMS *Clyde*, on the following day *Scharnhorst* was hurriedly assailed by six *Swordfish* that had been scrambled from the Orkneys in search of her. They had already covered around two hundred and forty miles when they overtook their prey and, if only because their flagging crews were more accustomed to conducting anti-submarine operations than torpedo strikes, the attack that ensued was a poorly coordinated, ineffectual one. Two planes were lost without any hits being scored.[120] Nevertheless, the *Kriegsmarine* had suffered terrible damage in the Norwegian campaign as a whole and yet more of its largest ships were destined to be ravaged or sunk in this theatre before the war's end. Among these were the *Lützow*, which, after having been crippled during 'Weserübung', ventured back into Norwegian waters in June 1941 only to be disabled once more by a *Beaufighter* torpedo-bomber. This was to put her out of action until the following May.

Long before this event the British had been compelled to ponder the impact of land-based aviation on their fleet's activities. As Admiral Forbes was to

---

117 Document 52, 'Letter from Commanding Officer, HMS *Ark Royal*, to Vice-Admiral Aircraft Carriers: . . . Attack on Warships at Trondheim, 13 June 1940', 15 June 1940, ibid., pp. 161–2; Document 56a, 'Letter from Vice-Admiral, Aircraft Carriers, . . . to Commander-in-Chief, Home Fleet: Attack on Warships at Trondheim, 13 June 1940', 18 June 1940, ibid., pp. 171–3; Document 56b, 'Report by HMS *Ark Royal*: Attack on Warships at Trondheim, 13 June 1940', 13 June 1940, ibid., pp. 173–6. Also see Till, *Air Power and the Royal Navy*, pp. 26–7.

118 See: Barnett, *Engage the Enemy*, p. 138.

119 For some details of these operations, see: Documents 40, 41 and 42 in *FAA in Second World War: Volume One*, pp. 109–12.

120 Till, *Air Power and the Royal Navy*, pp. 27–8.

observe in a letter of July 1940, the FAA had been assigned a limited role before the war, namely that of harassing enemy shipping at sea from carriers. Planes designed for that purpose and subject to the constraints imposed by operating from ships had always, he stressed, been likely to be at a disadvantage if pitted against modern, shore-based machines and sophisticated AA defences. Whereas the first seven months of the conflict had offered the FAA little opportunity to fulfil its preconceived mission, thereafter it was, Forbes pointed out:

> fully occupied but very frequently in tasks well beyond [its original] . . . terms of reference and nearly always in the face of enemy aircraft of superior performance. . . . This is not so much a criticism of the suitability of the aircraft for the particular conditions for which they were designed but rather a consequence of their employment in circumstances which go beyond these . . . conditions. . . . So far as the war in Home waters is concerned it has to be acknowledged that the occupation of Norway by Germany is a definite restriction on the useful operation of aircraft carriers.[121]

By this juncture, German air power dominated not just Norway but also the Baltic and the whole of Europe's Atlantic seaboard as far west as the Pyrenees. Indeed from mid-June onwards, by way of a prelude to the Battle of Britain, the *Luftwaffe*, operating from nearby airfields in France and the Low Countries, opened a campaign against shipping in the English Channel and its approaches, seeking to lure the RAF's precious fighters into protecting these maritime arteries. However, in the Mediterranean – through which ran Britain's links, via the Suez Canal, to India and Singapore – the military balance was even more precarious, insofar that here the threat on the Earth's surface was at least as ominous as that in the skies above it. France's defeat was of twofold significance in this respect. Her fleet – the world's fourth largest – had helped act as a counterweight to that of Italy. Now, under Vichy's dubious control, it might even aid the Axis against the overstretched Royal Navy. Britain's forces in the basin were, furthermore, dependent upon a clutch of isolated bases, notably Gibraltar and Malta, which respectively dominated chokepoints in the Mediterranean's western and central reaches, and Alexandria and Cyprus, which commanded the sea lanes of the Levant, including those radiating from the Suez Canal. Towards the end of 1940 Crete, too, was regarded as having swelling strategic potential, not least as an unsinkable aircraft carrier from which bombers might strike at Romania's Ploesti oilfields, Germany's main source of petroleum products. Airstrips were duly established on the island by the RAF at Máleme, Retimo (Réthímnon) and Heraklion (Iráklion).

---

121 Document 68, 'Letter from Commander-in-Chief, Home Fleet, to Secretary of Admiralty: Performance of existing aircraft types and possible replacements', in *FAA in Second World War: Volume One*, pp. 209–10.

Overlooked from the north by Italian aviation and naval units, the lines of communication and supply between these scattered outposts were colossal: from Gibraltar to Malta is approximately eleven hundred miles, with Alexandria situated a further thousand to the east of that island, the harbours, aerodromes and barracks of which were the main transit facilities for friendly forces travelling either westwards or eastwards. Its central location also made Malta an unsinkable platform from which aerial and nautical blows might be launched. Conversely, whilst Alexandria's comparative remoteness enhanced that outpost's security somewhat, Malta was overshadowed by RAI planes stationed on Sicily, which lies a mere sixty miles distant. Any advance towards Suez by the huge Italian army garrisoning Libya might also jeopardize Malta and Cyprus as much as Alexandria, insofar that this could bring more of the ports and airstrips along the African littoral under Axis control.

The fundamental strategic quandary confronting the British in the Mediterranean was that of balancing gain against risk. For much of the war the perils of suffering reverses here appeared greater than the benefits that might flow from conceivable victories. In the wake of France's capitulation, northern Africa was the one region where the soldiers of the British Commonwealth might still get to grips with those of the Axis. On the other hand, defeat on the seaboard could only jeopardize the bases upon which air and nautical power depended. The Royal Navy was in any event the hinge upon which so much turned. While the RAF's Bomber Command might molest metropolitan Germany and Italy from occupied Europe's periphery, the only way these territories might actually be invaded and conquered was through amphibious assault. Dominance of the maritime corridors was therefore essential for both defensive and offensive purposes. Yet first securing and then maintaining these proved very difficult, not least owing to the interplay of Axis surface forces and land-based aviation. If only because providing comprehensive air cover from their widely scattered outposts was physically impossible, the British had to look to carriers as the main means of projecting air power into many corners in and around the Mediterranean.

The numbers of pilots and planes aboard those ships were, however, at a low ebb for several months after the Battle of Britain had climaxed in September 1940.[122] At the time of Italy's entry into the war, there were some two to three hundred RAF and FAA machines scattered across the Mediterranean basin. Facing these were roughly twice as many Italian aircraft alone, many of which were from the outset pitchforked into assailing Allied shipping and bases that lay within range. Once France had capitulated, Malta and British naval flotillas were left as the primary targets of these operations, with the latter coming under intense pressure during the first half of July especially. Admiral Andrew

---

122 See, for instance: Document 124, 'Minute of First Sea Lord to Prime Minister: Availability of *Fulmar* Fighters', 26 May 1941, in ibid., pp. 412–13.

Cunningham, the Commander-in-Chief of the Mediterranean Fleet, noted that vessels plying between Crete and Cyrenaica in particular were 'bombed continuously' by planes based in Libya and on the Dodecanese, while other units were harassed by aircraft that sortied two hundred miles and more from Sicily.[123] These and other high-level attacks were embarrassingly ineffectual, however, the Italian high command concluding that the procurement of dive-bombers from Germany was an essential prerequisite for any improvement in the RAI's performance. The RMI – perhaps less shockingly but no less worryingly – also struggled to dent British sea power and, particularly after its lacklustre brush with the Royal Navy in the inconclusive clash off Punta Stilo, Calabria, on 9 July, focussed more on tying down as many enemy vessels as possible, much as the *Kaiserliche Marine* had done in the First World War.[124]

What the British remember as the 'Battle of Calabria' resulted from their and the Italians' simultaneous bids to conduct supply convoys to their forces in Alexandria and Libya respectively. *Swordfish* from the ageing carrier HMS *Eagle* participated in this running fight, as did some Italian aircraft, but neither side's planes managed to inflict much physical damage.[125] Nor did their battleships and cruisers: the encounter fizzled out without a single vessel having been sunk. Yet the engagement did have a significant psychological impact on the RMI, which became increasingly wary about confronting the British in pitched battles. This timidity was to be intensified by subsequent setbacks at Taranto and Cape Matapan, but was already palpable by the middle of July 1940. Certainly, the Royal Navy was more confident about its ability to contain if not defeat its Italian counterpart than it had been at the beginning of that month. For then it had seemed probable to Churchill for one that the French fleet would be commandeered by the Axis. Avoiding this was one of his most pressing and nerve-racking preoccupations. But how were France's warships to be kept out of Hitler's grasp without antagonizing the Vichy regime or, on the other hand, incurring losses Britain's armed services could ill afford?

Most sailors aboard the numerous French vessels that found themselves in British waters at the time of France's surrender proved pliable enough, including those at Alexandria. The sympathies of those in charge of the warships berthed further along Africa's northern littoral were less clear, however, while the squadron moored in Toulon was judged to be too close to home for proselytization. His attempts to negotiate a mutually satisfying settlement with his Gallic counterpart having proved fruitless, on 3 July 1940 London instructed Admiral

123 Document 64: 'Message from Commander-in-Chief, Mediterranean, to Admiralty: Air Attacks on Mediterranean Fleet', ibid., p. 64. Also see the continuation of this report: Document 65, ibid., pp. 197–8.

124 Greene and Massignani, *The Naval War in the Mediterranean*, pp. 80–1.

125 For details of the attacks by *Eagle*, see: 'Report by Commanding Officer: . . . Battle of Calabria, 9 July 1940', 16 July 1940, in *FAA in Second World War: Volume One*, pp. 198–202. Also see Document 66a, ibid., pp. 202–7.

James Somerville to open fire on the flotilla at Mers-El-Kébir, Algeria: 1,297 French sailors perished as the battleship *Bretagne* exploded, several other ships were heavily damaged and the battlecruiser *Dunkerque* ran aground. Her sister ship, *Strasbourg*, and some destroyers fled the scene, heading for Toulon. Passing dangerously close to *Ark Royal*, *Strasbourg* was attacked in the growing darkness by *Swordfish* from the carrier but managed to escape.[126] Five days later, the *Richelieu* was likewise assailed while moored in the Atlantic port of Dakar, French West Africa's cardinal harbour. Six torpedo-bombers pounced on her at first light. Although it is thought that their weapons all grounded amidst the shoals, their target had already sustained crippling damage in an impudent raid executed shortly beforehand: a motor boat from the carrier HMS *Hermes* had infiltrated the anchorage during the night and dropped four depth-charges directly under the battleship's stern.[127] (The vessels sheltering at Toulon – which came to include *Dunkerque* as well as the *Strasbourg* – all finished up being scuttled by their crews when Hitler's troops occupied Vichy France in November 1942.)

'*L'Affaire* Richelieu' and the bombardment of Mers-El-Kébir ineluctably embittered many French citizens. However, besides any military advantage that accrued from these ruthless actions, Churchill's determination to fight on was made apparent, not least to the USA, which was fearful that the British might be tempted to follow their vanquished European allies in making peace with the Axis powers.[128] At the end of the month, he sent President Roosevelt a somewhat peremptory message, urging him to expedite the handover of numerous old destroyers from the US Navy, which was gradually assuming *de facto* responsibility for the security of ever bigger swathes of the Atlantic.[129] More generally, Lend-Lease aid and other American measures short of war were now more essential than ever. The Battle of Britain was, Churchill realized, about to turn increasingly earnest, although it could at best achieve no more than to fend off the threat of immediate invasion. London's strategic planning had in any case to be global, not local, an imperative that inevitably generated a virtually insatiable demand for, among countless other items, planes, ships, tanks, trucks, munitions and fuel. Without assistance from overseas, simply replacing the mounds of materiel forsaken by the army at Dunkirk would have been a Herculean task for Britain's indigenous industries.

In September 1940, as the Battle of Britain neared its climax, Mussolini's armies in Albania and Libya slowly roused themselves. The latter lumbered across the Egyptian frontier, halting at Sidi Barrani to let its supply columns

126 See: Document 60, 'Report from Vice Admiral, Aircraft Carriers: . . . Attack on Oran, 3 July 1940', ibid., pp. 185–9.

127 'Report of Proceedings from Commanding Officer, HMAS *Australia*, to Commander-in-Chief, Home Fleet: Attack on *Richelieu* at Dakar, 8 July 1940', 31 August 1940, ibid., pp. 221–3.

128 For an insight into the political and diplomatic circumstances surrounding and arising from the attack on the French fleet, see: Jenkins, *Churchill*, pp. 620–6.

129 See: ibid., pp. 626–7; 641–2.

catch up. In the interim, reinforcements, notably more tanks, were rushed out to the few British troops safeguarding Cairo and the Suez Canal, while the Royal Navy, hoping to redress further the balance of sea power in the Mediterranean, struck at the Italian fleet much as it had the French, the FAA mounting a spectacular raid against the warships lurking in Taranto on 11 November – a feat about which more will be said elsewhere. Italy's soldiers, too, received a terrible beating: whereas those that had invaded Greece at the end of October were quickly routed by indigenous forces backed by a few RAF squadrons, the large but ponderous army in Egypt had, by January 1941, been comprehensively outmanoeuvred and outfought by its mercurial adversaries.

The Germans now intervened in this quarter, initially to secure their southern flank ahead of the invasion of the USSR. This was scheduled to take place on 22 June – a year to the day on which France had capitulated. While a small but doughty and innovative *Afrika Korps* was dispatched to bolster the Italians in Libya and *Luftwaffe* units reinforced the RAI on Sicily, an irresistible German army, supported by contingents of Hungarians and Bulgarians, began rampaging across neutral Yugoslavia towards Greece. The Italians had already renewed their efforts here on 9 March, fixing many of the available Greek forces in defensive positions. As these were progressively compromised by the advance of the other invaders, Athens, rather hesitantly, accepted an offer of substantial assistance from London. Much of the victorious army garrisoning Egypt was duly transferred to the Mediterranean's northern shore.

This undertaking precipitated a major clash between the Royal Navy and Italian warships on 28 March as the latter sought to intercept the troop convoys bound for the Aegean. As will be discussed elsewhere, the Battle of Cape Matapan was to prove another resounding success for the British, largely thanks to the contribution made by the FAA. However, by this juncture the *Luftwaffe* was already tilting the military balance within the Mediterranean in favour of the Axis powers. On 10 January a British flotilla just under two hundred miles west of Malta picked up, firstly, some Italian S.79 torpedo-bombers on its radars. The vessels were cocooned by a handful of *Fulmars* from the carrier *Illustrious*. These intercepted one of the Italian planes, shooting it down, but expending either all or much of their ammunition in the process. Before all of their replacements could take off and climb to their patrol altitudes, between 15 and 45 Ju.87s appeared. Falling on the ships, these concentrated their efforts on *Illustrious*.

It was, her captain conceded, a 'brilliantly executed' attack: six hits were scored with 1,000 lb bombs and there were several close shaves, splinters from which caused some damage and casualties; the flight-deck and one of the lifts were wrecked; there were fires fore and aft, including in parts of the hangar; and the carrier was no longer answering her helm properly. About an hour later, seven high-level bombers renewed the assault, but, spotted and engaged well short of their quarry, released their payloads too early to do any harm. A second flight of *Stukas*, escorted by a few fighters, had more success, dropping nine bombs, one of which hit the damaged elevator, killing and injuring men among the first-aid and fire-fighting teams. A fifth raid – made by a combination of high-altitude and dive-bombers – proved fruitless, however, as did another torpedo strike by Italian planes that approached the ships at twilight. Despite the tremendous punishment she had taken, *Illustrious* remained afloat and was able to dock for repairs at Malta.[130] A less robust ship would surely have foundered, as indeed proved the case with the modern, light cruiser HMS *Southampton* when she came under bombardment from numerous Ju.87s south-east of Malta the very next day. Gutted by fires, she had to be scuttled. Her sister ship, the *Gloucester*, which was damaged in this engagement, was also to succumb to what many British sailors had by now dubbed 'Stukaritis', being sunk off Crete on 22 May.[131]

Britain's triumph over the Italians in northern Africa was her first major victory on terra firma since the war's onset. But the subsequent weakening of Egypt's garrison and the vigour of the *Afrika Korps* soon led to a reversal of fortunes here. By May, Axis forces were again threatening Egypt's frontier, besieging the pivotal port of Tobruk and capturing airstrips from which they could strike seawards as well as along the coast towards Suez. The troops sent to stiffen the Greeks' resistance also recoiled as the German *Blitzkrieg* swept through the Balkans. The Royal Navy commenced shuttling them from coves and beaches in the Peloponnese and around Athens to Crete on 24 April. The evacuation fleet comprised several cruisers, around two dozen destroyers and a similar number of transport vessels. The available tonnage was insufficient to extricate the whole army in one fell swoop and so the embarkation was spread over seven nights, some of the faster ships, notably the destroyers, making several trips. Two of them and three transporters were bombed and sunk with numerous fatalities.[132] As at Dunkirk, most heavy equipment had to be left behind and several groups of soldiers – notably around seven thousand men waiting to be rescued at Kalamata – were compelled to surrender. Unlike at Dunkirk, however, the RAF could provide negligible cover for the withdrawal. This left the Allied shipping at the mercy of the *Luftwaffe*, whose accomplished engineering teams leapfrogged forward, repairing and revamping captured airfields. Confronted by some eight hundred German aircraft, the RAF's few surviving machines had to retreat to Crete to escape certain extermination.

Much the same pattern of events ensued when Crete in turn came under attack during the last days of May. But the problems faced here by British sea and air power were so much larger, the distance between Crete and the logistical hub of Alexandria being roughly twice that between that island and Athens and up to four times that between western Crete and parts of the Peloponnesian coast. Aviation could scarcely be sustained on Crete itself, if only because its three

---

130 Document 107b, 'Report from Commanding Officer, HMS *Illustrious*, to Rear Admiral Aircraft Carriers, Mediterranean: Air Attacks on HMS *Illustrious*, 10–19 January 1941', 26 January 1941, in *FAA in Second World War: Volume One*, pp. 344–56.

131 See: Costas N. Hadjipateras and Maria S. Fafalois, *Crete 1941: Eyewitnessed* (Athens, 1989), p. 120.

132 See: Stephen Prince, 'Air Power and Evacuations: Crete 1941', in I. Speller (ed.), *The Royal Navy and Maritime Power in the Twentieth Century* (London and New York, 2005), pp. 73–4.

aerodromes – especially that at Máleme on the island's north-western tip – were far too close to the mainland to be secure. The RAF and FAA were in any event hopelessly outnumbered. The Germans had amassed over four hundred bombers and nearly two hundred fighters for the invasion. They also had several hundred transport aircraft laden with paratroopers and engineers.[133] As in Norway and the Low Countries, these *Fallschirmjäger* were to form the attacking force's spearheads. Inserted and thereafter supported by air, over seventeen thousand of them started landing on 20 May, seeking to capture Máleme and to isolate its defenders from the rest of the island's garrison in their enclaves further east.

Since most of their transport vehicles had had, along with other heavy equipment, to be abandoned in Greece, the Allied troops were rather immobile, but they were far more numerous than the invaders had anticipated. Seizing Máleme was to prove a costly undertaking. Once completed, however, the aerial bridge allowed the invaders to build up their strength on the surface and, as elsewhere in the *Blitzkrieg*, closely combine land with air power. If the Allied soldiery concentrated to resist the former, they were vulnerable to the latter. If, alternatively, they dispersed to vitiate the impact of the Germans' combat aviation, they faced being defeated piecemeal on the ground. As General Archibald Wavell, the Commander-in-Chief, Middle East, warned London: 'Such continuous and unopposed air attack must drive [even the] stoutest troops from [their] positions sooner or later and makes administration practically impossible'.[134]

That the island's garrison did not face an even greater menace on the surface was because of the Royal Navy. 'Ultra' intelligence had revealed that the Axis forces were planning to ship heavy weaponry and equipment to Crete. It was assumed, moreover, that the RMI would emerge from its harbours to protect these convoys. Cunningham was urged by the Admiralty to try to disrupt if not block any such move: 'If the Fleet can prevent sea-borne reinforcements and supplies reaching the enemy until the Army has had time to deal successfully with all air-borne troops the Army may then be able to deal with sea-borne attacks'.[135] Cunningham's forces were accordingly and initially deployed to the west and south of Crete. Here, they could loiter out of sight and range of the *Luftwaffe* while staying sufficiently nearby to react should any hostile shipping materialize. Although the RMI remained puzzlingly quiescent, two flotillas of transport vessels did appear and were duly intercepted, the British inflicting appreciable losses on the first convoy and frightening off the second.[136] On the other hand, these forays took several cruisers and destroyers far into the Sea of Crete where they were woefully exposed to aerial counterstrokes. Thanks to the

133 Gerhard Schreiber, Bernd Stegemann and Detlef Vogel, *Germany and the Second World War: Volume Three: The Mediterranean, South-East Europe and North Africa, 1939–1941* (Oxford, 1995), p. 536.

134 Cipher telegram from General Wavell to War Office, 27 May 1941, TNA, CAB 121/537.

135 Message from Admiralty to Commander-in-Chief, Mediterranean Fleet, 23 May 1941, TNA, CAB 121/537.

136 See: Prince, 'Evacuations: Crete', pp. 75–6.

clear conditions and the many hours of daylight, Axis pilots were able to pursue the retreating ships for as long as their quarry remained in reach and their planes could be furnished with bombs and fuel. Stocks of AA ammunition aboard the flotillas were soon nearing exhaustion and some vessels, having sustained damage, began to lag behind the rest of their respective formation, slipping from under its protective umbrella of fire. The risks to individual ships then had to be weighed against those facing the fleet as a whole, just as the benefits from these sorties had to be set against their costs, which ultimately included the loss of *Gloucester* and another light cruiser.

Although the Royal Navy had denied Axis forces control of the waves in this theatre, the enemy's dominance of the skies was undermining British sea power's ability to help safeguard and sustain not only the beleaguered garrison of Crete but also that of Tobruk. Hostile aviation units could now subject passing maritime traffic to observation and bombardment from an arc of airstrips that curled from northern Africa through mainland Greece to the Dodecanese. Safe enough in mid-ocean, on entering many coastal areas vessels were running significant risks, General Wavell observing despairingly on 27 May that: 'No ship can approach within 100 miles of [Crete] during daylight without being detected by reconnaissance [planes] and then attacked by dive bombers. . . . Though our air force has made every effort and suffered considerable losses, it has been quite unable, without bases within fighter range, to deter [the] enemy'.[137]

Sure enough, all that was at hand to counter the *Luftwaffe* and RAI was a scratch force of long-range interceptors. Even this could be deployed 'for a day or two' at best over Crete, if only because it was not big enough to patrol simultaneously the skies around that island and those surrounding Tobruk. The need for such planes remained so 'acute', Cunningham cautioned the Admiralty, that 'until they are available in sufficient numbers we must continue to expect losses and damage . . . when the Fleet is at sea'.[138] Although his forces did include a carrier, so shrivelled had the complement of pilots and planes aboard HMS *Formidable* become that he was initially reluctant to risk employing her in either defensive or offensive operations, fearing that she would be damaged or even sunk to no avail.[139] His sombre expectations were indeed to be borne out on 26 May when she hurled a handful of *Albacores* and *Fulmars* against the aerodrome at Scarpanto (Karpathos) in the Dodecanese. This trifling attack, which

137 Message from Commander-in-Chief, Middle East, to Chief of the Imperial General Staff, 27 May 1941, TNA, CAB 121/537.

138 Message from Commander-in-Chief, Mediterranean Fleet, to Admiralty, 1 June 1940, TNA, CAB 121/537. Also see: Document 134, 'Letter from Secretary of Admiralty to Under-Secretary of State, Air Ministry: Shore-based fighter cover for naval forces', in *FAA in Second World War: Volume One*, pp. 442–6.

139 Document 123, 'Message from Commander-in-Chief, Mediterranean Fleet, to Admiralty', 23 May 1941, ibid., p. 412. Also see Cunningham's report on the Battle of Crete, 4 August 1941, Document 234 in *The Cunningham Papers: Volume I: The Mediterranean Fleet, 1939–1942*, Michael Simpson (ed.), (Aldershot, 1999), pp. 421–3.

was implemented at dawn, elicited a frenzied riposte from Axis airstrips on the African coast, dozens of Ju.87s and Ju.88s falling on the retreating carrier and her escort only hours later. With just two *Fulmars* immediately available to intercept them, *Formidable* was hit twice, sustaining damage to her waterline and other critical areas. After undergoing makeshift restoration work in Alexandria, she had to be sent to Norfolk, Virginia, for comprehensive repairs. Completing these was to take until December. In the interim Cunningham advised the Admiralty that he required two carriers with five squadrons of fighters if he was going to furnish his fleet with adequate protection against aerial attacks; maintaining a reasonable degree of security might well call for a minimum of 18 interceptors to be kept aloft, he opined.[140]

This could only be a guesstimate, for the belligerents were wrestling with a quandary that became so self-evident as to be infrequently articulated and, consequently, too often overlooked: whereas a single bomber unopposed might inflict fatal harm, whole squadrons of bombers might, if resisted, do no damage whatsoever. Even a handful of customized planes could have a disproportionate impact in maritime operations especially, as was adequately demonstrated in an air strike by six *Albacores* and two *Swordfish* during the Battle of Matapan in March 1941 and by a few torpedo-bombers stationed on Malta between 1940 and 1943. The sheer speed of aircraft relative to surface units meant that threats could develop in a trice at any point within the combat radius of the planes concerned, be their aerodromes static or mobile. Mustering and maintaining a countervailing concentration of firepower, whether it took the form of reactive or proactive defences or both, inevitably proved impracticable on occasion, not least in the Mediterranean where so much of the sea lay within the shadow of land-based air power. As Cunningham rather tetchily observed in a cypher marked 'hush: most secret' to the military and political hierarchy in London, for ships assigned to helping safeguard Crete the need to return to Alexandria intermittently for fuel, AA ammunition, repairs and other necessities was inescapable.[141] When plying between the island and this logistical haven, however, they could make no active contribution to Crete's protection yet were nonetheless exposed to aerial attack from Axis aviation in North Africa.

This often occurred, moreover, when their crews' psychological and material capacity for defence were at their nadir. Cunningham summarized the first days of the operations around Crete as 'nothing short of a trial of strength' between his fleet and the *Luftwaffe*. He had always feared that, under the prevailing circumstances, the odds against success would turn out to be too great unless he could catch his adversaries unawares while he himself employed the

---

140 Telegram from Commander-in-Chief, Mediterranean Fleet, to Admiralty, 20 August 1941, TNA, PREM 3/173/4.
141 Cypher from Commander-in-Chief, Mediterranean Fleet, to Admiralty, 24 May 1941, TNA, CAB 121/537.

'utmost circumspection'.[142] Neither of these hopes was realized sufficiently. On 27 May – the day the awesome *Bismarck* was sunk by British warships in the Atlantic – Wavell was reduced to telling London that, although he appreciated the 'grave effect' losing Crete would have on the defence of Malta and Alexandria and on public opinion in Egypt and (neutral) Turkey, any bid to prolong retention of the island would be 'likely so to exhaust Navy, Army and Air resources as to compromise [the wider] defence position in [the] Middle East . . .'.[143] Cunningham's warships were duly switched from cocooning the troops on Crete to evacuating them.

International politics as well as geography and strategy shaped the withdrawal, for much of the island's garrison comprised soldiers from Australia and New Zealand. Churchill and his senior military commanders were very mindful of the totemic and martial significance of the manpower recruited from the sprawling commonwealth.[144] As both an instrument and icon of that polity's unity, the Royal Navy especially was loath to forsake such close allies, although the price in blood and equipment of extricating the 'Anzacs' and their British comrades from Crete threatened to exceed that of the Dunkirk evacuation. Certainly, if the enterprise was going to have any chance of proving worthwhile, the scope for interference by the *Luftwaffe* would have to be minimized: embarkations would have to be crammed into the few hours of darkness available; warships rather than merchantmen would, if only because of their superior speed, have to be employed as ferries; as much long-range air cover as possible would have to be provided; and, ideally, the whole army – the exact strength and dispositions of which were now barely knowable – would facilitate its own rescue by first withdrawing into bridgeheads on the southern coast.

Mostly, these measures worked well enough. Although it lacked a sizeable harbour, Sphakia (Khóra Sfakíon) was relatively remote from the enemy's airstrips and several thousand men were to be plucked from its beaches with comparative ease. However, the only escape route for the troops in the north-east of the island lay through Heraklion. Whereas this port offered far better docking facilities for the evacuation flotilla, it lay at the end of a lengthier journey from Alexandria, not all of which could be completed under the cover of night. The ships involved endured protracted bombardment during both the outward and return legs; two destroyers were sunk and three cruisers damaged.[145]

In all, the Royal Navy had suffered 1,828 fatalities in the battle for Crete. Three cruisers and six destroyers had been sunk and numerous other vessels damaged. 'Thus at the close of May the Mediterranean Fleet was left in a sorry

142 Document 123, 'Message from Commander-in-Chief, Mediterranean Fleet, to Admiralty: Operations around Crete without air cover', 23 May 1941, in *FAA in Second World War: Volume One*, p. 412.

143 Message from Commander-in-Chief, Middle East, to Chief of the Imperial General Staff, 27 May 1941, TNA, CAB 121/537.

144 Prince, 'Evacuations: Crete', pp. 79–81.

145 See: ibid., pp. 77–80.

state', Cunningham lamented, 'though with the satisfaction of having achieved the almost insuperable task set it. . . . [W]ithout air support of any sort, [it had] had to be exposed to a scale of air attack which is believed to have exceeded anything of the kind yet experienced afloat'.[146]

Just three weeks after the British retreat from Crete, matters came full circle insofar that the Nazi-Soviet Pact was finally shattered by Hitler's invasion of the USSR. Left geopolitically isolated by his conquests in Scandinavia yet hoping to recover the lands wrenched from them by Stalin under the Treaty of Moscow in 1940, the Finns aligned themselves with the Axis, redirecting their principal exports – timber, nickel and copper, many consignments of which had traditionally gone to Britain – to Germany in return for grain and munitions. This economic dependency shackled Helsinki to Hitler's whims and the Finns were to remain his rather half-hearted allies until September 1944, when the Red Army's relentless advance westwards compelled them to conclude a separate peace. In the interim, squadrons of their and *Luftwaffe* aircraft based in the Arctic Circle in Petsamo (Pechengsky) – a region that was to be ceded to the USSR at the conclusion of hostilities – were able to molest Murmansk and Archangel and the sea lanes between them. Ineluctably, however, for the government in Helsinki the epicentre of what was portrayed as the continuation of the 'Winter War' of 1939–40 was always much further south, along the Karelian isthmus and the Gulf of Finland.

If Stalin can be pardoned for – like so many other onlookers – being taken aback by the rapidity and totality of France's collapse in 1940, his incredulity at Hitler's attack on the USSR only a year later remains harder to comprehend. Prior to the launching of Operation 'Barbarossa', the *Kriegsmarine* had established an outpost at Helsinki and, besides deploying numerous if small warships and U-boats in the Gulfs of Finland and Riga, had commenced mining the approaches to the abutting Soviet ports, notably the sea lanes between Libau (Liepaja), Tallinn and Kronstadt. Although the Soviet Baltic Fleet proposed to respond by fringing key installations and channels with more fields of its own while clearing those sown by the Germans and Finns, neither mines nor minesweepers were made available in anything like sufficient quantities. Much the same was true regarding AA artillery, with the result that many submarines and minor surface craft had to be dispersed in order to reduce their vulnerability to pre-emptive strikes. Similarly, the few capital vessels that made up the core of the 'Red Banner Fleet' were moved as far out of harm's way as possible, the revamped dreadnoughts *Oktyabrskaya Revolyutsiya* and *Marat* withdrawing as far as Kronstadt where the fixed air defences were denser. There were few plans for much cooperation between the fleet and proactive air power either. Of tremendous strategic importance though it was, interdicting Germany's imports of Swedish iron ore from Luleå was, for instance, a job that was to be left to submarines

---

146 Cunningham's report on the Battle of Crete, 4 August 1941, Document 234 in *Cunningham Papers: Volume I*, pp. 421–2.

alone.[147] The planes that nominally made up the Soviet maritime air arm were in any case almost entirely stationed at aerodromes that lay far to the east of the navy's forward bases and its most likely zone of operations. Once hostilities erupted, they were to be sucked into supporting the Red Army, particularly in the siege of Leningrad, which was destined to drag on for 900 days.

For months before the outright invasion of the USSR, German reconnaissance planes had been violating Soviet airspace on an almost daily basis and, moreover, had been doing so with impunity. Colonel-General Volonov, who was appointed as the head of the Air Defence Command in May, discovered with mystified dismay that his fighter units and AA batteries had no orders to try to restrain these interlopers. Even passive safeguards were done away with at Moscow's insistence: when Colonel-General Kuznetsov – who oversaw the Baltic Special Military District – instituted partial black-out measures at local air and sea bases, Voronov, who had praised his colleague's circumspection and recommended that such precautions be applied universally, found that the instructions were rescinded by the political authorities. More suffocating still was the contradictory guidance issued to commanders along the north-western frontier: on the one hand, they were to refuse to be provoked by German and Finnish actions; on the other, they were to react to any surprise attack with all available strength.[148] The resulting paralysis compounded the effectiveness of the ruses employed by the Axis powers. When the blow finally fell at 03:00 on 22 June, the whirlwind of operations mounted by the *Luftwaffe* alone inflicted appalling damage in the space of a few hours. Hundreds of Soviet combat planes were destroyed, many of them on the ground, leaving between 700 and 1,000 German machines free to attack further targets more or less at will. In fact so unchallengeable did the aggressors' control of the skies over this sector of the front become that, by July, their bombers were confidently dispensing with fighter escorts.[149]

As the jaws of the Axis closed around Leningrad, other German columns thrust through Lithuania, Latvia and Estonia, quashing pockets of resistance. The Red Army recoiled, leaving the naval bases here open to attack on their landward side. After a few weeks only Tallinn remained in Soviet hands and, with their backs to the coast, the only way out for its defenders and inhabitants lay in evacuation by sea.[150] Many of the troops and thousands of civilian refugees duly squeezed themselves onto an armada of assorted warships and commercial vessels that, led by the cruiser *Kirov*, sailed for Kronstadt and Leningrad on the night of 27 August. There were no combat planes to protect the embarkation but, amidst the fading light and billowing smoke from countless fires, the prowling *Luftwaffe* had difficulty picking out viable targets. There were other dangers aplenty, however, not least the minefields that littered the Gulf of Finland and the

147 Erickson, *Road to Stalingrad*, pp. 100, 121.
148 Ibid., pp. 120, 160.
149 Ibid., pp. 162, 170, 208.
150 Ibid., pp. 171, 202, 263–5, 270.

heavy artillery batteries that were dotted along its shores. Barely had the flotilla left Tallinn than the first losses occurred, the rate increasing as Finnish torpedo boats joined in the massacre. At dawn the *Luftwaffe* returned, Ju.88s and *Stukas* chasing the remaining vessels as they plodded eastwards. In all, over forty were lost, among them five destroyers, ten smaller warships and three submarines. Although the final death-toll remains incalculable, this was surely the goriest of the evacuations seen in Europe since the war began.[151]

Those warships that made it to Kronstadt quickly discovered that the anchorage was no longer much of a refuge. The Germans and Finns had all but encircled nearby Leningrad and the bigger vessels' guns were now trained on the besiegers' lines, provoking retaliatory blows from bombers especially. Indeed, echoing Cunningham's remark about the Royal Navy's 'trial of strength' off Crete, Admiral Vladimir Tributs, the Commander-in-Chief of the 'Red Banner Fleet', described what ensued as 'their duel' against the *Luftwaffe*: the *Kirov* was seriously damaged, the *Oktyabrskaya Revolyutsiya* even more so, while the *Marat*, struck by bombs that detonated the contents of the magazine under her front turret, was reduced to a semi-submerged hulk as she slumped onto the seabed, her entire bow section pulverized by the explosions.[152] Refloated and partially refitted, what remained of her was used for the rest of the siege as a static platform for guns, including AA artillery.

Operation 'Barbarossa' went some way to redress the military balance in Britain's favour insofar that it gave her a continental ally in place of those she had lost. However, the USSR was unable to assist much directly. This was partly owing to her geographic remoteness – which effectively increased as the Germans and their vassals occupied huge swathes of Soviet territory – and partly to her own strategic situation, which was to remain dire until the end of 1942, despite or because of Japan's failure to assail her from behind. Indeed the British, though extremely hard-pressed themselves, now had to forward substantial amounts of materiel to their new partners, much of which could only originate in the USA. Thus was born the need for the Arctic convoys to Murmansk and Archangel.

Along with linguistic and technical complexities, the weakness and preoccupations of the Soviet fleet and maritime aviation precluded much in the way of operations in concert with British forces. Even active cooperation regarding the security of the Arctic convoys proved rather limited. Until such time that she could muster the strength – which would perforce have to come from further alliances – to risk putting an army onto the continent again, the principal instruments of Britain's assistance to the USSR could only be, firstly, material aid and, secondly, aerial operations, notably the relentless strategic campaign mounted by RAF Bomber Command against the war machines of Germany in particular and the Axis powers in general.

151 See: Harrison E. Salisbury, *The 900 Days: The Siege of Leningrad* (Cambridge, MA, 2003), pp. 221–42.
152 Erickson, *Road to Stalingrad*, p. 270; Jürgen Rohwer, *Chronology of the War at Sea, 1939–1945* (Annapolis, MD, 2005), pp. 94–5.

The carrier-based planes of the FAA were, however, chosen as the best if only tools for a mission that had rather more political than strategic significance: synchronized raids on the Norwegian harbour of Kirkenes and the port of Liinahamari across the frontier in the Finnish region of Petsamo.[153] The journey involved was a very lengthy one, Kirkenes lying all of fifteen hundred miles north of Oslo and over seven hundred from Helsinki. After refuelling in Iceland, a task force including HMS *Furious* and *Victorious* came within striking distance a week later on 30 July 1941, 24 aircraft from the first of these carriers setting out for Liinahamari while 29 took off from *Victorious* for Kirkenes.

Things went awry from the outset. In the 'Land of the Midnight Sun' achieving surprise in July was always going to be a challenge, but the flotilla was spotted by a surveillance aircraft almost immediately. Whereas matters were too advanced to abort the operation, the Germans could hardly have been given more warning of it.[154] Coast-watchers and radar also detected and monitored some of the British machines as they headed landwards, several of the crews of which betrayed their own movements to eavesdropping foes through recklessly lax use of radio transmissions.[155] Furthermore, since conducting timely, prior reconnaissance of such remote objectives was hardly feasible and could even prove counterproductive, the attacking planes were sent in with little knowledge of what actually awaited them, either in terms of defences or potential targets.[156] Certainly, the nine *Fulmars* from *Victorious* that were supposed to serve as fighter cover for her *Albacores* struggled to provide the bombers with much protection once German combat aircraft started swarming into the skies at news of the impending raid.

The *Swordfish*, *Albacores* and bomb-armed *Fulmars* that assailed Liinahamari found the port and its approaches virtually devoid of shipping but studded with static AA weapons that were soon reinforced by fighters. A couple of little craft were holed, some jetties were wrecked and an oil tank destroyed. Two *Fulmars* and an *Albacore* were lost and four other planes damaged.[157] At Kirkenes, meanwhile, the *Albacores* from *Victorious* – all of which were armed with torpedoes – had been greeted by clouds of Messerschmitts and AA fire as they crested the

---

153 See: Document 166, 'Letter from Secretary of Admiralty to Rear Admiral, Naval Air Stations: . . . Attack on Kirkenes-Petsamo area, 30 July 1941', 10 December 1941, in *FAA in Second World War: Volume One*, p. 525.

154 Document 166d, 'Letter from Commanding Officer, HMS *Victorious* to Rear Admiral, First Cruiser Squadron: . . . Attack on Kirkenes-Petsamo area, 30 July 1941', 7 August 1941, ibid., p. 539.

155 Document 166c, 'Report from Commanding Officer, HMS *Furious* to Rear Admiral Commanding, First Cruiser Squadron: . . . Attack on Kirkenes-Petsamo area, 30 July 1941', 5 August 1941, ibid., p. 536.

156 Document 166, 'Letter from Secretary of Admiralty to Rear Admiral, Naval Air Stations: . . . Attack on Kirkenes-Petsamo area, 30 July 1941', 10 December 1941, ibid., p. 525.

157 Document 166c, 'Report from Commanding Officer, HMS *Furious* to Rear Admiral Commanding, First Cruiser Squadron: . . . Attack on Kirkenes-Petsamo area, 30 July 1941', 5 August 1941, ibid., pp. 536–9.

mountaintops and descended towards the anchorage. Here, several merchantmen were moored or underway, most of which were targeted. Many of the bombers were, however, driven off line by flak or fighters that streamed at them straight out of dazzling sunshine. Although a couple of vessels were seen to be hit, all but one of the *Albacores* were either damaged or downed. Two *Fulmars* were also lost. In all, of the 29 machines that had been committed to the raid, 13 failed to return.[158]

When it came to reporting on the outcome of the strikes, Admiral Sir John Tovey – who had succeeded Forbes as the Commander-in-Chief, Home Fleet, in October 1940 – wrote that: 'The material results of this operation were small and the losses heavy. This had been expected . . . [S]ome of the survivors felt that an attack on such poor targets against heavy opposition was not justified and their morale was rather shaken until they appreciated the political necessity for the operation'.[159] Sandwiched between the Axis and Soviet forces that wrestled for control of northern Scandinavia, Kirkenes was fated to become one of the most frequently bombed places in Europe, almost rivalling Valetta, Malta. In October 1944 it became the first Norwegian town to be liberated by the Red Army.

Some asymmetrical engagements

Ships that were lacking air cover were that much more vulnerable to attack. This was especially true of those caught without the mutual support furnished by a defensive formation that, ideally, comprised vessels with differing but complementary capabilities. By exploiting the synergy between these, a flotilla's protection could – as was seen in several clashes off Norway and Crete – be made to amount to significantly more than the sum total of its parts. Further, the inherent limitations of adversaries' armaments have always acted as something of a shield for combatants. For instance, land-based aircraft employed in maritime strikes during the Second World War – such as the American B.17, Japan's G4M, Italian S.79s and the German *Stuka* and *Kondor* – had characteristics that varied appreciably across their assorted designs, with some of these machines proving in the main to be more dangerous to opponents than others.

Yet even a handful of bespoke maritime planes could have a disproportionate effect. Whilst strike aircraft occasionally lacked adequate fighter escorts, which left them vulnerable to enemy interceptors, they could seldom be thwarted by AA fire alone.

The threat to carriers in this respect was as great as it was to other vessels if not worse. Early experience in the Mediterranean suggested that insufficient fighter

158 Document 166b, 'Report from Commanding Officer, HMS *Victorious* to Commander-in-Chief, Home Fleet: . . . Attack on Kirkenes-Petsamo area, 30 July 1941', 30 July 1941, ibid., pp. 528–30; Document 166d, 'Letter from Commanding Officer, HMS *Victorious* to Rear Admiral, First Cruiser Squadron: . . . Attack on Kirkenes-Petsamo area, 30 July 1941', 7 August 1941, ibid., pp. 539–40.
159 Document 166a, 'Letter from Commander-in-Chief, Home Fleet, to Secretary of Admiralty: . . . Attack on Kirkenes-Petsamo area, 30 July 1941', 12 September 1941, ibid., p. 526.

protection could lead to a carrier being disabled if not destroyed outright, HMS *Illustrious* and the *Formidable* alike being rendered *hors de combat* in such circumstances during January and May 1941 respectively. Agglomerating sufficient ships and aircraft was only one aspect of a wider problem, however, challenging enough though this was for the overstretched British especially, partly because of early losses among their few carriers and partly because those built since 1937 accommodated roughly half as many planes as some of their forefathers. Replicating something akin to the integrated air-defence network that had foiled the *Luftwaffe* during the Battle of Britain was what was called for. This meshed firepower, sensors and command and control cells with one another. Yet erecting and maintaining a similar system around aerodromes that moved across the seamless cloth of the world's oceans was always going to demand a lot of time, thought and innovation as well as material resources.

Although the sinking of the *Yamato* (one of the two largest battleships ever constructed) in April 1945 was to highlight the overwhelming might that the US Navy's carrier groups above all were ultimately to dispose of, this event has to be seen in context. On 18 November 1942 – barely a year after Japan's attack on Pearl Harbor plunged America into the Second World War – the Future Building Committee of the Royal Navy was emphatic that: 'The . . . Carrier must be the core of the Fleet, and the deployment of aircraft, both Carrier-borne and shore-based, must be the king pin of Naval strategy'. Noting that a prominent feature of the conflict so far had been the growing exposure of both military and commercial vessels to observation and fire from the skies, the panel suggested that 'revolutionary' change was afoot and needed to be addressed as a matter of the highest priority. It was clear, the committee argued, that:

> with the increasing performance and range of aircraft, all ships at sea, whether friendly, neutral or enemy, will become ever more liable to air attack. . . . The effectiveness of ships' AA gunfire depends on many factors, but it is probably safe to say . . . that the adequate defence of ships against air attack cannot be achieved by gunfire alone . . . [and] that fighter defence will be a necessity. The two forms of defence are complementary and in proper combination each will enhance the effectiveness of the other. . . . It is not possible to dissociate ships, carrier-borne aircraft and shore-based aircraft from each other as they all have an important part to play in the Control of Sea Communications. . . .
>
> Attacks on enemy warships and merchant ships will in future be carried out most economically and efficiently by carrier-borne aircraft when the distance of the enemy exceeds about 500 miles from our nearest aerodrome. . . . Reconnaissance must be shared by shore-based and carrier-borne aircraft. . . . [T]he duties of shore-based and carrier-borne aircraft are complimentary to each other and . . . on many occasions they will have to work in the closest co-operation.
>
> It is apparent . . . that as the aircraft carrier provides the primary means of striking the enemy and an essential means of defeating enemy strikes, . . .

other vessels are in a sense ancillaries to . . . [it]. The possibility of developing a ship combining the functions of a battleship and an aircraft carrier has been recently . . . examined but discarded . . . . [Yet] there will be conditions when aircraft cannot be effectively used and when the gun and torpedoes from surface vessels can . . . . This implies that in the foreseeable future battleships and cruisers will continue to have a definite though more limited function, . . . [the] main role [of the battleship becoming] that of AIRCRAFT CARRIER HEAVY SUPPORT SHIP.[160]

Although this transition was to be accelerated by events in the Indian and Pacific Oceans, it was primarily a reaction to the transformed geostrategic situation that prevailed across Europe in the aftermath of Germany's *Blitzkrieg* of 1940–41. Until such time that aircraft carriers were available in sufficient numbers to universally supplant battleships at the heart of fleets, the latter would retain some importance, not least as stable, robust platforms for large amounts of AA weaponry. The battleship's ability to pound coastal targets and other vessels from afar also gave it an advantage over combat planes that had to close to relatively short distances before releasing their payloads. Indeed there were those in the British Admiralty who were especially fearful that the *Kriegsmarine* might deploy the *Bismarck* or other big-gun vessels alongside Germany's embryonic carrier, the *Graf Zeppelin*, in the spiralling *guerre de course*, the Director of Plans noting that: 'In good weather the aircraft carrier could reconnoitre some 20,000 square miles in one day and could hardly fail to locate some of our large convoys. Her reconnaissance would serve equally to defend the attackers from our hunting groups. [This] . . . is going to turn the scales in favour of any raider'.[161]

Whereas *Bismarck* was to be sunk fairly early on in the conflict, many of the grandest of her kind managed to survive for much of the war, among them the *Tirpitz* (her twin) and Japan's gigantic *Yamato* and *Musashi*. All three of these battleships were eventually destroyed by planes, although the hounding of *Tirpitz* by air power dragged on for years, during which time raids by carrier-based machines gave way to attacks by large, land-based bombers that packed a commensurately heftier punch. One of the most imposing attacks against her occurred on the 3 April 1944 while she was in the Kaa Fjord, Norway. To try to ensure that she could not interfere with the impending D-Day landings, no fewer than 40 *Barracuda* bombers escorted by 81 fighters were launched against her from six British carriers of varying classes. Approaching their quarry over the neighbouring mountains in two waves, the planes achieved complete surprise and encountered comparatively light resistance; only two were lost. It is thought that the *Barracudas* scored 14 hits in all, including four with 1,600 lb

---

160 Future Building Committee, Royal Navy, 'First Interim Report: Future Composition of the Royal Navy', 18 November 1942, TNA, ADM 116/5150.
161 Document 29, 'Minute by Director of Plans: Threat posed by *Bismarck* and *Graf Zeppelin*, 29 January 1940', in *FAA in Second World War: Volume One*, p. 80.

armour-piercing bombs. Despite the fact that one of the latter failed to detonate, damage to the ship was very extensive and took three months to repair. Around 400 of her crew were also killed or injured.[162]

Thereafter RAF Bomber Command rather than the FAA spearheaded attempts to molest the leviathan. Five months later a force of *Lancasters* flew to Yagodnik near Archangel, whence, having been refuelled and armed with Soviet assistance, they set out on 15 September to bombard *Tirpitz* at her moorings in the Kåfjord, a branch of the Altafjord, far in the north of Norway. The battleship suffered at least one direct hit, while several other bombs landed sufficiently close to inflict further damage. Long since reduced to an immoveable hulk, she eventually succumbed to her nemesis near Tromso on 12 November 1944. Assailed by *Lancasters* dropping 'Tallboy' bombs that had originally been developed for the demolition of massive, static targets such as U-boat pens and bunkers fashioned from reinforced concrete, she sustained two direct hits and capsized.

So it was that the Norwegian fjords where *Tirpitz* had loitered for most of her existence became her last resting place, too. *Musashi* and *Yamato*, by contrast, met their respective ends on the high seas at the hands of carrier-based torpedo- and dive-bombers. By this stage of the contest in the Pacific especially, the big-gun ship – traditionally the mainstay of fleets – had been relegated by aviation's superior flexibility to essentially auxiliary roles, notably that of escorting carriers. Battleships in general and *Yamato* and *Musashi* in particular were, however, far too big to manoeuvre in tight formations with other vessels and also presented prime targets in themselves. Destroyers and light cruisers were better suited as close escorts to carriers, although they were commensurately inferior to capital ships in terms of seakeeping and their capacity to accommodate AA weaponry in particular.

Such subtleties formed part of the wider conundrum of aerial defence in maritime environments. By the Second World War's end, planes had had a hand in the demise of a great many vessels, more than a few of them aircraft carriers, whereby the failings of one side's air power were part and parcel of the other's success.

Matapan, 1941

The first noteworthy, asymmetrical engagement of the war between nautical forces took place off Cape Matapan on 28 March 1941. It occurred when a formidable Italian flotilla under Admiral Angelo Iachino endeavoured to harass the British convoys ferrying troops and supplies from Egypt to Greece. The Royal Navy had been alerted by 'Ultra' intercepts as to Iachino's intentions and, having diverted the vulnerable convoys out of harm's way, Admiral Cunningham

---

162 See: Roskill, *The War at Sea 1939–1945, Volume Three: The Offensive, Part One*, p. 277; and Report on Operation 'Tungsten' from Vice Admiral, Second-in-Command, Home Fleet, to Commander-in-Chief, Home Fleet, 10 April 1944, TNA, ADM 199/844.

concentrated elements of the Mediterranean Fleet to confront him. Among the foremost of these vessels was the carrier *Formidable*, although, owing to the replenishment problems besetting the FAA in the wake of the Battle of Britain, she had only 14 serviceable strike planes and 13 fighters on board.

Nevertheless, just a clutch of these few machines was to pull off one of the most impressive feats of arms seen in the Mediterranean theatre, one of Cunningham's senior subordinates going so far as to suggest that it would 'remain a standard example of the operations of an aircraft carrier in battle'.[163] Besides maintaining a constant combat air patrol and shadowing the Italian vessels, *Formidable* struck three times at the opposing flotilla, which was provided with next to no support by the RAI. A fruitless attack by two S.79 torpedo-bombers on *Formidable* aside, no Italian aircraft were seen throughout the fighting. (Both torpedoes were released at 2,000 yards from the carrier and were easily dodged.[164]) By contrast, a strike undertaken at dusk by just six *Albacores* and two *Swordfish* started a sequence of events that ended in calamity for Iachino.

At the Italian fleet's heart was the battleship *Vittorio Veneto*, with three cruisers to either side of her. When the planes got within two miles of the outlying destroyer pickets, 'a smoke screen was put up and an intense AA barrage of all calibres was started, which the aircraft were unable to penetrate [in formation]. They then . . . scattered, and carried out individual attacks . . . from different bearings . . .'.[165] The captain of one of the cruisers, the *Pola*, remarked: 'I have never seen such courage as was displayed by the aircraft attacking me. It came in about 5 feet above the water under a withering fire to within very short range. . . . I can only describe it as an act of God'.[166] Amidst the AA barrage, flares, smoke, searchlight beams and fading light, few of the crews were able to observe precisely what harm they were inflicting and several of the planes almost collided with one another. The *Vittorio Veneto* and at least one of the cruisers were believed, however, to have been hit.

The *Pola* certainly had been and was drifting in the night. By 'fixing' parts of the enemy's battle-line, the *Swordfish* and *Albacores* had fulfilled one of the principal roles for which they had always been envisaged. Indeed unbeknown to his adversary – who believed that Cunningham's battleships were far away in Alexandria – some of the Mediterranean Fleet's 'heavies' were plodding up behind the lighter vessels of the vanguard, seeking to finish off any stragglers from Iachino's fleeing squadrons. They soon located the *Pola*, if only because some of the British ships were equipped with radar suites and were better able

---

163 Document 114, 'Letter from Rear Admiral, Mediterranean Aircraft Carriers, to Commander-in-Chief: . . . Battle of Cape Matapan, 28–29 March 1941', 15 April 1941, in *FAA in Second World War: Volume One*, p. 374.

164 Document 114a, 'Report from Commanding Officer, HMS *Formidable*, to Rear Admiral, Mediterranean Aircraft Carriers: . . . Matapan, 28–29 March 1941', 10 April 1941, ibid., pp. 378–9.

165 Ibid., p. 387.

166 Ibid., p. 381.

to discern what was happening in the darkness than the Italians; like the RAI, the RMI had fallen behind in the development and introduction of electronic sensors and in the refinement or supersession of mechanical engineering by mechatronics. Four destroyers and two heavy cruisers – the *Zara* and the *Fiume* – that had been detached to assist the paralyzed *Pola* promptly blundered into the *Warspite* and two other battleships, the *Barham* and the *Valiant*, which raked them at short range. All of the Italian vessels save two of the destroyers were sunk in due course, with casualties exceeding two thousand men.

Inconclusive though it was, the clash off Cape Matapan did substantial material and psychological damage to the RMI. Whilst *Vittorio Veneto* and her remaining escorts managed to elude destruction, Cunningham's task group slipped away virtually unscathed, having lost just one *Albacore* and its crew of three. Predictably, the following day the *Luftwaffe* lunged after the British as they fell back on Alexandria, a dozen Ju.88s endeavouring to sink *Formidable* in particular. But this riposte was parried; none of the bombs struck home and a couple of planes were seen to be shredded by the barrage of AA fire thrown up by the other ships. A solitary S.79 that tried to keep tabs on the fleet was also chased off by patrolling fighters and shot down some forty miles away.[167]

Arguably experience at Matapan did more than any other single event to persuade the RMI that it needed aircraft carriers of its own. Work was duly started on transforming the liners *Roma* and *Augustus* into such ships. However, resources, not least time, were running low. Neither vessel was to be completed before Italy was drubbed into surrender in 1943.

The sinking of the Bismarck, *1941*

On 20 May 1941 – the same day that *Fallschirmjäger* first assailed Máleme Aerodrome on Crete – the German battleship *Bismarck* was spotted steaming northwards through the Kattegat. As new as she was intimidating, she had spent some months conducting final sea trials in the Baltic, her activities monitored by, among others, Sweden's intelligence bureau.

Forewarned by confidants in Stockholm and Norway of her approach, the hard-pressed British had long been pondering how best to respond to such an ominous eventuality. For over a year the Admiralty had feared that one or more of the capital ships of the *Kriegsmarine* might make a foray into the Atlantic accompanied by the fledgling *Graf Zeppelin* – the first of four carriers envisaged under 'Plan Z', the blueprint for the evolution of Germany's battle-fleet during the period 1938–48. But early in 1941 *Graf Zeppelin* was not yet serviceable. Nor was she destined to become so. Always the neglected sibling of air power that could support his armies directly, to Hitler ocean-based aviation was actually diminishing in importance at this stage of the war. After all, the *Luftwaffe*

---

167 Ibid., pp. 382–3.

was now ideally placed to contain any fresh efforts by the Royal Navy to strike landwards from its own carriers, both in northern waters and the Mediterranean. Malta and metropolitan Britain alike also lay within easy reach of land-based bombers and, sure enough, were experiencing an escalation of the aerial bombardment to which they had both been subjected since June 1940; the 'Blitz' against population centres the length and breadth of the UK was to climax in the middle of May 1941, just as raids on Valetta and other Maltese towns were intensified prior to the Axis shifting much of its deadliest firepower eastwards in readiness for 'Barbarossa'. Indeed the subjugation of the USSR and the British Isles – by invasion and blockade respectively – was Hitler's primary strategic goal at this juncture. The marginalization if not defeat of the USA – which, despite the entreaties of the French prime minister for one, had remained neutral throughout Germany's victorious campaigns of 1940[168] – was something to be accomplished over the longer term, not least because it would necessitate major investments in naval power, including carriers and overseas bases. This would call for immense resources that, for all his conquests, Hitler did not yet have at his disposal, although he did anticipate their acquisition through his impending annihilation of the British Empire and his erstwhile partner in crime Comrade Stalin, which would bring him hegemony over what Halford Mackinder had in 1919 termed the 'Heartland' of the 'World-Island'. Certainly, Churchill did not hesitate to draw President Roosevelt's attention to the implications for the USA's maritime security that would stem from any German seizure of not merely the French fleet but also the Royal Navy and its global network of harbours.[169] In the event, however, Hitler's bid to safeguard his gains – particularly from those who struck at 'Fortress Europe' from the seamless skies and oceans and thereby had more choice as to the exact timing and place of any attack – would prove a ruinously costly and ultimately futile exercise that contributed to the Axis being defeated on every front and by land, sea and in the air.

So it was that *Bismarck* left the relative safety of the Baltic accompanied by, not *Graf Zeppelin*, as the British Admiralty had dreaded, but another big-gun ship, the heavy cruiser *Prinz Eugen*. Nevertheless, this was not much consolation for the Royal Navy, which was simultaneously attempting to fulfil several commitments that were as demanding as they were widespread. Besides the operations around Crete, it was imperative that the *Scharnhorst* and *Gneisenau* be contained. These powerful battlecruisers had recently taken refuge in Brest on France's western shore after sinking 22 merchantmen in a wrecking spree that had lasted from January into March. With these and other surface raiders as well as U-boats at loose, sizeable escorts had to be scraped together for the numerous convoys plying the Atlantic especially. Whilst it was presumed that *Prinz Eugen* and *Bismarck*

---

168 See: Horne, *France 1940*, pp. 640–41. For Churchill's views on the likelihood of direct American intervention, see Jenkins: *Churchill*, pp. 609, 612–13.

169 Ibid., pp. 608, 614–15.

would attempt to slip into that ocean, too, judgements as to when and how they were going to do this were more speculative still.

Simply keeping track of the two ships as they progressed along the Norwegian coast proved difficult enough, if only because of the prevailing weather. This blighted reconnaissance sweeps and, thereby, stalled impromptu bombing raids, such as one the FAA strove to improvise on the evening of 22 May after both *Prinz Eugen* and *Bismarck* had been caught on camera at Bergen by a prying *Spitfire*. The experience of a surveillance plane that was dispatched at 17:00 from the Orkneys to confirm the targets' positions is sufficient illustration of the obstacles encountered. In the course of this lengthy journey, the pilot twice risked taking his Martin *Maryland* down to just 100 feet, yet the occupants were still unable to glimpse the sea below, so dense and extensive were the intervening layers of cloud and mist. On making landfall, where, fortuitously, the sky turned out to be clear up to around two thousand feet, the aircraft was finally rewarded with a view of the harbour but was equally exposed to the AA batteries surrounding it. Although these inflicted some damage, the *Maryland* managed to get away, its mission accomplished.[170]

There was to be no torpedo strike, however, as by this time the German birds had flown. Their commanding admiral, Günther Lütjens, doubtless conscious that Bergen had proved the grave of the *Königsberg*, seems to have been spooked by the appearance of the *Spitfire* photographic reconnaissance plane and by the *Whitley* and *Hudson* bombers that came searching for him not long after, for he had ordered his ships to leave the area in some haste. Certainly, whereas *Prinz Eugen* had topped up her fuel stocks from a tanker while at anchor, *Bismarck* herself had neglected to take the precaution of doing so. This was a curious oversight for a ship bent on venturing far from friendly waters.

The relationship between fuel consumption and time and distance were among the factors that Lütjens's opponents in the Royal Navy also had to bear in mind. For as long as there was any possibility that large German surface raiders might sortie into the Atlantic, the Admiralty was obliged to keep a reserve of suitably powerful ships at Scapa Flow, from where they might have a good chance of heading off any interloper. This force frequently included at least one carrier. After all, in British doctrine the primary role of such vessels was to assist in countering an enemy's capital ships. But in May 1941 the overstretched Royal Navy had to switch carriers from other theatres and roles to contain if not defeat *Bismarck* and *Prinz Eugen*; *Victorious* was taken off convoy escort duties and *Ark Royal* was to be summoned from the Mediterranean. Admiral Tovey's Home Fleet also had to judge the right moment to leave Scapa Flow in order to intercept Lütjens's flotilla, the ultimate destination of which was uncertain. Intelligence reports had noted numerous *Luftwaffe* reconnaissance flights that

170 Document 126, 'Letter from Commanding Officer, RNAS Hatston, to Admiral Commanding Orkneys and Shetlands: 771 Squadron reconnaissance flight to Bergen, 22 May 1941', 3 June 1941, in *FAA in Second World War: Volume One*, pp. 419–20.

were evidently charting the extent of the ice sheet in the Greenland and Norwegian Seas. Would either one or both of the German vessels remain on the Norwegian coast, or were they bound for Icelandic waters or some other corner of the Atlantic or Arctic? Tovey surmised, correctly, that their target was Britain's transatlantic trade.

Lütjens, however, had three or four viable routes to choose from when it came to breaking out into that ocean. This obliged his immediate antagonists – Tovey's vessels operating out of Scapa Flow and long-range planes of RAF Coastal Command – to disperse themselves to cover the Denmark Strait (between Greenland and Iceland) as well as the passages to the north and south of the Faroes. Lütjens selected the first of these channels so as to minimize his exposure to snooping British aircraft, but his approach was first detected and then monitored from a respectful distance by the radars of the cruisers *Suffolk* and *Norfolk*. At one point *Bismarck* opened a rapid fire on the latter as she glimpsed her amid the freezing fog. No hits were scored, but the reverberations from her massive guns dislocated her own forward radar system. Thereafter, *Prinz Eugen* preceded Lütjens's flagship as they groped their way along the ice floe's edge. Meanwhile, the battleship *Prince of Wales* and the battlecruiser *Hood*, under Vice Admiral Lancelot Holland, were vectored onto a course that, all things staying equal, promised to bring them into range of their quarry at dawn on 24 May. Holland anticipated catching his prey unawares, if only because, whereas the German ships would appear silhouetted against the skyline, his would still be masked somewhat by the receding shadows.

It was not to be. The climactic manoeuvres were not executed as well as they might have been and Holland's force found itself slightly but significantly behind schedule, maldistributed and downwind from *Bismarck* and *Prinz Eugen* when the shooting started. In the prevailing conditions *Prince of Wales* could not get her *Walrus* spotting plane aloft, while the seakeeping of the destroyers accompanying her and *Hood* was inadequate for them to get within striking distance of Lütjens's ships in time; in the absence of aerial reconnaissance they had had to be spread fairly widely along Holland's axis of advance to help in locating the enemy. Likewise, the cruisers that had been shadowing the German vessels were a little too far off to engage and distract *Prinz Eugen* – as Holland had anticipated they would – from the battle's outset. In a bid to retain the element of surprise, the British admiral had maintained radio silence, leaving *Suffolk* and *Norfolk* without detailed instructions as to their part in the incipient clash. Holland assumed, reasonably enough, that the cruisers would attack the rearmost German ship, which he mistakenly believed to be *Prinz Eugen*. Neither could *Prince of Wales* and *Hood* bring all their turrets to bear immediately; they were obliged to advance under fire, with wind-blown spray clouding the vision of their gun-directors, before they could wheel broadside to the enemy.

Holland was in any case anxious to reduce the range between his and Lütjens's ships as quickly as possible, for he knew the resilience of the *Hood* especially to be precarious. Laid down in the same year that Jutland was fought, she was as elderly as she was venerable. More of a fast battleship than a traditional battlecruiser, she was the Royal Navy's largest vessel and, through her numerous flag-flying voyages around the world, had long since become the pre-eminent icon of British martial might. But although her eight 15" guns endowed her with a colossal punch, the plating protecting her vitals varied in depth, creating some very dangerous chinks. Whereas her sides were sheathed in slabs of armour that were set at a slight angle so as to increase its thickness from 12" to as much as 15" without adding to the overall weight, her decks were scantily shielded. This made her vulnerable to plunging shot especially, as a series of interwar experiments on replica targets confirmed.[171] Projectiles fired from relatively short distances – and thereby along correspondingly flat trajectories – were less perilous to such a ship than those that followed more of a parabolic path, much like bombs dropped from an aircraft. Indeed the ageing *Hood* was scheduled to undergo a major modernization programme in 1941, if only because of the perceived need to counter the growing danger posed by planes. For want of anything better, she had been fitted with a UP AA system, for instance, which in many combat situations was likely to prove as much of a liability as an asset.

Hydrophones aboard Lütjens's ships had been picking up the sound of approaching vessels for some time when, at about 05:30, the converging forces came in sight of one another. Among the witnesses to what ensued was the crew of an RAF *Sunderland* of 201 Squadron.[172] Initially exchanging fire with the foremost enemy vessel, which was actually *Prinz Eugen*, *Hood* was quickly straddled with shots from both. What seems to have happened is that a high-explosive round from the German cruiser struck *Hood* near her mainmast, igniting a conflagration amidst the stowage lockers where primary reserves of 4" shells and 7" UP projectiles were kept close to hand. Shortly after, *Bismarck* unleashed a (fifth) salvo of 15" armour-piercing shells at *Hood*. One of the plunging rounds was seen to plummet into the battlecruiser in much the same area. A split second later, the contents of an underlying magazine erupted.[173]

The resulting chain of tremendous explosions demolished the ship's centre instantaneously. *Hood*, her hull broken in two, slid beneath the frigid waves, taking Holland and all but three of her 1,418 crew with her. The *Prince of Wales* – so fresh from the slipway that she still had teething problems, most notably with her Type 284 gunnery radar and the operation of her turrets – now found herself in a very unequal contest. After firing a few salvoes and being straddled by rather more, she shrouded herself in smoke and turned away, an unexploded 15" shell

171 See: David K. Brown, *The Grand Fleet: Warship Design and Development, 1906–1922* (London, 1999), p. 99.

172 The report from this flight and details of other aerial operations in the hunt for *Bismarck* can be found in TNA, AIR 14/415.

173 See: 'Loss of HMS *Hood* in Action with German Battleship *Bismarck*: Boards of Enquiry', 2 June and 12 September 1941, TNA, ADM 116/4351–2.

lodged in her side. Another had passed clean through her compass platform, killing or injuring several key personnel.[174]

Bismarck had been damaged as well, mostly around her forward fuel tanks, which were haemorrhaging irreplaceable oil. Detaching *Prinz Eugen*, Lütjens decided to head for Brest. Once here, besides getting her repaired and replenished, he might unite his fearsome battleship with *Gneisenau* and *Scharnhorst* to form a uniquely puissant squadron. A last-ditch bid to 'fix' their quarry – a torpedo attack by *Swordfish* from the carrier *Victorious* – having failed, the remaining capital vessels of Tovey's Home Fleet could not hope to overtake *Bismarck*. By dawn on 25 May Lütjens had even shaken off the cruisers that had been tracking his movements with radar during the night. Yet he betrayed his whereabouts intermittently by breaking radio silence and, on the morning of 26 May, came under surveillance from a *Catalina* dispatched by Coastal Command from Lough Erne, Ireland. With no aircraft to drive off this flying boat, which loitered at a respectful distance beyond the reach of most AA guns, *Bismarck* could not escape scrutiny.

Later that day *Swordfish* from the *Ark Royal* – which had hurried north from Gibraltar – attacked the cruiser HMS *Sheffield*, having mistaken her for the *Bismarck*, which was actually a few miles distant. Fortunately for all concerned, the torpedoes employed had been armed with novel, magnetic fuses that malfunctioned, the warheads detonating harmlessly on impact with the water. But this tragicomical incident wasted precious time. By now, dwindling fuel stocks were also as much of a problem for the hunters as they were for the hunted; unless *Bismarck* was slowed down in the next few hours, the 'heavies' of the pursuing Home Fleet would have to head homewards.

Having regained the carrier and re-armed themselves with traditional torpedoes, in fast-fading light the biplanes again took off from *Ark Royal*. Almost two hours later, they sighted *Bismarck* about twelve miles beyond *Sheffield*. Several of the 15 *Swordfish* were able to descend towards their prey under the cover of low banks of cloud. It is also conceivable that some of the cutting-edge, automated AA weapons affixed to *Bismarck* pivoted too quickly to remain pinpointed on such slow-moving targets. Certainly, despite very heavy and well-directed defensive fire none of the aircraft was downed, although four were damaged, one of them being riddled with 175 holes from shell fragments. Indeed so intense was the flak buffeting his aircraft that one pilot wavered, aborting his attack and jettisoning his 'Fish'. Torpedoes from two other machines struck home, however. By accident or design one of these hit the jinking battleship in the steering compartment, jamming her rudder.[175]

Unable to maintain a straight course, *Bismarck* was left steaming unwillingly but inexorably towards the Home Fleet and away from Brest. She had almost

---

174 See: 'Battle Damage Sustained by HMS *Prince of Wales*, 24 May 1941', TNA, ADM 267/111.
175 Document 127, 'Report from Commanding Officer, HMS *Ark Royal*, to Secretary of Admiralty: Torpedo Attack on *Bismarck*, 26 May 1941', 6 June 1941, in *FAA in Second World War: Volume One*, pp. 421–4.

come within range of *Luftwaffe* aircraft based in France and U-boats that were hastening to her aid, but now she lay beyond all help. During the night she was observed and harassed by the Fourth Destroyer Flotilla, the *Piorun* (a Free Polish ship) going so far as to trade shells with the leviathan. However, with the dawn came two British vessels that individually possessed just as much firepower as Lütjens's stricken command and, fully manoeuvrable, were better placed to exploit it, low though their fuel stocks had become. After being raked by salvoes from the battleships *King George V* and *Rodney*, *Bismarck* was torpedoed by the cruiser *Dorsetshire*. A mere 115 out of her 2,222 crewmen could be saved.

The sinking of HMS **Repulse** and HMS **Prince of Wales**, *1942*

Five months after her baptism of fire at the hands of *Bismarck* in the Denmark Strait, HMS *Prince of Wales* was ordered to sail for the Far East. Although at this juncture the Admiralty wished to retain all of its latest battleships in the axial, occidental theatre, Churchill and the war cabinet's Defence Committee were persuaded that the dispatch of a sizeable flotilla to Ceylon (Sri Lanka) or Singapore might help deter Tokyo from resorting to war and, failing that, would constrain the IJN's scope for action in both the Indian and Pacific Oceans.[176] The deployment of one of the Royal Navy's most modern battleships to distant, oriental waters at a time when the situation on Britain's doorstep was so disquieting would, it was hoped, be taken as a sign of London's earnestness by Australia, New Zealand, India and potential allies and foes alike. Simultaneously, however, it was recognized that *Prince of Wales* and the few vessels that might be spared to accompany her could only be more influential diplomatically than militarily.[177] Whilst what was dubbed 'Force Z' was envisaged as the mere kernel of an armada that was expected to come to comprise no fewer than seven battleships, ten cruisers, a carrier and two dozen destroyers, it was realized that bringing this plan to fruition would remain impracticable until March 1942 at the earliest.[178] Again, the British were struggling to reconcile their peculiar need for a global strategy with resources that were unavoidably insufficient to meet each and every actual or possible commitment. Should hostilities with Japan break out, the capacity of 'Force Z' to strike at the enemy and parry his blows would be very limited, if only because the flotilla lacked the versatility afforded by a spectrum of maritime capabilities, including proactive air power. Although the admiral selected to lead it, Sir Tom Phillips, was officially designated as 'The Commander-in-Chief, Eastern Fleet', this title was somewhat grandiose under the prevailing circumstances. Certainly, the martial assets immediately at his disposal were fewer in number and far less sophisticated than it might

176 'Report on the Loss of HMS *Prince of Wales* and HMS *Repulse*', by First Sea Lord, 25 January 1942, TNA, PREM 3/163/2, paragraphs 9 and 14.
177 Ibid., paragraph 18.
178 Ibid., paragraph 1, 'Remarks on the Formation of the Eastern Fleet'; and paragraphs 8 and 10.

have suggested. Phillips's command was all but monofunctional and, largely as a consequence of that, proved fatally ill-matched to its operating environment.

Phillips arrived at Singapore on 2 December 1941, a few days after the largest carrier task force the world had yet seen set out surreptitiously from Japan. Commanded by Vice Admiral Chūichi Nagumo, its destination was Oahu in the distant Hawaiian Islands – the location of Pearl Harbor, the anchorage of the US Navy's Pacific Fleet. There were already some cruisers and destroyers at Singapore, among them the Dutch light cruiser *Java*, but most of these were undergoing refits and repairs. Four American destroyers were also en route to the British base, essentially as a gesture of solidarity. Phillips, too, strove to reassure key partners in the wider region by sending the battlecruiser *Repulse* and two of his destroyers on a quick, flag-waving visit to Port Darwin in Australia. However, the endeavours of London and Washington to forestall any resort by Tokyo to violence were already being overtaken by events. Pearl Harbor came under attack from Nagumo's aircraft around 08:00 on 7 December and, shortly after this, high-altitude bombers stationed in Indo-China raided Singapore as well, although they did not succeed in inflicting any significant damage. What was very much more concerning for the garrison was the prospect of the entire enclave, naval dockyards and all, being seized by the enemy by means of an overland offensive – a fate that could not befall Oahu. Even before America's battleships were assailed in their berths at Pearl Harbor some seven thousand miles to the east, Japanese soldiers had started infiltrating the Isthmus of Kra from Thailand and through amphibious landings at Singora (Songkhla), Pattani and Kota Baharu. Within just a few hours they had made appreciable progress towards Singapore and, by the end of January, were pounding on its gates.

Admiral Phillips was thus dragged into the whirlpool of war while Britain's 'Eastern Fleet' was still embryonic and Singapore's landward defences were already crumbling. *Repulse* and her companions were immediately recalled, if only because the few ships immediately available for any interdiction of the Japanese invasion convoys were essentially those comprising Force Z. Besides *Prince of Wales* and *Repulse*, this consisted of four destroyers, all but one of which had yet to be fully 'worked up'. There were commensurate doubts about the mechanical endurance of these escorts and – somewhat ironically as things turned out – Phillips, mindful of the fate that had befallen so many destroyers off Crete, reasoned that they were the most vulnerable of all his vessels to aerial attacks.[179]

Against such threats his flotilla had no proactive defences whatsoever. Whereas the handful of reconnaissance seaplanes housed aboard *Repulse* and *Prince of Wales* promised to provide glimpses of whatever lay beyond the horizon, since he did not yet have an aircraft carrier at his disposal Phillips would be obliged to manoeuvre his ships without fighter escorts alongside them. That said, ten RAF

179 Ibid., paragraph 43.

Buffaloes stationed at Sembawang, Singapore, were specifically earmarked for fleet protection duties and, subject to being given adequate notice, might be able to come to the aid of Force Z, if only spasmodically and ephemerally. Although this sort of arrangement had worked adequately enough around Sphakia during the choreographed evacuation of Crete, with a few long-range fighters from Alexandria chaperoning convoys en route to Egypt, furnishing aerial cover from distant aerodromes on an impromptu basis plainly left much more to chance. The Brewster *Buffalo* was, moreover, far from being the ideal instrument for such an undertaking. This early American-made monoplane had only been acquired by the British to make up for the widespread shortage of better alternatives, notably *Hurricanes* and *Spitfires*. Its engine did not function at its best in humid, tropical climes and was insufficiently powerful to endow this stubby, rather weighty machine with much agility. At the same time neither was its range comparable to that of the main Japanese interceptor, the Mitsubishi A6M 'Zero'. Whilst London had readily approved in principle to the sending of four squadrons of long-range fighters to the Far East primarily for maritime support operations, in practice this had not proved feasible.[180] Demand for such machines far outstripped supply for much of 1941, not least in the eastern Mediterranean, where both the FAA and the RAF were thinly spread.

On the other hand, if viewed as a single entity Force Z did have appreciable reactive defences with which it might hope to thwart attacks from the skies, not least the AA weaponry affixed to the ships' superstructures. Although the flotilla did not include a specialist AA cruiser and the elderly *Repulse* had comparatively few AA armaments for a vessel of her size, *Prince of Wales*, having had hers overhauled and upgraded in the summer of 1941, boasted several batteries of such weaponry, including pom-pom guns. Again, however, qualitative factors threatened to make this shield more apparent than real should it be put to the test: the 5.25" dual-purpose guns were just too ponderous and their rate of fire rather too slow for them to pose much of a threat to fleeting aerial adversaries, whereas the pom-pom – which was more nimble and generated a greater volume of shot – lacked the range to engage many targets in a sufficiently timely fashion.[181]

Phillips's ships might also have justifiably expected to glean some protection from the limitations of the opposition's armaments, too, which is after all a common aspect of combat. In the climactic (and secret) report on the loss of *Prince of Wales* and *Repulse* compiled in January 1942 by the First Sea Lord, Admiral Sir Dudley Pound, it was stressed that up to this juncture all the available empirical evidence had indicated that, whereas aerial torpedo strikes were not to be anticipated far out to sea and that dive-bombers also had a maximum reach of approximately two hundred miles from their bases, high-altitude, horizontal bombing by suitably large aircraft remained a possibility at considerably greater

180 Ibid., paragraph 37.
181 See: Brown, *Nelson to Vanguard*, p. 32; and J.A. Clements, 'Royal Navy Ship-Based Air Defence, 1939–1984', *RUSI Journal*, 129, (1984) p. 19.

distances from the shore. Nevertheless, events to date suggested that 'against modern capital ships [this technique] was not likely to cause vital damage'. The case of *Repulse* and the *Prince of Wales* was to prove the first significant exception to this apparent rule insofar that they were sunk by land-based torpedo planes that were operating almost four hundred miles from their aerodromes near Saigon (Bho Minh) in Indo-China. It is worth noting that this distance greatly exceeds that between Scapa Flow in the Orkneys and Bergen in Norway, for instance, and is roughly equivalent to that between Alexandria in Egypt and Crete. In view of these factors and the lack of actual combat experience among Japanese aviators operating over the oceans, it is difficult to disagree with Pound's observation that Admiral Phillips had 'had no cause to rate the Japanese Air Forces on a par with . . . [the *Luftwaffe*]'.[182]

Indeed it is probable that at this juncture Phillips believed he knew as much about Japan's aerial capabilities as anybody within the British Admiralty. He had been Vice Chief of the Naval Staff until just a few weeks before the outbreak of the Pacific war and was consequently more familiar than most of his colleagues with the details of the evolving situation in the Far East. Alas, Admiral Pound's report conceded, what intelligence had been obtainable regarding Japanese air power especially was patchy and 'left much to be desired. It was in the main based upon the most recent [observable] actions . . . in China generally'. There had been next to no maritime engagements per se in the protracted Sino-Japanese conflict, although IJN aviation units such as the Kanoya Air Group had played a significant part in the support of operations on terra firma. Whilst Tokyo was known to have some impressive carrier-based planes at its disposal, little had been gleaned about shore-based units that suggested they were anything extraordinary. Neither Phillips nor any of the other theatre commanders, Pound concluded, 'appear to have had reasonably accurate information of the strength, disposition, types of aircraft and efficiencies of Japanese Air Forces in [the] Indo China area. . . . [I]t may be that the consistently adverse reports of the capabilities of Japanese . . . [aviation] may have caused . . . [Phillips] to underestimate the . . . threat and discount the possibility of . . . a heavy scale of attack at long range'.[183]

On the other hand Phillips seems to have been far from complacent. Whilst concluding that the imperative of securing Singapore dictated that the Royal Navy had to try to disrupt the Japanese landings to the north, he recognized that, without at least some air support, this would be a risky enterprise. Yet the foe's offensive was such that the RAF would be hard-pressed to resist it and simultaneously furnish Phillips with much assistance. Before setting out with Force Z on the evening of 8 December, he asked for surveillance sweeps to be performed at least a hundred miles ahead of his flotilla starting at dawn the very next morning. He also requested that the area around Singora be reconnoitred

182 'Report on the Loss of HMS *Prince of Wales* and HMS *Repulse*', paragraphs 30–32.
183 Ibid., paragraphs 24–9.

at first light on 10 December and that fighter patrols then rendezvous with his ships off that port and be maintained over them for the rest of the day. Shortly before sailing, however, he was warned that it was increasingly doubtful that the last of these measures would prove practicable, essentially because of the deteriorating situation ashore. Barely had his little fleet steamed out of Singapore than Japanese counterforce operations precipitated the evacuation of the aerodrome at Kota Bharu and plunged the RAF's other northern airfields into chaos and jeopardy. Indeed a message Phillips received less than eight hours after departing for Singora confirmed that no interceptors would be waiting to meet him there, although the reconnaissance flights he had asked for were expected to go ahead.[184]

Phillips had hoped that, with the advantage of surprise, Force Z could reach and ravage the enemy's most southerly beachheads and lines of communication before the Japanese could even begin to react. There were, he was aware, hostile warships in the region, including a battleship and at least seven cruisers, but no carriers were reported to be among them. He doubtless anticipated that any enemy aircraft that did materialize would have been diverted abruptly from missions inland to meet the looming threat posed by his flotilla. As a consequence, they would almost certainly be armed with general purpose ordnance rather than with bespoke anti-shipping weaponry, notably torpedoes, that might pose a serious threat to Force Z. Although he knew that no fighter cover would be waiting for him off Singora, on the other hand the prevailing weather – low cloud and numerous rain squalls – promised to vitiate any surveillance or combat operations that Japanese planes might try to undertake against his fleet, should it be discovered.

Nevertheless, Phillips erred on the side of caution, particularly when the skies began to clear late in the afternoon of 9 December. Unbeknown to him, a submarine, the *I-65*, had already spotted his ships at 14:00 and had duly alerted other Japanese units in the area. The troop transporters and supply vessels in the Gulf of Siam (Gulf of Thailand) were promptly ushered out of harm's way while elements of the Second Fleet rushed to head off the British flotilla. Naval aviation groups stationed around Saigon in Indo-China were mobilized, too, their presence being noted by intelligence staff officers in Singapore, who were also to report an aircraft carrier off Singora some twelve hours later.[185] When three seaplanes – which had actually been launched by cruisers accompanying the invasion convoys – were sighted by *Prince of Wales*, Phillips concluded that he no longer had any chance of catching the enemy off their guard and was now in danger of being attacked by large numbers of aircraft. He duly wheeled Force Z about and, after shaking off the shadowing planes, steamed off at high speed into the growing darkness, bound for Singapore.[186]

184 Ibid., paragraphs 37–9.
185 Ibid., paragraph 28.
186 Ibid., paragraphs 42–4.

The destroyer HMS *Tenedos* – which, running low on fuel, had already been ordered to proceed homewards – was well ahead of the rest of Force Z when the bulk of the flotilla altered course once more. Advised at around midnight that Kuantan, which lies midway between Kota Bharu and Singapore, had also been seized, at 00:52 on 10 December Phillips – who would otherwise have retraced his steps by pivoting on the Anambas Islands, over two hundred miles to the north-east of Singapore – increased speed and swerved back towards the Malaysian Peninsula, hoping to nip this landing in the bud.[187] He was unaware, however, that a submarine, the *I-58*, had again sighted his ships, reporting their position and heading at 03:40 and even unleashing some torpedoes at the British vessels as they sped across the horizon at 25 knots. As Force Z neared the coast three hours later, a Japanese plane duly appeared, although the flotilla now lay all of four hundred miles from the nearest aerodrome known to be under the opposition's control.[188] There was no sign of the foe at Kuantan, on the other hand, which was found to be completely peaceful. Accordingly Phillips swivelled his ships about again, steaming eastwards with a view to resuming his journey around the Anambas Islands.

Roughly two hours later, *Tenedos* – which was then about a hundred and fifty miles to the southeast of the rest of Force Z – reported that she was being assailed by high-level bombers, none of which managed to hit her. The warm, humid conditions of the Far East had spawned some gremlins within the radar systems aboard Phillips's ships, but just before 11:00 the antennae of HMS *Repulse* picked up formations of aircraft approaching from the south, the first of which were sighted within a few minutes.[189]

The planes concerned were from the Genzan, Kanoya and Mihoro Groups of the Twenty-Second Air Flotilla stationed near Saigon. Each of these three units had been sent hunting for Force Z along slightly differing axes and had probed well to the south of the Anambas Islands in search of Phillips's ships. As they received confirmation of their quarry's position each of the groups veered to the northwest, encountering the British flotilla east of Kuantan and nearly four hundred miles from Singora. The Kanoya's G4M1 torpedo-bombers especially were at the end of their tether. They had by this stage been aloft for six hours and were within a few minutes of having to head homewards for lack of fuel.[190]

In assailing *Prince of Wales* and *Repulse* the seasoned occupants of the Japanese aircraft followed the tactical procedures that had been drilled into them over recent years. The bomber squadrons bombarded the warships from around 5,000 feet, releasing bombs in compact, clover-leafed patterns. The ordnance employed was not the novel, armour-piercing variety that, adapted from huge naval shells, had wrought such havoc at Pearl Harbor, if only because these

---

187 Ibid., paragraph 45.
188 Ibid., paragraph 46.
189 Ibid., paragraph 49.
190 Peattie, *Sunburst*, p. 169.

weapons were in short supply and had all been allocated to the operation against Oahu.[191] Rather, the aim of these attacks was to suppress the British vessels' AA defences by strafing their superstructures and, above all, by drawing the gun crews' attention away from the torpedo planes. Flying in at right angles to their targets and, if possible, from both sides simultaneously, these unleashed their weapons from about a hundred feet above the water and at a distance of some three thousand to six thousand feet from their quarry.

Up until this point in the war the Royal Navy had seldom faced the challenge posed by aerial torpedoes. Their use in conjunction with horizontal bombardment proved particularly perplexing, although the initial lunges of the Japanese planes towards Phillips's capital ships were poorly synchronized. There were seven attacks in all, the first of which commenced at 11:18 and was executed by nine aircraft dropping bombs. Whereas just one of these projectiles hit *Repulse* (and did so without inflicting any real harm), one of the nine torpedo planes in the following wave dealt *Prince of Wales* a near fatal blow. As had happened with her old nemesis the *Bismarck*, the battleship was struck astern by a single torpedo. Many hundreds of tons of water promptly surged up a damaged propeller shaft and inundated several compartments, including one of the engine rooms, a dynamo chamber and an auxiliary machinery room. With her pumps struggling to stem the flooding and many areas below deck plunged into total darkness – a complexity that grossly hampered her damage-control teams – *Prince of Wales* slowed to 15 knots and slumped 13° to port.

This list, the flooding and the attendant loss of speed and manoeuvrability greatly diminished the battleship's capacity to evade further blows. The loss of current in a large, modern vessel, where so much mechanical engineering had been supplanted or made more convoluted by electrical components, was also a crippling handicap. Her crew never managed to regain control of the stricken leviathan. Meanwhile, with her starboard side tilted skywards and parts of the network shorn of power, the ability to focus the AA armaments that studded *Prince of Wales* was seriously impaired as well. Several of her 5.25" gun turrets alone had been rendered *hors de combat* at one fell swoop and the muzzles of the remainder, like those of some of the other weaponry that was still in action, could not be depressed sufficiently to engage low-flying aircraft to the right of the ship's line of travel. An attack from this quarter by torpedo planes would be especially difficult to thwart.

At 11:56 nine such aircraft made for *Repulse*, but she dodged all of the weapons that they dropped. Two minutes later, she was assailed by nine high-altitude bombers, none of which managed to hit her. However, attacking at 12:22 the next wave of planes succeeded in striking the battlecruiser with one torpedo and in putting no less than three into the vulnerable starboard side of the *Prince of Wales*. One of these twisted the outer propeller shaft against the inner, bringing the battleship to a virtual standstill.

191 Ibid.

Just before noon *Repulse* notified Singapore that Force Z was being bombed. If any specific request for fighter cover had been made prior to this, for one reason or another it had not been received. Neither had the RAF been made aware that Phillips had abandoned his sortie to Singora and Pattani and was now east of Kuantan. Although it remains a matter for conjecture, it is highly probable that the admiral had maintained radio silence so as not to betray his presence to his prey or other hostile forces he understood to be at large in the region. Yet if that was indeed the case any assumption on his part that he would be met at Kuantan by friendly aircraft was as odd as it proved misplaced. *Prince of Wales* and *Repulse* had, moreover, been steaming away from Kuantan for the best part of three hours before they came under attack.

The most plausible explanation for the chain of events is that Phillips genuinely believed Force Z to be too far from the enemy's aerodromes to be molested by anything other than high-altitude bombers. Should any locate his flotilla, combat experience gained off Norway and in the Mediterranean earlier in the war suggested that capital ships especially would be able to ride out any ensuing storm. The ineffectiveness of the initial bombing attack doubtlessly seemed to vindicate this expectation. The torpedo strikes that commenced nearly half an hour later were, by contrast, both surprising and potentially more dangerous. Nevertheless, *Prince of Wales* was hit by just one – albeit one that happened to do extraordinary harm – out of the nine weapons released around 11:44, while *Repulse* managed to evade all nine of the torpedoes directed against her at 11:56.[192] It is probable that it was only at this point that the gravity of the situation, not least the seriousness of the wounds sustained by Phillips's flagship, became fully apparent. This turn of events precipitated the call for help that was issued by the captain of the *Repulse* at 11:50 and which was responded to immediately, despite the RAF having its hands full trying to contain the invaders around Singora, Pattani and Kota Bharu. In the interim, having plainly disabled the largest of the British vessels, the Japanese aircraft redoubled their efforts against the limping *Repulse*. A coordinated attack by nine torpedo planes at 12:25 scored four hits, inflicting terminal damage. Finally, at 12:46, a last flight of nine machines swooped over the floundering *Prince of Wales* disgorging bombs. One of these penetrated a deck while others landed close to her hull, buckling plates and loosening rivets.

Phillips was acutely aware that the RAF was desperately overstretched and might not have been in a position to assist Force Z at all.[193] Nevertheless, it is surely striking that no call for help was issued until a full hour after the leading Japanese combat planes were sighted by his capital ships. The request proved sadly belated. Whilst it is conceivable that the handful of *Buffalo* fighters available

192 Admiral Pound noted in his report that *Prince of Wales* might possibly have been struck by two torpedoes during this attack rather than one. This could not be ascertained under the prevailing circumstances. See: ibid., paragraph 50, 'Aircraft Attacks on Force "Z"'.
193 Ibid., paragraph 61, sub-section b.

might have spoilt some of the Japanese torpedo attacks especially, it is apparent that they could not have reached the scene of the action promptly enough to intervene. On learning at noon that *Repulse* was being bombed, these sluggish interceptors were on the way to the rescue within a quarter of an hour but did not arrive until 13:10, just in time to see *Prince of Wales* slip beneath the waves. The *Repulse* had preceded her at 12:33, only a few minutes after the fighters had taken off in response to her report that Force Z was being bombarded. The planes subsequently watched over the flotilla's destroyers – all of which had emerged unscathed from the engagement – while they finished gathering survivors from the lost ships. They then escorted the remnants of Force Z back to Singapore. In all, 840 British sailors, including Admiral Phillips, had perished in the disaster. The Japanese lost four aircraft.

The destruction of two capital ships at a cost of just four planes represented a tactical triumph for the Japanese and was another illustration of the emergent axiom that, if inadequately opposed, even a few aircraft might do tremendous harm, especially in maritime environments. Admiral Pound concluded that the risks that Phillips took were 'fair and reasonable in the light of the knowledge he had of the enemy when compared with the very urgent and vital issues at stake and on which the whole safety of Malaya may have depended. . . . [H]owever, . . . in the light of after events it would have been better if he had asked for fighter protection at least when the attack was known to be developing'.[194]

A few days after the loss of *Prince of Wales* and *Repulse*, Rear Admiral Lyster, the Fifth Sea Lord, wrote:

> [T]he lesson now being driven home to us by bitter experience . . . is that the dominating factor in naval warfare is no longer the big gun but the air striking force whether shore based or carrier borne . . . . I see no reason why a fleet of carriers only escorted by cruisers and destroyers should not be able to deal effectively with what is usually termed a 'well balanced' fleet. . . . [W]e would be well advised to divert a great deal of our [resources] . . . into the production of carriers, aircraft and torpedoes as quickly as can be – particularly if we hope to compete with the Japanese fleet. . . . If we do not provide our fleet with ample aircraft both for offense and defence we are liable to get a caning.[195]

The Japanese claimed a total of 21 torpedo hits on Phillips's vessels. Arthur Marder's study of the engagement suggests that *Repulse* was struck by five

---

194 Ibid., paragraph 63. For fuller narratives of the events surrounding the destruction of Force Z, see: Martin Middlebrook and Patrick Mahoney, *Battleship: The Loss of the Prince of Wales and the Repulse* (London, 1977); and Arthur Marder, *Old Friends, New Enemies: The Royal Navy and the Imperial Japanese Navy: Volume One* (New York, NY, 1981), pp. 465–90.
195 Document 168, 'Minute by Fifth Sea Lord: Requirements for Aircraft Carriers', 27 December 1941, in *FAA in Second World War: Volume One*, p. 541.

torpedoes and *Prince of Wales* by six.[196] Yet the findings of the First Sea Lord's official investigation were evidently much nearer the mark. Forensic surveys of the wrecks undertaken by divers in 2007 and 2008 were conclusive insofar that they confirmed that each of the ships was hit by four torpedoes. (Pound's insistence that *Repulse* was struck by a fifth is presumably correct, too, although it was not possible to corroborate this because of the rusting battlecruiser's posture on the seabed.)[197]

It is clear from all this that, whereas as many as three-quarters of the torpedoes dropped by the Japanese bombers missed their targets, sufficient did strike home to inflict what ultimately amounted to lethal damage. It is conceivable that a submarine could have achieved as much – and the *I-58* did try to do so. Indeed on 25 November 1941, just days before the dismemberment of Force Z by the Japanese, HMS *Barham*, the sister ship of the *Warspite*, was sunk by the *U-331* in the Mediterranean.

Submarines, like aircraft, exploit the third dimension. Yet, together with what occurred at Pearl Harbor, the sinking of HMS *Prince of Wales* and *Repulse* is often depicted as a watershed, as the exemplary triumph of air over sea power in general and over battleships in particular. The simple truth is that it had always been possible to destroy these and any other design of vessel through the concentration of adequate firepower in time and space, just as any platform's vulnerability had always been and continued to be partially determined by extraneous factors, not least its operational environment. The battleship diminished in utility not because of its vulnerability but because of the comparative inflexibility of its weapon systems. In fact such ships, along with heavy cruisers, remained the most resilient of ocean-going craft in many respects and were retained in service by all the principal belligerents for the war's duration. Notwithstanding her ground-breaking development of nautical aviation, Japan was pre-eminent in this regard, going so far as to produce the largest vessels of this kind ever seen, the *Yamato* and *Musashi*, which had a displacement of nearly twice that of the *Prince of Wales*. But like Germany's *Tirpitz*, neither of these colossi ever really got an opportunity to fulfil the role for which they had been designed and, as such, posed a threat that remained more passive than active. Although their 18" guns had a range of over twenty-three miles, this paled into insignificance when set against that of bombers and torpedo planes the reach of which was limited only by meteorological conditions, the capacity of their fuel tanks and the physical and psychological endurance of their crews. Once control of the skies around them was compromised, as had occurred with Force Z, capital ships were unable to roam

196 Marder, *Old Friends, New Enemies*, p. 470.
197 See: William H. Garzke, Kevin V. Denlay and Robert O. Dulin, 'Death of a Battleship: The Loss of HMS *Prince of Wales*, December 10, 1941: A Marine Forensic Analysis of the Sinking', (Society of Naval Architects and Marine Engineers, International Marine Forensics Symposium, National Harbor, MD, 2012).

the seas without fear of being observed and bombarded from afar. *Musashi*, *Yamato* and *Tirpitz* all eventually succumbed to poundings meted out by planes that originated from bases the ships never even glimpsed. The *Tirpitz* – which spent most of her service career skulking in Norwegian fjords, where she was bombed on more than twenty separate occasions, among them a raid by Soviet aircraft – never actually got to train her 15" guns on another vessel, although she did once join *Scharnhorst* in shelling the Allies' meteorological station and other installations on the remote outcrop of Spitzbergen (Svalbard) in the Arctic during September 1943.

The circumstances that made this possible arose with the extension of air power's reach across ever larger areas of the waters that make up so much of the Earth's surface. Overlapping patrols by shore-based machines operating from abutting land masses covered, not just chunks of the Mediterranean and Baltic, but also swathes of the Atlantic and the Pacific especially. Regions of the high seas that were too remote for even long-range bombers such as the *Liberator* and *Sunderland* to penetrate could be monitored, too, but only by machines housed aboard vessels sent into the areas concerned. Here, the fact that any concentration of power is relative could be particularly noticeable. A couple of *Swordfish* flying from the smallest of platforms might well thwart any foe that was likely to be encountered. Elsewhere, several big carriers with large complements of fighter and strike aircraft might turn out to be a poor match for the opposition, if only because quantity can have a quality all of its own.

This quandary bedevilled the provision of a passable degree of air cover on many fronts, with the Japanese especially failing to make good their losses, let alone keep up in the race to improve the overall quantity and quality of naval aviation to hand. Indeed during the opening years of the Second World War the rate of attrition among carriers was generally worse than that among battleships, which were increasingly employed as their escorts. Before 1942 was out Britain had already endured the sinking of five of the six front-line carriers she had possessed when the conflict commenced; the US Navy had, after just twelve months of hostilities, lost all those it had entered the war with save *Saratoga*, *Enterprise* and the diminutive *Ranger*; while the Japanese, besides losing the small carrier *Shōhō* and many planes in the Battle of the Coral Sea, had seen four of their premier carriers destroyed in the debacle of Midway. Although the Americans' building programmes especially were eventually to give the Allies relatively enormous numbers of carriers, at the moment when the Royal Navy's Future Building Committee was urging that maritime aviation be made 'the king pin of Naval strategy', the First Lord of the Admiralty had to confess that: '[T]he number of aircraft carriers available is still totally insufficient for the deployment of adequate aircraft to support large combined operations'.[198]

---

198 Memorandum, D.C. (S) (42) 88: 'The Needs of the Navy', by First Lord of the Admiralty, 5 October 1942, TNA, CAB 70/5.

The sinking of HMS Cornwall, HMS Dorsetshire and HMS Hermes, 1942

Following the crippling of much of the US Pacific Fleet and the destruction of Force Z in December 1941, the IJN could, as Admiral Yamamoto had predicted, run amok for a time. Wake Island – a small but important American outpost roughly half way between the Philippines and Pearl Harbor – succumbed to a Japanese task group before the month was out, as did the pivotal port of Rabaul on the large island of New Britain, part of the Australian mandate of New Guinea. The Japanese invasion of the Netherlands East Indies proceeded apace as well. Java, Sumatra, the Moluccas and Timor had all been conquered by the end of March, while areas of Borneo, Celebes (Sulawesi) and New Guinea were in the process of being overrun. Similarly, Singapore was left increasingly beleaguered as enemy troops pushed down the Malaysian Peninsula. Isolated and demoralized, the garrison surrendered – unconditionally if somewhat controversially – in the middle of February. Although widely regarded as a fortress, their island base had never really amounted to more than an anchorage and logistical hub, its fixed defences being designed to protect the facility from attack by naval rather than land forces.

With Singapore captured, in March 1942 the Japanese turned their attention to British strategic outposts further west. Jubilant in the wake of its triumphs at Pearl Harbor and elsewhere, Nagumo's awesome *Kidō Butai* moved to within striking distance of Ceylon. Simultaneously, submarines and a flotilla including cruisers and the carrier *Ryūjō* under Vice Admiral Jizaburō Ozawa descended on the Indian shoreline to the north of the island. Here, they were to spread panic and inflict appreciable damage: 23 merchantmen were destroyed in the Bay of Bengal, mostly in the space of a few hours, and several others were torpedoed off the sub-continent's western coast.

By this time Britain's Eastern Fleet was commanded by Admiral James Somerville and included five ageing battleships, five cruisers and three carriers, *Hermes*, *Indomitable* and *Formidable*. Somerville had been alerted to Nagumo's approach by 'Ultra' intercepts and was planning to surprise the Japanese carriers with a pre-emptive nocturnal attack in which his forces' electronic eyes promised to offset some of their other tactical disadvantages. At this stage of the war, the IJN was highly adept at fighting by the glare of star shells and searchlights but, like the RMI, was trailing behind the Allies in the harnessing of radar and mechatronics. However, come the start of April there was still no sign of Somerville's quarry and he was obliged to withdraw most of his ships westwards to a secret base in the Maldives to refuel and take on fresh water. Meanwhile the veteran *Hermes* – the Royal Navy's very first custom-built carrier – together with the Australian destroyer *Vampire* and the heavy cruisers HMS *Dorsetshire* and *Cornwall* lingered in Ceylonese waters to undertake escort duties or undergo repairs.

On the evening of 4 April an RAF *Catalina* spotted Nagumo's fleet – which included five carriers with some three hundred aircraft – about three hundred and fifty miles to the south of the island. The British reconnaissance plane was

shot down by a patrolling 'Zero' but not before it had got off a report of its findings. Somerville began gingerly retracing his steps. In the meantime, however,
Nagumo struck at Colombo's anchorage and the airfield at nearby Ratmalana.
Approaching from inland, the massive attack caught some of the defenders on
the hop. Fortunately, as a precaution the port had already been cleared of vessels
for the most part, leaving few worthwhile targets other than the rickety destroyer
Tenedos and an armed merchant cruiser, the *Hector*. Riddled by heavy AA fire,
nine of the Japanese planes were destroyed, although a score and more of RAF
machines were also wrecked, most of them on the ground.

Another group of Japanese aviators found far richer pickings in the waters
southwest of Ceylon. Here, the cruisers *Cornwall* and *Dorsetshire* were on
their way from Colombo to rendezvous with Somerville's main force near his
makeshift base at Addu Atoll, the southern tip of the Maldives. They had left
Colombo at 22:00 on 4 April and, on receiving reports of Nagumo's approach
at 08:00 the next day, were zigzagging as fast as they could about a mile apart.
With virtually no cloud cover to mask their movements, they were spotted by
surveillance planes at 11:30 and again at 13:00, after which several flights of
dive-bombers appeared. These attacked with the sun behind them, approaching
their prey head-on.

Partly because of the ships' architecture and partly because of the dazzling
sunshine, this quarter was something of a blind spot for the cruisers' AA batteries. *Cornwall* had no radar suite and that aboard *Dorsetshire* seems to have
provided little if any warning.[199] Although fired on, a trio of aircraft swooped
to less than a thousand feet over *Cornwall* before releasing their payloads; one
bomb struck her hangar deck and another narrowly missed her port side.[200]
Simultaneously, three other dive-bombers descended on *Dorsetshire*, disgorging
the first in a cascade of projectiles nearly all of which caused damage. Despite
the cruiser's evasive lurch to starboard, all of the first three bombs hit home:
one disabled the steering; the second severed much of the internal communication network; and the third struck amidships on the port side, knocking
out all the neighbouring AA defences except for a solitary pom-pom gun.
After just four minutes during which several more blows rained down on her,
Dorsetshire was out of control, losing speed and listing heavily to port; all communications between the bridge and other parts of the cruiser had been cut, as
had her radio link to the outside world; her entire complement of AA weaponry save a couple of machine guns was no longer functioning; and flames
and smoke had completely engulfed everything aft of the bridge as far as the
stern. At least four more direct hits were sustained in the next two minutes,

---

199 'HMS *Cornwall* and HMS *Dorsetshire*: Bomb Damage: Report DNC 4B/R.158 on sinking of Corn-
    wall and Dorsetshire by C.S. Lillicrap (for DNC [Director of Naval Construction])', 13 Novem-
    ber 1942, TNA, ADM 267/84, Subsection I, paragraph 4 and subsection III, paragraph 1.
200 Ibid., subsection I, paragraph 3 and subsection III, paragraph 2.

making a total of ten, in addition to several near misses. Six minutes after the attack's onset, orders were passed by mouth for the crew to abandon ship via the starboard side.[201] She sank moments later, her bows pointing skywards.[202]

Meanwhile, much the same story had unfolded where *Cornwall* was concerned. She, too, had swerved to starboard in an unsuccessful bid to dodge the first three weapons hurled at her, one of which had evidently done serious damage to her port waterline and wrecked a dynamo compartment, dislocating the electricity supply to much of the ship. Close on their heels came more bombs. With a nigh continuous stream of projectiles landing on and adjacent to her, in less than five minutes *Cornwall* was all but stationary and helpless: her power and internal communication systems had failed; her pom-pom guns could only be operated manually; several fires were raging, belching ominous clouds of thick, black smoke that obscured the attacking aircraft from the remaining machine gunners and filled the damage-control centre; many of the engineers and medical and fire-fighting personnel had been killed or wounded; both boiler and both engine rooms were flooding rapidly; and the stricken cruiser's bows were already so low in the water as to make the starboard outer propeller break the surface. Barely eight minutes after the Japanese had commenced their onslaught, her captain judged *Cornwall* to be beyond redemption and ordered her abandonment. She foundered four minutes later.[203]

The planes' attack was described as having been 'one of remarkable accuracy and efficiency'.[204] Both cruisers were lethally pummelled in a trice, with the dive-bombers achieving an extraordinary success rate of 90 per cent.[205] *Cornwall* was subjected to so many blows and near misses in such rapid succession that her surviving crew were left uncertain just how many times she had been struck. However, nine direct hits were logged, as were six close-shaves that inflicted varying degrees of damage.[206] *Dorsetshire* is likewise known to have experienced several near misses – one of the first of which did fatal harm – as well as sustaining ten direct hits.[207] As had occurred in the case of HMS *Prince of Wales* four months earlier, the loss of electrical power had a disastrous effect on the defensibility and survivability of HMS *Cornwall* for one, her commanding officer commenting that: 'One bomb destroyed the two forward dynamos, another flooded the engine room containing the remaining two. Those two bombs alone would have virtually put the ship out of action, even if she had suffered no other damage'.[208]

201 Ibid., subsection II, paragraphs 1–4.
202 Ibid., subsection II, paragraph 5.
203 Ibid., subsection III, paragraphs 2–6.
204 Ibid., subsection I, paragraph 4.
205 Ibid., paragraph 8.
206 Ibid., subsection III, paragraph 7.
207 Ibid., subsection II, paragraph 4.
208 Ibid., subsection III, paragraph 12. Also see: Ministry of Defence (Navy), *War With Japan, Volume II: Defensive Phase*, p. 126.

After sinking *Cornwall* and *Dorsetshire*, Nagumo wheeled his *Kidō Butai* north-eastwards before venturing back towards Ceylon. By dawn on 9 April his fleet was lurking off the coast, preparing to attack the port of Trincomalee. On the previous evening, HMS *Hermes*, accompanied by the destroyer HMAS *Vampire*, had slipped out of that harbour temporarily, presuming that an air strike from the Japanese carriers was imminent. The two Allied ships steamed down the coast during the night and were well to the south when Nagumo's planes fell on Trincomalee at 07:00.[209] Alerted by radar, the garrison was somewhat bet-ter prepared for them than that of Colombo had been four days earlier. Several *Fulmars* and *Hurricanes* were already aloft and assisted numerous AA batteries in harassing the attacking swarms.

Persuaded that any danger had passed, *Hermes* and *Vampire* turned about at 09:00. At 10:15, however, they were urged to proceed with 'the utmost despatch' to Trincomalee, for a report from a Japanese surveillance plane detailing their position had been intercepted. As the ships increased speed to 27 knots, some fighters hastened to their aid from Ratmalana, but within ten minutes scores of enemy dive-bombers were lining up above the two vessels. Concentrating on *Hermes* initially, they came out of the sun, which was nearing its zenith, at extremely steep angles, presenting particularly difficult targets for the AA bat-teries to engage. The elderly carrier, which had first put into Trincomalee for maintenance work, had long since disembarked her own planes and thus had no aerial shield to hand whatsoever. Like *Cornwall* and *Dorsetshire* before her, she was hit by bomb after bomb as over forty aircraft fell upon her in quick succession. Irrecoverably flooding and ablaze, she heeled over and sank at 10:45.[210]

Fifteen planes then turned on *Vampire*, the first three bombs landing very close to the ship and causing some damage to gun mountings especially. The sixth projectile to be dropped penetrated a boiler room, severing the main and auxiliary steam lines. The destroyer all but stopped moving and sagged to port as further blows rained down on her. One of these actually broke her in two, the vessel's forward section vanishing beneath the waves almost instantaneously. The after magazine then blew up and the rear portion of the ship sank as well. The AA weaponry had continued in action until most of the available ammuni-tion had been expended and the crews had been blown or washed from their battle stations. They downed at least one of the attacking bombers and damaged others.[211]

The RAF had, meanwhile, glimpsed an opportunity to give Nagumo a dose of his own medicine. The raid on Trimcomalee by carrier-based planes con-firmed that his *Kidō Butai* had come within striking range of *Blenheim* bombers

209 'Report of Proceedings from Senior Surviving Officer, HMAS *Vampire*, to Commander-in-Chief, Eastern Fleet: Loss of HMS *Hermes* and HMAS *Vampire*, 9 April 1942', 10 April 1942. TNA, ADM 267/106.

210 Ibid.

211 Ibid.

stationed on Ceylon. Nine duly set out to catch the interlopers while they were still preoccupied in mounting their own offensive operations. The British planes eventually found Nagumo's fleet and penetrated the surrounding fighter cordon without the alarm being raised. (The crew of the carrier *Hiryū* apparently noticed the intruders but did not issue any warning, perhaps because they mistook them for friendly aircraft.) Certainly, it was only when ordnance hurtled into the water off her starboard bow that those aboard the flagship – the large but unarmoured carrier *Akagi* – realized that anything was awry. This proved to be just another instance of high-level bombing that missed its mark, however. The *Blenheims* were soon being assailed by embarrassed, vengeful 'Zero' interceptors that quickly destroyed four of them. Another was shot down as it blundered into more enemy planes returning to their ships. A sixth crashed on landing.[212] Still, had the British machines been *Skuas* or some other type of dive-bombers, in such a setting it is likely that they would have inflicted severe – perhaps fatal – damage on at least one of Nagumo's precious carriers, which had formidable proactive but perilously limited passive and reactive defences. In fact it was to be under much the same circumstances that *Kidō Butai* was dismembered by American dive-bombers off Midway Island just two months later.

In the interim, however, the IJN proved virtually unstoppable, certainly in the Indian Ocean. In the face of Nagumo's aerial dominance, Admiral Somerville was reduced to withdrawing his battleships as far as Kilindini (Mombasa) on the Kenyan coast of British East Africa. This corner of the empire had, unlike in the First World War, remained a comparatively quiet front and, particularly since the annihilation during the spring of 1941 of Mussolini's large army in Italian East Africa – the combined provinces of Abyssinia, Eritrea and Italian Somaliland – had been all but immune from invasion. Indeed once Axis forces had been ousted from the littoral of the Horn of Africa, unaligned powers, not least the USA, had ceased classifying the Red Sea and the Gulf of Aden as combat zones. With the Suez Canal's southern approaches deemed safe for neutral shipping, the colossal strain on the Allies' merchant tonnage was alleviated and, thereby, Britain's logistical difficulties in the eastern Mediterranean especially began to ease.

Although for a time it was feared that the raid by Nagumo and Ozawa in the spring of 1942 might presage an amphibious landing on Ceylon, the Japanese flotillas had neither the intention nor the wherewithal to linger in this quarter. They soon realized that whatever remained of Somerville's Eastern Fleet had managed to elude them. Ineffectual though it had proved, the strike mounted by the RAF's *Blenheims* on 9 April had, furthermore, highlighted their own exposure to counterblows from land-based planes and other forces in a region where they had neither harbours nor airstrips of their own. Above all, however, Nagumo's *Kidō Butai* especially was required elsewhere, for the immense task of consolidating the territorial gains of the past four months was just commencing.

212 See Parshall and Tully, *Shattered Sword*, p. 145.

It was to get off to an ominous start: just three weeks after the sinking of the *Hermes*, in a battle fought by aircraft from ships that never even came within sight of one another, the IJN was to be checked in the Coral Sea and was to be dealt a crippling blow at Midway at the very beginning of June.[213]

The sinking of the Yamato, 1945

Japan's naval aviators accounted for the most memorable tactical triumphs witnessed during the opening phase of the conflict in the Indian and Pacific Oceans. Notwithstanding this, the IJN could not and did not succumb to the temptation to place all of its eggs in the seductive basket of air power. Whereas even prior to the mortifying experiences of December 1941 and April 1942 the Royal Navy was pondering transforming the carrier into the core of its fleet and maritime aviation into the lodestar of its strategy, the Japanese more than any of the other leading belligerents persisted in anticipating a Jutland-like clash in which aircraft would play an essentially subordinate role to that of surface units, notably battleships. The US Navy – which had long been regarded by Tokyo as the IJN's nemesis in the Pacific – was, somewhat ironically, unshackled from any such mindset by the very success of Japan's air power at Pearl Harbor in December 1941. With so many of their battlewagons out of action, if only for the time being, the Americans' principal hope and means of containing Japanese expansionism initially lay in their carriers and long-range strategic bombers.

Since Tokyo's whole strategy was geared to keeping hostile aviation at a safe distance from metropolitan Japan, not least by denying any adversaries the use of forward bases in the Philippines, the Marianas and on Wake Island in particular, floating aerodromes became of pivotal importance to both sides. The more remote they were, the more dependent the outlying Japanese garrisons became on carriers and other shipping for fuel, replacement aircraft and personnel, spare parts and other necessities. On the other hand, the very paucity of fixed bases amidst the colossal expanse of the Pacific had long since encouraged the US Navy especially to develop the doctrine and wherewithal to maintain forces on the high seas. Whilst American carrier groups with their supporting fleet train were thus capable of conducting protracted undertakings, the IJN's *Kidō Butai* was better suited to delivering the sort of short, sharp stab seen at Pearl Harbor, Trincomalee and Colombo. Once the operational emphasis swung from the infliction of pre-emptive blows to the retention of territorial gains, however, the Japanese inevitably struggled against opponents who enjoyed appreciable leeway as to the timing and location of their counterstrokes in a theatre of war that was truly enormous. Admiral Yamamoto's vision of a flexible defence rooted in long-range naval aviation based on island fortresses proved impracticable. After

---

213 For details of the Battle of the Coral Sea, see: John B. Lundstrom, *The First Team: Pacific Naval Air Combat from Pearl Harbor to Midway* (Annapolis, MD, 1984), pp. 157–305. For details of Midway see: ibid., pp. 309–447; and Parshall and Tully, *Shattered Sword*.

all, any concentration of force is relative and, if only in the longer run, the Allies had many more resources at their disposal than Japan could hope to muster. Just as her rambling defensive perimeter could only be as strong as its weakest link, bolstering individual outposts if and when they came under threat relied on the security of the interlinking lines of communication. It was a comparatively easy matter for the US Navy to sever these vital conduits. Other platforms, notably submarines, played their part, but time and again planes from numerous carriers helped achieve an adequate if localized dominance of both the skies and the surface. Isolated from the rest of Japan's defensive glacis – which the Allies could safely ignore until they felt it necessary to tackle it – one key bastion after another faced subjugation by amphibious armies that were inserted under an umbrella of fire from the heavens and the ocean.

Some of the most destructive of bombardments came from big-gun ships, without which the conquest of many of the Japanese-occupied islands in the Pacific and undertakings such as the D-Day landings in France could only have proved more difficult. But here these vessels were functioning as integral parts of forces that combined land, sea and air power. This reduced their vulnerability by making potential threats more manageable and thereby providing the ships with a more favourable operating environment. Sending them unaided against forces they themselves were not primarily designed to counter was always likely to turn out more risky than worthwhile.

Whereas asymmetrical engagements were an almost day-to-day feature of warfare by land – with thousands of tanks, for instance, being lost while fighting with inadequate support against better balanced forces – the destruction of seemingly mighty battleships by dissimilar opponents has proved a more eye-catching and controversial aspect of the Second World War. This is presumably because such occurrences were comparatively rare and because such vessels were once widely regarded as *the* icon of martial might. As one of the two greatest battleships ever built, the *Yamato* is a particularly good case in point. She, however, was as much a victim of inimical geostrategic and geopolitical circumstances as she was of the combat aircraft that actually sank her.

Yamato was a superlative battleship. She was designed to take on vessels of her own kind and make short work of them, ideally before they could even get close enough to return fire. Her nine 18" guns could hurl 3,000 lb projectiles nearly twenty-five miles and penetrate the armour of any foe. Moreover, though almost half as big again as any ship built up to this point in time, she was capable of steaming at 27 knots, partly because of her innovative bulbous bow. Recognizing that they could not match their putative adversaries quantitatively speaking, the Japanese had invested in quality, hoping that *Yamato* and her sister ship, *Musashi*, would give them a decisive edge in any Jutland-like confrontation.[214]

That type of battle never came, however. What Japan had hoped would prove a brief conflict turned into a lengthy, total one that she could not end

---

214 For further details regarding *Yamato* and *Musashi*, see: Evans and Peattie, *Kaigun*, pp. 370–83.

through a compromise peace. Neither could she sustain hostilities against as formidable and implacable an opponent as the USA, if only because her industrial and manufacturing capacity was so much smaller and too much of the technology she could turn out was significantly inferior. After all, even Nazi Germany – which for much of the war was in a much stronger position in these respects than Japan ever was – ultimately failed to contend with the Allies in producing the entire gamut of essential equipment, the skilled personnel to operate and maintain it, and the fuels without which mechanized warfare was impracticable, regardless of the quality of the technology concerned. Japan's resources, both human and material, ended up being fatally overstretched, not least because of the sheer size and complexity of the theatre of operations, the competing demands exerted by her army and navy and by different schools of tactical thought within those services.[215] With her meagre means so thinly spread across such a wide spectrum of requirements, capability gaps and imbalances in force structures were bound to emerge at some crucial point or other. Providing naval flotillas with adequate protection against aerial attacks was one imperative that became steadily harder to fulfil.

By the autumn of 1944 the Allies had made such inroads into Japan's strategic glacis that even her links to the Philippines – the fountainhead of so many of her remaining natural resources – were in jeopardy. This precipitated the Battle of Leyte Gulf, to which the IJN committed virtually all of its disposable warships, including *Yamato* and *Musashi*. The resultant clash was arguably the nearest the protagonists came to a Jutland-like contest, if only insofar that it was to prove the forlorn hope of Japanese maritime power. The conflict the IJN had started with a sudden blow from a carrier task group was now being ended by such forces, with those of its adversary having swollen to overwhelming dimensions. By this stage of the war many of the US Navy's pilots had accumulated more combat experience than those of the IJN. Their aircraft were, moreover, that much more capable than either their forebears or most of their Japanese counterparts, if only because of their powerful radial engines. These enabled machines to bear ever heavier loads of fuel, armour and armaments without much if any loss of aerodynamic performance. Whilst the Japanese enjoyed some success at Leyte Gulf, this proved ephemeral and their fleet subsequently took a severe beating at the hands of American aviation especially. *Musashi* was among the many vessels lost, although it predictably required an exceptionally large amount of ordnance to sink her.[216]

It was against this backcloth that, early in April 1945, *Yamato* set sail for beleaguered Okinawa. She was accompanied by a light cruiser and eight destroyers, which was virtually all of the serviceable vessels the IJN still possessed. There was no air cover available whatsoever and the escorts that made up the bulk of the flotilla had very few AA defences between them.

215 See, for instance, Peattie, *Sunburst*, pp. 159–60, 166–7, 192, 194–5.
216 For an analysis of the Battle of Leyte Gulf, see: H.P. Willmott, *The Battle of Leyte Gulf: The Last Fleet Action* (Bloomington, IN, 2005).

But then the battleship's mission was regarded from the outset as a suicidal one: with Japan's few remaining aviators immolating themselves in *Kamikaze* attacks, the political and cultural pressure on the IJN to sacrifice its last great asset in the motherland's defence was irresistible. On the other hand the prospects for much if any tangible success were extremely bleak. It was hoped that *Yamato* might get close enough to the American invasion convoys to shell them. After this, she would be run aground at best, if only because she had insufficient fuel to return home. However, spotted in the East China Sea by a reconnaissance plane at 08:23 on 7 April, the Japanese flotilla was assailed by over three hundred strike aircraft from enemy carriers well before it got within range of its quarry. Without fighter support to keep the threat at anything like a manageable level, the ships' AA batteries and passive defences proved fatally inadequate. The first bombs and torpedoes rocked *Yamato* at 12:41 but she did not founder until almost two hours later. The light cruiser was sunk as well, as were half of the destroyers, three of them by their own crews who judged them too badly damaged to be redeemable.

Attacks on fleet anchorages: Taranto and Pearl Harbor

Even prior to the First World War the proliferation of increasingly sophisticated mines and submarines especially had rendered the old technique of close blockade too hazardous to be applicable in many situations. Certainly, during the Eurocentric conflict of 1914–18 the Allied battle-fleets had found themselves obliged to try to contain their adversaries' nautical power from a suitably safe distance. Whilst aviation's maturation had further complicated matters in this respect, it had also revitalized the old notion of 'Copenhagening' warships when they believed themselves to be at their most secure. The ability of carrier forces in particular to initiate engagements at times and places of their choice gave them a significant advantage over static, shore-based defences and other relatively immobile opponents. Principal harbours were among the most tempting of targets, for a successful attack here might go far in tilting the wider strategic balance, if only for a time.

On 16 March 1940 a force of Ju.88s crossed the North Sea from Germany to bombard Britain's Home Fleet in its main anchorage at Scapa Flow. This raid inflicted negligible damage and was subsequently derided by Winston Churchill as 'ill-directed'. Nevertheless, along with the sinking of the *Königsberg* at Bergen by the FAA and their torpedo strike on the *Richelieu* at Dakar, the affair did highlight the exposure of vessels in harbour to aerial attack. Unlike those of fleets manoeuvring at sea, the geographical coordinates of known ports were confirmed easily enough and, just like aerodromes, harbours serving as bases were susceptible to a variety of threats. Shipping periodically needed dry docks and other repair and maintenance facilities, as well as access to fuel, clean water and other day-to-day essentials. If deprived of these, vessels would be immobilized almost as surely as if they had been sunk.

Eight months after the raid by the *Luftwaffe* on Scapa Flow, Churchill was able to report to the House of Commons that:

> The Royal Navy has struck a crippling blow at the Italian fleet. . . . This . . . was, of course, considerably more powerful on paper than our Mediter-ranean Fleet, but it had consistently refused to accept battle. On the night of 11th–12th November, when the main units . . . were lying behind their shore defences in their . . . base at Taranto, . . . the Fleet Air Arm attacked them . . . . [O]nly three Italian battleships now remain effective. This result, while it affects decisively the balance of naval power in the Mediterranean, also carries with it reactions upon the naval situation in every quarter of the globe.[217]

The raid on Taranto indubitably had appreciable strategic consequences insofar that it slashed the number of battleships available to the RMI – a navy that, if only because it had shunned carriers, retained the big-gun ship as the bedrock of its doctrine. The FAA's use of torpedo strikes against vessels moored in a relatively shallow anchorage was also something of a seminal case, although – contrary to widespread belief – it is improbable that it directly inspired the IJN's use of this tactic at Pearl Harbor.[218] As we have noted elsewhere, the Japanese had been developing suitable techniques (and not just with Hawaiian shoals in mind) since 1939, while the FAA had already had to grapple with the intrica-cies involved under actual combat conditions in Norwegian fjords and Dakar's harbour. In any case, the IJN was faced with a rather different set of strategic problems from those confronting the Royal Navy in European waters, espe-cially in the Mediterranean. Whereas the attack on Taranto did promise to ease Britain's worries in a war that had long since begun, Japan's disabling of the American battleships at Pearl Harbor could only bring far more difficulties in its wake than tangible benefits.

The Taranto raid ineluctably had its tactical and other distinctions. Planning for a torpedo strike against the RMI had been underway for months, but it was not until HMS *Illustrious* joined the British forces in the central Mediterranean during the autumn of 1940 that sufficient aircraft were to hand to make such an undertaking a feasible one. With several machines from *Eagle* added to her own, *Illustrious* duly took up position approximately two hundred miles from Taranto. Divided into two waves an hour apart, 21 *Swordfish* were to be employed in all. Eleven of these were armed with torpedoes while the rest carried bombs. As this was to be a nocturnal operation, two planes were also designated as pathfinders; equipped with flares, they were to mark the cardinal targets in the main anchor-age before staging diversions around the inner harbour. Of vital importance, too, was the last-minute aerial reconnaissance of the port prior to the mounting of

217 *Commons Debates*, 13 November 1940, vol. 365, cols 1,712–13.
218 See: Peattie, *Sunburst*, p. 144.

the strike; photographs taken by RAF planes based on Malta confirmed the exact positions of the target vessels and the locations of protective barrage balloons and torpedo nets. Forecasts from Malta's meteorologists also helped the aviators select a moment when the weather conditions would favour an attack.[219]

Complete surprise was achieved. The defenders had neither radar nor night fighters and, although their AA batteries put up an extremely heavy barrage, most of this fire was poorly directed. At the cost of just two *Swordfish*, five hits were scored on three battleships. A cruiser was also damaged, as were parts of the dockyard. As Admiral Cunningham was to say of the raid: 'As an example of "economy of force" it is probably unsurpassed'.[220]

Had the weather not changed, the FAA would have resumed its assault the very next night. Few worthwhile targets would have been found, however. Within hours of the attack, all of the ships that were still seaworthy had hurriedly departed for ports on Italy's western coastline. Here, in theory at least they would be somewhat safer, but they would also be that much further from the sea lanes upon which the survival of Britain's Mediterranean outposts depended. By virtue of the fact that they were lying in an anchorage rather than in an ocean's depths, even the vessels left awash in Taranto's shallows might also come to pose a threat again. Indeed as was to happen with all of the battleships that foundered at Pearl Harbor save the *Arizona* and the *Oklahoma*, those crippled at Taranto were eventually refloated. Whilst this painstaking, laborious procedure was essentially an optional one for a US Navy that would go on to wage a conflict in which the carrier was the final arbiter, the battleship's centrality to the RMI's strategy dictated that the vessels disabled at Taranto be brought back into service if that were at all possible.

The IJN's attack on Pearl Harbor on 7 December 1941 has become such a familiar story as to merit little discussion here. Nevertheless, some details are deserving of mention, if only as illustrations of manoeuvre warfare's fundamental constituents – the often challenging processes of finding, fixing and striking opponents – and how even the most glittering of tactical triumphs does not necessarily translate into strategic success.

The operation's scale and stealthiness were unprecedented. Bit by bit Vice Admiral Nagumo's *Kidō Butai*, which included six carriers, slipped out of Kure naval base over a period of several days. Having rendezvoused unobserved off the Kurile Islands and with its true movements cloaked by, on the one hand, the maintenance of radio silence and, on the other, a smokescreen of specious signals, Nagumo's strike force steamed undetected for almost three and a half

219 See: Document 101, 'Letter from Commander-in-Chief, Mediterranean, to Secretary of Admiralty: Operation "Judgement" – attack on Taranto, 11–12 November 1940', 16 January 1941, in *FAA in Second World War: Volume One*, pp. 314–16.

220 Ibid., p. 316. For details of the attack, see: Document 101a, 'Report from Commanding Officer, HMS *Illustrious* to Rear Admiral, Aircraft Carriers, Mediterranean: Operation "Judgement" . . . 11–12 November 1940', 13 November 1940, ibid., pp. 316–29.

thousand miles to bring its aircraft within reach of Oahu. The ensuing attack was to be the first occasion on which carriers were employed *en masse* under authentic combat conditions.

Whereas Rear Admiral Husband Kimmel, the Commander-in-Chief, US Pacific Fleet, reasoned that any danger would emanate from the Marshall Islands – the nearest Japanese enclave, but still all of two thousand miles away – and pardonably focussed his few long-range reconnaissance resources on the seas to the south-west of Hawaii, Nagumo approached from the north. Kimmel's military and political masters in Washington were in any event persuaded that the archipelago's very remoteness made it virtually invulnerable to aerial attack especially; the Philippines were surely that much more exposed in this regard. (Accordingly, the moment hostilities appeared imminent, instructions were issued for some of the P.40 fighters stationed on Oahu to be ferried to the more westerly outposts of Wake and Midway from where they might cover reinforcements bound for the Philippines.) Although it was conceivable that Japanese submarines might venture as far as Hawaii, the anchorage at Pearl Harbor was barely deep enough for ocean-going types to dive in and, not least for the same reason, was adjudged to be an unlikely setting for an attack by torpedo-bombers. For sure, Kimmel was provided with neither barrage balloons nor anti-torpedo nets with which to shield his battle-fleet. Nor were the anchorage's AA defences kept at a high state of readiness: batteries were left unmanned or with purely token ammunition stocks to hand; fighters, most of them kept unfuelled and unarmed, were crammed together on runway aprons where they were thought less accessible to saboteurs – the greatest threat perceived by both the local commanders and the authorities in distant Washington; and, although the surrounding skies were monitored by a radar network, this vigil was only maintained for a few hours each day. Air-defence was, furthermore, just one of several security spheres in which allotting responsibility between the autonomous services that made up Oahu's garrison was inescapable. The contrasting concerns of the USAAF and the US Navy could only complicate the pooling of their assets and the division of responsibilities. Information dissemination proved to be one of the weaker links in the Hawaiian command and control chain that was forged by Kimmel and his army colleague, Lieutenant-General Walter Short, both of whom were subject to the whims of their respective superiors within a compartmentalized military and political hierarchy.

Washington's preoccupation with bolstering the Philippines' garrison did fortuitously mar the IJN's plans for the raid on Pearl Harbor, however. Kimmel's carriers – which numbered among Nagumo's foremost targets – had still to return from transporting fighters to Midway and Wake when the first Japanese aircraft appeared near Oahu. Whilst a floatplane was picked up by radar at about 06:45, this single, slow-moving signature evidently seemed innocuous enough, for no action was taken. Even when the approach of the first wave of strike aircraft was detected by a solitary radar station – a turn of events that, again, owed more to luck than design – it was mistaken for a flight of B.17s that was scheduled to arrive from the continental USA. Still, even if an alarm had been sounded at this juncture, it is grossly improbable that in the few minutes available the island's proactive and active AA defences could have been mobilized sufficiently to make much difference.

In all, 351 aircraft were committed to the raid. They struck in two waves, the first of which comprised no fewer than 183 machines including 36 'Zero' fighters. Although tremendous damage was inflicted on both Kimmel's ships and the USAAF's planes, Pearl Harbor's facilities as a base were essentially left intact.

The failure of the Japanese to consolidate their success by destroying the port's fuel depot, dry docks and repair facilities remains somewhat controversial. For instance, it has been suggested that the explanation for this apparent oversight largely lies in Nagumo's professional background: had a pedigree aviator rather than a graduate of the big-gun school of tactics been entrusted with the *Kidō Butai*, he might have possessed sufficient imagination to launch a third strike immediately.[221] But regardless of his particular *métier*, Nagumo was surely as aware as any sailor that naval power stems from installations and processes on terra firma, as does aviation. In all the Second World War's combat environments, nowhere was this fundamental factor more noticeable than in the Pacific.

'The selection of objectives . . . and determining the order in which they are to be destroyed', Douhet had observed in *Il Dominio dell'aria*, 'is the most difficult and delicate task in aerial warfare, constituting what may be defined as aerial strategy'.[222] At Pearl Harbor, Nagumo was confronted with an abundance of potential targets relative to the number of planes at his disposal, large though that was in absolute terms. His first goal had perforce to be the suppression of the Americans' AA defences so that the wider operation could proceed. Although it promised to have broader, strategic benefits, disabling any enemy carriers in the locality was a vital element of this tactical prerequisite. In the event, however, this was not achievable for the simple reason that the whereabouts of Kimmel's carriers were unknown at the moment the Japanese assault commenced.

The date for this had after all been set by Tokyo, not the IJN. Indeed Admiral Yamamoto had voiced grave reservations about any resort to violence against the USA as an instrument of policy. Similarly, once the bombardment of Pearl Harbor had actually begun, Nagumo's actions were dictated more by the perennial problems of air power's impermanence than by his professional instincts. Tempting though it might have been to prolong the assault with a third round of attacks, to do so the Japanese carriers would first have had to retrieve their aircraft and refuel and rearm them, a process that would have taken two to three hours at least. It would have been mid-afternoon at the earliest before a third wave of planes could have reached Pearl Harbor and it is doubtful whether many of them could possibly have regained their ships before the sun set at 17:12.[223]

221 See, for instance, Peattie, *Sunburst*, pp. 196–7.
222 Quoted in Gates, *Sky Wars*, p. 43.
223 See: Zimm, *Attack on Pearl Harbor*, pp. 95–6; and H.P. Willmott, *Pearl Harbor* (London, 2001), pp. 142–57.

Increasingly poor visibility might well have caused difficulties in the immediate combat zone, too. Whilst setting their fuel stockpiles ablaze could only have hamstrung Kimmel's remaining vessels, the resultant billows of thick smoke would most likely have obscured other targets from the bombers, rendering pinpoint attacks that much harder to perform.

In any event, no amount of physical or psychological damage inflicted here could realistically have been expected to end the bitter conflict that the raid had ignited. Nagumo might have chosen to postpone any resumption of the bombardment until the next morning. However, running low on fuel and with surprise no longer on his side, he was forgivably unwilling to risk snatching defeat from the jaws of victory.[224] Anticipating a counterstroke by the American carriers and submarines, he headed homewards.

As he did so, Wake was assailed by, firstly, long-range bombers stationed in the Marshall Islands and, thereafter, amphibious forces. The defenders held out until two of Nagumo's carriers joined the battle.[225] Wake was physically occupied by the Japanese for the remainder of the war, which was the only reliable way of preventing their adversaries from exploiting it for their own ends. However, like so many of the bastions within Japan's defensive perimeter, it was one that the besiegers could afford to ignore; circumvented and isolated, the garrison became more of a liability than an asset. In the interim, not least thanks to the efforts of the 'Seabees' – the construction battalions formed by the USA to build, renovate and improve aerodromes, naval facilities and overland communication networks – the Allies' ships, submarines and aircraft enjoyed the support of a swiftly growing web of depots and operating bases.

Engagements in littoral waters: The 'Channel Dash', 1942

Land-based aircraft – some of them ferried into the theatre concerned by carriers – were occasionally used to cover or attack naval forces operating in littoral waters, such as at Guadalcanal and Rabaul in the Pacific and off Malta in the Mediterranean. Of all the instances one could explore, there is arguably no better case study than the so-called 'Channel Dash', which also highlights some of the complexities that could be encountered in finding, 'fixing' and striking adversaries, however exposed and vulnerable they might have appeared to be.

Long before the Battle of the Atlantic had entered its third year the British had compelled the battlecruisers *Scharnhorst* and *Gneisenau* as well as the heavy cruiser *Prinz Eugen* to take refuge in the French port of Brest. Although these ships were conveniently placed to molest the Allies' transatlantic trade, in the event that they ventured back into the sea lanes they risked – like the *Bismarck* before them – being cut off and sunk by a combination of warships and air power, including carrier-based planes. Indeed Brest proved as seductive a

---

224 See: ibid., p. 149.
225 See: Lundstrom, *First Team*, pp. 32–44.

sanctuary for the surface raiders as it did a thorn in the Royal Navy's side especially: while the numerous U-boats stationed there wrought havoc among the Allied convoys, months and seasons ticked by without the German capital ships sallying forth. In the interim, however, the British had to do what they could to contain if not destroy the enemy cruisers – an exigency that threatened to tie up masses of resources in essentially fruitless operations.

This frustrating situation was to change in February 1942 when Hitler, persuaded that the Allies were about to invade Norway, insisted that all three ships be recalled to the North Sea. The shortest route open to them seemed, however, the most dangerous: they could either risk the lengthy voyage round the British Isles – which would give their opponents more time and opportunities to find and hunt them down – or head eastwards up the English Channel, running the gauntlet of enemy radar, mines, destroyers, submarines and land-based aviation. Despite the misgivings of *Großadmiral* Raeder, it was decided to try the latter.

Brest lay within comfortable reach of strategic and even torpedo-bombers stationed in England. If only because of this consideration and the port's pivotal position among the hubs used by the *Kriegsmarine*, the city's AA defences had been repeatedly upgraded. By 1942 they were quite formidable. In any case RAF Bomber Command's strategy generally shunned direct attacks on Germany's military might (including her maritime strength) in favour of inflicting blows against her industrial heartlands and transport webs; these, it was reasoned, would eventually wear her forces down on every front. But the bombing campaign had proved far less virulent than had been hoped, essentially because of an inability to hit the selected nodes with sufficient accuracy and firepower. By early 1942 a new approach – area bombing – was increasingly being resorted to and, by mid-February, had become standard procedure.

This had tremendous ramifications for the training and capabilities of the bulk of Bomber Command's squadrons. Nevertheless, the RAF remained willing in principle to continue mounting raids on some discrete targets, notably capital ships. Among these was the *Tirpitz*, which, alarmingly, had just quit Kiel for Trondheim. Despite the disappointing results of earlier forays against those at Brest, a memorandum of 9 February proposed 'a light but sustained scale of attack [here] with a relatively small number of heavy bombers'.[226] This was evidently acceptable to the Admiralty, providing that, as the First Lord rejoined, 'certain immediate naval requirements of long-range . . . [reconnaissance] aircraft are met'.[227] By this juncture, however, the cruisers that had been at Brest had escaped.

The Germans had done all they could to ensure the success of the dash for the North Sea – Operation 'Cerebus'. While the cruisers themselves were rotated

---

226 Memorandum DO(42)14, 'Bombing Policy' by Sir Archibald Sinclair, Secretary of State for Air, 9 February 1942, TNA, CAB 69/4.

227 Memorandum DO(42)15, 'Bombing Policy' by First Lord of the Admiralty, 14 February 1942, TNA, CAB 69/4.

through Brest's dry dock to maximize their seaworthiness, an escort of torpedo boats, minesweepers and destroyers was assembled in the harbour. Flotillas of E-boats and other small craft likewise began gathering in ports between Le Havre and the Hook of Holland. Meanwhile, jamming equipment was aligned with the radars dotting England's south coast, the French side of the Channel was combed for mines and the *Luftwaffe* began preparing numerous fighters to cover the convoy as it moved eastwards.

Some at least of these measures caught the attention of the British and they guessed quite correctly what was afoot. However, they remained uncertain as to when exactly the German cruisers would try to break out and what direction they would take. Since it was improbable – but not inconceivable – that they would make for either home waters via a northerly route or turn south-west towards an Italian port, it was deduced that they would seek to progress up the Channel – a voyage of roughly four hundred miles – under the cover of darkness, spreading their journey over two nights with an intervening pause in Cherbourg.

The British Admiralty and Air Ministry had first compiled plans – Operation 'Fuller' – for use in such a contingency as early as April 1941.[228] It was acknowledged from the outset that aviation would have to take the lead in 'fixing' and striking any German cruisers that tried to force their way through the Straits of Dover. Unwilling to deploy its own capital ships where they might too easily fall prey to the *Luftwaffe*, the Royal Navy earmarked nothing more than six destroyers and eight torpedo boats for use should 'Fuller' have to be enacted. Air power's relative speed was also a critical consideration, for any window of opportunity that did open would surely remain so only fleetingly. All unit commanders duly concluded that they should seek to 'delay the enemy in any way possible by immediate attacks with any forces available [rather] than to risk losing the . . . [chance in an] effort to arrange co-ordinated attacks'.[229]

The fundamental problem here was that the Germans held the initiative. The British simply could not afford to tie substantial resources up indefinitely in anticipation of an engagement that might never actually occur. The Admiralty, sensing that the enemy cruisers would sail on 10 February, started making dispositions some days beforehand: six FAA *Swordfish* were transferred from Manston in Kent to Lee-on-Solent in Hampshire (opposite Cherbourg); around a thousand mines were laid in six new fields between Ushant and Boulogne; and three submarines took up position along the Channel. In the meantime the RAF sowed yet more mines in five zones off the Dutch coast, not least in areas the Germans had just swept, and warned units of torpedo-bombers stationed as far afield as Cornwall and Scotland that they might be required in south-eastern England at

228 'Escape of German battlecruisers *Scharnhorst* and *Gneisenau* and heavy cruiser *Prinz Eugen* up the Channel, [12 February 1942]: Operation 'Fuller' and the Board of Enquiry, 2 March 1942', TNA, ADM 116/4528, paragraphs 5–6.

229 Ibid., paragraph 111.

short notice. Bomber Command also diverted scores of its planes from training and other duties and put them on two hours' standby to mount attacks with armour-piercing bombs. Similarly, Fighter Command allotted numerous squadrons to providing support to both the bombers and the naval surface forces.[230]

Things were already starting to go awry, however. Firstly, the cruisers did not leave Brest until as late as 22:45 on 11 February, by which time some of the British units awaiting their appearance had had to drop to a lower state of readiness. Secondly, the German ships were not planning to interrupt their voyage at all but to steam as fast as possible right through to the North Sea. Their initial movements cloaked in darkness, they had calculated that their progress up the Channel would coincide with the arrival of a weather front that was slipping southwards from Iceland.[231] Come daybreak, visibility would be low at best and, moreover, aerodromes across England might well be fogbound and covered in ice and snow.

Nocturnal surveillance of the Channel was largely entrusted to interlocking patrols by *Hudson* aircraft equipped with Air-to-Surface-Vessel (ASV) radar that in theory could pick out a large ship from up to thirty miles away. However, the secret report into the 'Dash' conceded that: 'In the best circumstances the reliability and efficiency of these instruments in their present state of development and application in *Hudson* aircraft cannot be assessed higher than approximately 50 per cent'.[232] Certainly, the most westerly patrol – which covered the area around Brest – was suspended for three, critical hours as a result of technical problems with the ASV set aboard one of the planes. Then, by an unfortunate coincidence, the *Hudson* monitoring the adjacent sector of the Channel to the east also suffered an equipment malfunction and had to be recalled. Having, as a result of these mishaps, slipped into the Channel unobserved, *Prinz Eugen*, *Scharnhorst* and *Gneisenau* were, on the other hand, insufficiently far to the east to be detected by the planes circling between Le Havre and Boulogne, the last of which ended its reconnaissance prematurely for fear of fog blanketing its home base.[233] The deteriorating weather also encouraged radar stations along the English coastline to blame atmospheric interference – rather than subtle German jamming – for the growing 'clutter' on their screens.[234]

So persistent did this problem become, however, that by mid-morning the phenomenon was being attributed to enemy action. Numerous sorties by *Luftwaffe* aircraft were also being noted. Nevertheless, it was not until after a special reconnaissance foray by two *Spitfires* caught a glimpse of a German flotilla including at least one large ship that the British could be at all certain what was happening out in the Channel. Even news of this sighting was not disseminated immediately, for the fighters had orders to maintain radio silence and submit any contact reports on returning to their base. Debriefing of the pilots was just concluding when news arrived that other planes had overflown a cluster of capital

230 Ibid., paragraphs 7–16.
231 Ibid., paragraph 16.
232 Ibid., paragraph 97.
233 Ibid., paragraphs 17–22, 91–7.
234 Ibid., paragraph 27.

ships while engaging German aircraft near Boulogne. This intelligence had been passed up the chain to Fighter Command and thence to the Air Ministry and Admiralty, reaching them at around 11:30.[235]

The British promptly started pitchforking aviation into an assault on the cruisers. If only because hardly any of the strategic bomber crews had had any training in attacking fast, armoured ships on the high seas, the most potent forces available were a few FAA and RAF torpedo-bombers, many of which were not immediately to hand. Fourteen *Beauforts* that had been keeping an eye on *Tirpitz* from Scotland had belatedly headed south, planning to refuel and take on ordnance at North Coates in Lincolnshire. However, that aerodrome had been closed by the recent snowfalls. The planes headed for Coltishall, Norfolk, instead, arriving at around noon. Here, there were insufficient torpedoes for all the machines, two of which had in any case developed engine trouble. It was well into the afternoon before the others could resume their journey southwards. Similarly, twelve *Beauforts* that had been left at St. Eval in Cornwall in case the German ships were spotted heading westwards were summoned to Coltishall at 12:20. They did not arrive there until 17:00.[236]

As a result of all this just six *Swordfish* and seven *Beauforts* were initially within striking distance of the German flotilla. All of the former and four of the latter were supposed to rendezvous over Manston at 12:25 with not only one another but also with five squadrons of RAF fighters. Agglomerating so many contrasting machines from different aerodromes was a complex undertaking, however, and few were ready on schedule. At 12:28 the *Swordfish* tired of waiting for the *Beauforts* – which were in any case far swifter than 'string bags' – and set off accompanied by only ten *Spitfires*. These, together with two more squadrons that came hurrying after them, soon outpaced their plodding charges and even lost sight of the camouflaged *Swordfish* among the leaden clouds. Engaged by both interceptors and the warships' numerous AA batteries, all of the torpedo-bombers were shot down. Although most are believed to have launched their weapons, none of them struck home.[237]

The four *Beauforts* had, meanwhile, arrived over Manston, where they not only failed to find their fighter escort among the traffic thronging the airfield but also got split up. At 15:40, having espied the German flotilla with ASV, two approached under the cover of rain and low cloud to within a thousand yards of *Prinz Eugen* and unleashed their weapons. Similar strikes were executed by the third and fourth *Beauforts* at 16:40 and 18:00 respectively. One plane was damaged by flak and all of the torpedoes evidently missed their marks.

A second cluster of three *Beauforts* had in the interim been prepared for action, assembling over Manston at 15:00. These aircraft also attacked singly: one was brought down by fighters and AA fire before it could release its torpedo; the

235 Ibid., paragraphs 24, 27–31, 88.
236 Ibid., paragraphs 9, 12, 53–7.
237 Ibid., paragraphs 32–41.

others unleashed theirs but without success.[238] Barely had this raid subsided, however, than other British planes entered the fray as well, namely the nine *Beauforts* that had travelled south via Coltishall and five *Hudsons* that had joined them from Manston. The latter tried to distract the opposition with bombing runs while the former, divided into two flights, closed in on their prey from both sides: one *Beaufort* lost its bearings in the mist and never attacked; another's torpedo failed to release; and two of the *Hudsons* were destroyed by the intense AA fire. Visibility was so poor that nobody knew whether any of the seven torpedoes that were launched struck the target.[239]

Fading light also put paid to any hope of using the *Beauforts* that had hurried from St. Eval to Coltishall, where they were supposed to rendezvous with some fighters. When these did not materialize, the torpedo-bombers set out unescorted for the German flotilla's estimated position, arriving there at 18:05. Amidst the rain and twilight they glimpsed some minesweepers but could not find the cruisers. Even if they had, the chances of mounting an effective attack in such conditions would have been negligible. The planes, which had now lost sight of one another, duly turned for home. Two never made it back.[240]

The atmospheric conditions had also played havoc with RAF Bomber Command's plans to bombard the German ships with over two hundred aircraft.[241] Not only had some *Wellingtons* been grounded by snowfalls but also the dense, low cloud and wet weather ruled out the use of armour-piercing projectiles, for these had to be dropped from a considerable height. Most planes were duly re-armed with general purpose bombs at the last minute, although a few armour-piercing weapons were retained 'in the hope that possible breaks in the cloud might enable a high-altitude attack to be carried out'. In all, 242 sorties were mounted, from which 15 bombers failed to return. Just 39 of the aircraft actually unleashed their payloads against shipping; all of 188 bombers were either unable to locate their quarry or see it clearly enough to bombard it.[242] Even if more strikes had proved feasible, however, it is very questionable whether significant damage would have been inflicted on the cruisers, if only because, as the inquiry into the 'Dash' noted, 'the training of the greater part of Bomber Command is not designed for effective attack on fast-moving warships by day'.[243] Indeed, it was acknowledged that 'owing mainly to the continuous employment of these units on other operational duties, some of the crews of [Coastal Command's] *Beauforts* had not had . . . sufficient training and experience to render them fully efficient against such a target as was presented to them . . .'.[244]

238 Ibid., paragraphs 45–52.
239 Ibid., paragraphs 53–5.
240 Ibid., paragraphs 56–7.
241 See: ibid., paragraphs 67–75.
242 Ibid., paragraph 74.
243 Ibid., paragraph 129.
244 Ibid., paragraph 126.

4 The defence of trade and the sea lanes

The military theorist Julian Corbett once observed that: 'So vital indeed is financial vigour in war than more often than not the maintenance of the flow of trade has been . . . a paramount consideration'.[1] In deciding the outcome of the conflicts of both 1914–18 and 1939–45 the protagonists' capacity to procure, process and distribute sufficient raw materials to satisfy their burgeoning needs was indubitably a fundamental factor. The reciprocal nautical blockades witnessed during the First World War were, together with the economic 'Crash' of 1929 and its global ramifications, to have a profound impact on the pattern of international relations in the course of the 1930s. Indeed it could be argued that the Second World War was in the main a product of these events, insofar that the hardships and deprivations that attended and followed them not only helped to alienate and radicalize many individual citizens within established polities but also impelled Germany, Italy and Japan above all to seek substantial alterations to the world's geopolitical order. While, much to the Americans' chagrin, the Japanese especially strove to dismember the tottering Chinese Empire, Hitler wanted his new *Reich* to acquire *Lebensraum* at the USSR's expense in particular. This was so largely because expansion eastwards would simultaneously render Germany less susceptible to any repetition of the maritime strangulation that had proved so debilitating between 1914 and 1918. As the *Wehrmacht* and *Luftwaffe* had perforce to be allotted the vast majority of the available resources, in the mid-1930s Berlin did not even anticipate possessing a battle-fleet that would be able to contend directly with those of the major maritime nations for at least another decade. On the other hand, the Germans could not afford to neglect control of the sea entirely if only because the reorientation of trade was an essential prerequisite for their wider strategic goals and because so much of the world's trade was transoceanic. Although Halford Mackinder had claimed in 1919 that continental powers enjoyed a growing advantage over maritime ones in this respect, this reasoning was suspect, not least because the movement of bulk cargoes remained more efficient by water than by alternative means. As

1 Corbett, *Principles*, p. 160.

Charles Callwell remarked as long ago as 1897: 'It is a broad economic principle well known in the commercial world that, as a general rule, the conveyance of goods by ships is cheaper than their conveyance by road or rail'.[2] The fledgling *Kriegsmarine* was duly conceived and configured for a *guerre de course*, with costly surface raiders rather than relatively inexpensive U-boats being seen as its most promising weapons.

Similarly, Mussolini's ambitions in Africa and the Balkans were a manifestation of a vicious circle whereby some states' quests for autarky spawned rearmament programmes both at home and among their putative foes. This, in turn, stoked demand for strategic commodities and the concomitant desire for further geo-political aggrandizement. Rome's bid to gain access to more natural resources through violence ineluctably alarmed the very powers – France and Britain – that dominated the trade routes criss-crossing the Mediterranean, the great maritime barrier separating metropolitan Italy from her enclaves further afield. Sure enough, when Mussolini invaded Abyssinia in 1935, the League of Nations imposed an economic boycott on Italy that was enforced by above all French and British warships. Somewhat incongruously, however, these sanctions did not encompass fundamental strategic materials such as oil, iron, coal and steel since it was feared that any ban on the importation of these might ignite a wider conflict, not least over freedom of the seas. Mussolini was thus left unfettered but antagonized, while the League was utterly discredited as an instrument of collective security.

Introduced at a juncture when the wider international order was undergoing immense change, the economic sanctions inflicted on Tokyo in the summer of 1941 could only have appeared to many Japanese as a mortal threat to their country's bid to remain among the great powers. Just as Italy had shared in the partition of metropolitan France in June 1940, Japan had also exacted concessions from Vichy with regard to Indo-China, notably the right to establish air bases and army encampments on what was still nominally French sovereign territory. The enclave thereby became a *de facto* part of Tokyo's 'Greater East Asia Co-prosperity Sphere' – Japan's New Order for a vast swathe of the Earth's surface and one which, she protested, was a progressive, mutually beneficial concept that would liberate the region's indigenous peoples from colonialism and imperialism. Actually, for most of the Second World War Tokyo used the Vichy administration in French Indo-China as a facade behind which ever more of the region's natural resources, notably its large rice and rubber harvests, were siphoned off by Japan to the growing detriment of the local inhabitants. With no realistic prospect of resisting, the Vichyite governor-general, Vice Admiral Jean Decoux, was bullied and otherwise manipulated into collaborating. However, when Japanese troops occupied Saigon and moved into Cambodia in July 1941, the Netherlands East Indies (at the behest of the Free Dutch cabinet in London), Britain and the USA retaliated by imposing a choking commercial boycott on Tokyo. The sanctions excluded Japan from key markets in the USA especially, froze many of her overseas assets and severely truncated her ability to import commodities that were strategically indispensable, such as petroleum products (including kerosene), metals and rubber. The oil embargo alone was worrying enough, for at least two-thirds of her demand was satisfied through imports, nearly all of which emanated from the Netherlands East Indies. Indeed so paralysing were the measures in their entirety that the Japanese government calculated that it had just four months in which to either submit to its opponents by withdrawing from China and Indo-China, or to resort to war against them in an effort to gain new markets and access to raw materials.[3]

The USA, too, prised some undertakings out of Vichy as Washington and London increasingly made common cause with one another. Admiral Jean François Darlan – who, by the summer of 1941, had risen so far and so fast within the Vichy cabinet as to have accumulated several of its most important portfolios, including those of the vice-premiership, the foreign ministry and both the marine and defence departments – assured the Americans that France's battle-fleet and her possessions in the Caribbean – Guadeloupe, Martinique and French Guiana – would not be allowed to fall into Germany's hands. However, Darlan – who was something of an Anglophobe even before Mers-El-Kébir and Dakar – simultaneously favoured close collaboration with Hitler and even went so far as to negotiate the so-called Paris Protocols of May 1941, which, among other concessions, granted Berlin military installations in French West Africa, Tunisia and Syria. Although this particular diplomatic initiative ultimately withered on the vine, Darlan's attempts to run with both the hare and the hounds precipitated a pre-emptive invasion of Syria by British, Australian and Free French forces in June 1941, by which time Washington had likewise taken the precaution of bolstering the active defences of the Caribbean basin.

The eastern gateway to the Panama Canal and a major source of oil and bauxite in particular, that region was of great strategic significance to the Americans and British especially and most of the leases proffered to the former by the latter under the destroyers-for-bases deal of September 1940 related to sites that were located on Jamaica and within the Lesser Antilles. In the wake of that accord numerous new military installations sprang up and the USA's existing web of facilities – which comprised nodes in Cuba, the Dominican Republic, Haiti, Puerto Rico and the Canal Zone – was simultaneously enlarged and upgraded. Around the time of the Paris Protocols the USA also set up a Caribbean Defense Command (CDC) to oversee the sprawling network of outposts within the basin and to harmonize their activities. This step accelerated and consolidated a trend that had been underway in this corner of the globe since the end of the 1800s and across the northern reaches of the Western Hemisphere since the dawn of the Second World War, namely the transfer of responsibility for security

3 See: Morison et al., *American Republic*, p. 630.

from European armed forces, notably the Royal Navy, to those of the USA. By February 1942, as had happened in Iceland the previous year, American units had relieved British ones that had been safeguarding the Free Dutch enclaves of Aruba, Curaçao and Guiana. The CDC also organized a convoy system for commercial shipping and the methodical scouring of the Caribbean by numerous, coordinated sea and aerial patrols.

Although for much of 1942 U-boats intermittently prowled the sea lanes of the Caribbean, sinking nearly three hundred merchantmen between them, Germany's capacity to prosecute a *guerre de course* in this quarter proved comparatively ephemeral, not least because of her adversaries' ability to exploit air power in protecting the region's trade routes. Whereas their numerous aerodromes, harbours and other facilities enabled the Allies to remain extraordinarily vigilant and active, the Caribbean's relative remoteness from the bases available to the *Kriegsmarine* posed operating problems that, even when other difficulties were left aside, were more than a little daunting. Whilst the wider Battle of the Atlantic continued to rage, by the end of 1942 that for the Caribbean was all but over, the U-boats having chosen to focus their efforts on the one environment where, if only for the time being, they still enjoyed the upper hand.

In 1939 almost half of the UK's crude oil supplies originated from wells on the Caribbean's rim. At this juncture oil was one of roughly thirty substances that states required in order to be able to wage war, for these commodities were essential ingredients for one or more of the primary, secondary and tertiary processes of production. By way of an illustration, fossil fuels – notably coal – were needed for the generation of the heat, light and motive power that made countless other activities at all practicable. For example, together with the appropriate ores, huge quantities of coal were vital for the smelting of metals, ingots of which could then be fashioned into all manner of artefacts including weapon platforms and actual armaments. Principally derived from bauxite, aluminium, for instance, was increasingly prized by, among others, aircraft manufacturers; resistant to corrosion, sturdy and yet relatively light, aluminium alloys especially were ideal materials from which to make airframes and components, including parts of engines. Similarly, the consumption of tungsten – which has an exceptionally high melting point – and molybdenum grew considerably as mechatronics came to play an ever bigger part in aerial warfare, not least air-defence, since these comparatively rare metals were needed for the production of filaments in thermionic valves. Demand for other specialist metals such as titanium and vanadium increased as well: titanium possesses similar qualities to aluminium but is even stronger, while vanadium pentoxide, for example, was employed as a catalyst in the Haber Process for the creation of ammonia, the raw material for most nitrogen compounds.

In addition to serving as crop fertilizers that boosted the production of foodstuffs – one of the most basic of necessities for armed services and civilians alike – many of these chemicals were highly explosive and could be readily adapted as propellants and charges for munitions. Other substances, notably petrochemicals and rubber, were absolutely indispensable in manoeuvre warfare especially, since the former served as fuels and lubricants for mechanical machinery while the latter was the main constituent of the tyres supporting aircraft undercarriages and almost all wheeled vehicles. In 1941 virtually all of the rubber consumed by the USA and the British Empire was imported from South-East Asia. The sudden seizure by the Japanese of much of that region – and, with it, the bulk of the world's rubber output – underscored the dangers of banking on outside sources and triggered the recasting of the patterns of international trade in this particular commodity. It also intensified and broadened an ongoing quest for man-made alternatives to a clutch of materials that were traditionally obtained from Mother Nature. Besides Buna – an artificial rubber made by the polymerization of butadiene – synthetic oil and textiles were developed and used by several states to reduce their reliance on external supplies that, especially in wartime, might prove perilously volatile. However, the manufacture of ersatz substances could be a relatively costly process and its feasibility depended in many cases on the sophistication of the economic base of the country concerned; the churning out of Buna, for instance, was but one tiny detail of a wider transformation whereby in the space of a couple of years America's industrial output soared to twice that of all the Axis nations combined.[4] Similarly, there were practical limits to just how far any state might go in maximizing the production of sought-after materials within its own borders, just as there were to the peacetime procurement and stockpiling of strategic commodities. Whilst then as now autarky might have been a desirable goal, it was essentially an unattainable one.

Although the demand for certain strategic commodities could be alleviated through such processes as the judicious rationalization of production and consumption, recycling programmes and the development of synthetic substitutes, none of the major belligerents in the Second World War managed to do away entirely with the need to import significant quantities of raw materials. For all her natural riches, even the USA relied on trade with the British Empire to satisfy her need for cobalt, chrome and nickel, for example, and the importance of the Arctic convoys that were to help supply the USSR at a point when so much of her indigenous wealth had fallen into German hands should not be underestimated. That said, on the whole the powers that were destined to become the opponents of the Axis embarked on hostilities with larger endowments of raw materials and easier access to additional sources, not least neutral states with which they could do business by sea. The Allies were also that much readier and able to pool and share resources, whereas the European and Asian wings of the Axis waged war all but independently from one another. Until such time that Germany and Japan could mobilize the assets of the regions that they conquered in the first flush of their expansionist campaigns, they and Italy alike trailed their adversaries either in terms of the quantities of critical commodities that

4 See: ibid., pp. 632–5.

they managed to muster, or in the efficiency with which they processed them, or both. Between 1941 and 1943 the USSR, for instance, squeezed appreciably more armaments out of its economic base – grossly emaciated though it was after the German invasion – than Hitler succeeded in garnering from his bloated *Reich* and its various vassals. Similarly, until 1943 the far smaller British economy outproduced that of Germany in almost every major category of weapons.[5]

Whilst some of these disparities can be attributed to the Allies' superiority when it came to the mobilization, organization and allocation of their labour forces and other resources, the relative efficacy of the economic warfare waged by the belligerents explains a great deal as well. In the final analysis shortages of raw materials and finished products proved fatal to, not the Allies, but the Axis, with many of the critical deficiencies being spawned or exacerbated by strategic bombing and maritime blockades. The attempts by the naval and air forces of the Axis to disrupt the Allies' trade and industrial activities were, on the other hand, essentially a failure, although the *guerre de course* prosecuted by the *Kriegsmarine* especially did enjoy sufficient success to cause grave concern from time to time. In March 1942, for instance, Admiral Sir Dudley Pound, Britain's First Sea Lord, dismayed by the growing losses of commercial vessels in the Atlantic especially, warned the War Cabinet that: 'If we lose the war at sea we lose the war. We lose the war when we can no longer maintain those sea communications which are essential to us'.[6] During June 1942 no fewer than 141 merchantmen were sunk – a casualty rate that simply could not be absorbed for long if only because it outpaced Britain's ability to find replacement ships and crews.

On entering the Second World War, first Italy and then Japan followed Germany in coming under blockade from Allied naval forces. Whereas the economic infrastructure of the USA – 'the arsenal of democracy' – lay at too great a distance from enemy territory to be vulnerable to attack by land-based aircraft, the Axis powers also fell victim to strategic bombing offensives that steadily grew in scale and effectiveness. From the outset, Germany's steel and synthetic oil plants were, along with docks and railway marshalling yards, among RAF Bomber Command's top targets. By the same token, however, most of Britain was exposed to raids by the *Luftwaffe*, especially after the fall of France and Norway in 1940. Tremendous damage was sustained, particularly in the so-called Blitz that dragged on from the August of that year until the following May, when Hitler started shifting forces eastwards for his invasion of the USSR. Still, neither this nor the second Blitz of 1944 – which saw the use of indiscriminate '*Vergeltungswaffen*', namely V-1 flying bombs and V-2 ballistic rockets – had sufficient impact to undermine Britain's determination and ability to sustain the conflict. By contrast, the Allies' Combined Bomber Offensive against Italy and Germany and the fire- and atomic bombing of Japanese cities did much

---

5   See: Richard Overy, *Why The Allies Won* (London, 1995), pp. 182–205.

6   Memorandum DO(42)23 by the First Sea Lord, 'Air requirements for the successful prosecution of the war at sea', 5 March 1942, TNA, CAB 69/4.

to decide the contest's outcome, not least by ravaging the industrial heartlands of the Axis powers.

The impact of this approach was twofold insofar that it simultaneously impelled the Germans in particular to divert an ever bigger proportion of their disposable resources from acquisitive, offensive operations to defensive ones. Huge numbers of personnel, artillery pieces, munitions and fighter planes especially were tied up in trying to safeguard population and industrial centres within the *Reich* from aerial raids that might be anticipated but were by no means inevitable. This was so if only because the whole notion of strategic bombardment went hand in hand with that of distinguishing an opponent's Achilles heel from a plethora of other potential targets. At various junctures in the Combined Bomber Offensive either the USAAF's Eighth Air Force, or RAF Bomber Command, or both, anticipated that the destruction of Germany's hydroelectric dams, ball-bearing factories, petroleum plants or transport hubs would bring much of her economy to a standstill. Similarly, there were moments when the Germans – despite the glaring failure of the *Luftwaffe* in this regard during the summer of 1940 – hoped that, by the targeting of discrete bits of infrastructure upon which so much more appeared to depend, air power might yet undermine Britain's resolve and capacity to go on resisting. Expecting that her harbours would prove rather more susceptible to attack than the merchant shipping that linked them with those of the wider world, early in February 1941 *Großadmiral* Raeder persuaded Hitler to refocus the Blitz on Britain's foremost entrepôts, namely Avonmouth, Belfast, Bristol, Clydeside, Hull, Liverpool, Newcastle, Portsmouth and Plymouth. The *Luftwaffe* duly mounted dozens of nocturnal raids on these port complexes in a period of less than twelve weeks. The strength and sophistication of Britain's integrated air-defence system had, however, continued to grow, not least through the development of electronics that jammed the German bombers' navigation mechanisms and through the advent of night-fighters equipped with tactical radar sets. Although appreciable damage and disruption was inflicted in places, the size and intricacy of the targets were sufficient by themselves to confound the attackers, who became steadily more vulnerable as their quarry's defences improved in quantity and quality. By the time she embarked on her quest for *Lebensraum* at the expense of the USSR, the epicentre of Germany's campaign against British maritime power was already being pushed and pulled onto the high seas.

Land-based aviation was nonetheless playing a major role in the fighting. During April 1941 all of 644,000 tons of British shipping were sunk. Roughly half of this figure is to be attributed to aerial attacks, with planes destroying over a million tons of vessels in the course of that calendar year.[7] Besides the Focke-Wulf *Kondor* aircraft that prowled the Atlantic approaches from bases

---

7 For a statistical breakdown of the *guerre de course* in the Atlantic and elsewhere, see: S.W. Roskill, *The War at Sea 1939–1945, Volume Three: The Offensive, Part Two, June 1944 – August 1945* (London, 1961), Appendix Y, p. 472 and Appendix ZZ, p. 479.

in France, German and Italian machines harassed ships in the Mediterranean's eastern reaches especially. If only because of environmental peculiarities, many of the convoys that, beginning in August 1941, were to ferry vehicles, planes, clothing, radios, strategic commodities and other materiel to the beleaguered USSR also sailed under the shadow of the bomber for long stretches of both their outward and homeward journeys. Whereas the darkness that prevailed during the polar winter might conceal their movements and thereby afford them some protection, on the whole the intertwined phenomena of weather and geography helped create a situation that was perilous in the extreme, not least because of the numbingly low temperatures that affected both men and equipment. Barred by ice floes from giving the coastline of northern Scandinavia a wide berth, ships were more often than not funnelled onto courses that were both fairly predictable and within comfortable reach of *Luftwaffe* and Finnish aircraft stationed along the littoral. U-boats and surface raiders compounded the dangers. Convoy PQ13, for instance, which set out in March 1942, was assailed by destroyers that sortied from Kirkenes as well as by submarines and planes.[8] Apart from around Murmansk and Archangel, where Soviet fighters might occasionally intervene, most of the theatre was far too remote for the Allies' own shore-based aircraft to penetrate, leaving their Axis counterparts free to manoeuvre almost at will. Moreover, once it became clear that the *Blitzkrieg* of Operation 'Barbarossa' was being superseded by a protracted, attritional struggle, the *Luftwaffe* and *Kriegsmarine* began intensifying their efforts to interdict the flow of aid to the USSR. On 27 May 1942 alone, Convoy PQ16 was molested by in excess of a hundred planes.

The pattern of events within the protracted Battle of the Atlantic especially was shaped by a number of interlocking factors, the cardinal ones being as follows. Firstly, the shape and size of the theatre of operations were influenced as much by changes to the geopolitical environment as by the reach and endurance of particular forms of sea and air power. Just as Germany's conquest of Norway and France in 1940 had a major impact in this respect, so too did Washington's decision to assume responsibility for the security of an expanding swathe of the Atlantic basin. Well before the USA entered the war, this zone had been extended as far as 26° West and American troops were garrisoning Iceland, which, by April 1941, was serving as a forward operating base for British escort vessels and maritime patrol planes. The USA's embroilment in the conflict at the end of that year was likewise to impact on the fighting's geometry and intensity, with, for instance, U-boats ravaging comparatively vulnerable merchantmen along America's eastern seaboard in the last weeks of 1941 and the first of 1942. This helped spawn a leap in Allied shipping losses from 104,000 tons in November to no less than 583,000 in December.[9]

---

8   See: Barnett, *Engage the Enemy*, pp. 702–3.
9   Roskill, *The War at Sea 1939–1945, Volume One: The Defensive*, Appendix R, p. 616.

Then as now, the ability to engage targets at all was as dependent upon reliable information and accurate navigation as it was upon tactical capabilities and prowess. Locating opponents amidst the enormity of the seas was a hurdle that had to be overcome if any combat was to occur. Although tapping the wealth of knowledge that both sides transmitted over the airwaves promised to reduce the scale of this challenge, gleaning useful information and then exploiting it to a degree that did not risk proving counterproductive were formidable tasks in themselves. During the first eighteen months or so of the war Nazi cryptanalysts made considerable inroads into some British maritime ciphers, but this only accelerated the introduction of refinements that made them appreciably harder to crack. Meanwhile Bletchley Park began unravelling the 'Enigma' and 'Lorenz' codes that, the Germans remained persuaded, were impenetrable. In May 1941, for example, a boarding party from HMS *Bulldog* spirited away a three-rotor 'Enigma' machine and concomitant documents from the sinking *U-110*, which, holed by the British destroyer, had been forsaken by her crew. Whilst, through the addition of a fourth rotor to the encryption devices of its submarines, the *Kriegsmarine* managed to foil Bletchley's boffins for much of 1942, in the December of that year the security of the 'Enigma' network was undermined afresh when HMS *Petard* crippled and cornered another U-boat, the *U-559*. As had been the case with that of *U-110*, her crew had abandoned her, confident that she was doomed and would take her secrets with her to the bottom of the Mediterranean. Rescued by the British and escorted below deck, the survivors could not suspect that a handful of their captors had boldly ventured into the stricken submarine, whence they snatched a four-rotor encoding console and related manuals just moments before *U-559* finally foundered.

This coup and ensuing breakthroughs in the intelligence war helped the Allies gain the upper hand in the *guerre de course*, with May 1943 proving the climax of the drawn-out Battle of the Atlantic. But quantitative factors as well as qualitative ones were bound to be key determinants in what was essentially an attritional struggle that was ultimately all about the control of resources. At the outbreak of hostilities Germany had 39 serviceable U-boats, of which only 25 were capable of roaming the Atlantic. This was an insufficient force for the 'wolf pack' tactics favoured later on in the conflict. Even when significantly more submarines became available, there were times when requirements far exceeded supply, compelling the *Kriegsmarine* to focus on some sea lanes at the expense of others. During autumn 1941, for instance, 20 U-boats were transferred from the North Atlantic to the Mediterranean and 6 to the Norwegian Sea in the hope that they might hamper the Allies' efforts to, respectively, relieve Tobruk and ship aid to the USSR. At this juncture many transatlantic convoys were in any case proving difficult to target, not least because the Allies, forewarned by deciphered signal intercepts, were simply steering them round known agglomerations of U-boats. Their skippers exasperated, following America's entry into the war in December 1941 some German submarines ventured as far as the coastal waters of her eastern seaboard in search of prey. Here they found abundant, easy pickings in the form of freighters that were sailing independently. Operating so far from home, relatively few raiders could be kept astride this distant sea lane, but the damage they were able to inflict belied their numerical strength. If only because it was abruptly confronted with spiralling requirements both here and in the Pacific, the US Navy was short of escort craft and did not institute a convoy system until May 1942. For the U-boats, the intervening weeks proved to be *'glückliche Zeiten'* – 'lucky times'.[10]

The Royal Navy, too, intermittently struggled to find enough escort vessels to fulfil its copious commitments, while, owing to the competing demands on the RAF and FAA, air cover for commercial shipping was frequently sparse and spasmodic as well as patchy. In the Atlantic and Pacific theatres alike, numerous machines and extensive webs of convenient aerodromes were vital prerequisites for any aerial operations over the high seas. This was particularly true of anti-submarine campaigns, since, generally speaking, all but the smallest of submersibles had greater reach and endurance than even the biggest, long-range planes. Periodically, the threat was multi-faceted as well, compounding the need for safeguards. During February and March 1941, by way of an illustration, shipping in parts of the Atlantic faced attacks from not only U-boats but also from *Kondor* and Heinkel He.111 bombers and from the surface raiders *Scharnhorst* and *Gneisenau*. Inherent in such circumstances was the danger that convoys would merely present better targets to predators than solitary merchantmen and that, unless sufficiently supported by aircraft, no number of traditional escort vessels would necessarily afford adequate protection to either.

By mid-1941, however, the pendulum was swinging back in the Allies' favour. Not only had progress been made on the intelligence front but also more material assets were entering the fray. Early in 1941 the Royal Navy hastily reconfigured a couple of suitable freighters and an ageing seaplane tender as Fighter Catapult Ships (FCS). Intermittently, other commercial vessels were also called on to accommodate handfuls of planes as well as their usual cargoes. These initially took the form of Catapult Aircraft Merchantmen (CAM), with the first Merchant Aircraft Carrier (MAC) not entering service until May 1943. The former had no means of recovering machines, which had to ditch in the sea after use; the latter were fashioned from grain transporters and small tankers through the addition of crude flight-decks and, where feasible, small hangars. Such vessels normally catered for three or four *Swordfish* aircraft. In all, 18 were to be produced specifically for the defence of convoys in the North Atlantic. In the interim, the few FCS and CAM planes available helped suppress the threat posed by hostile bombers and shadowing aircraft especially. In July 1941, for instance, out of the 43 vessels (120,000 tons) lost, only 11 (9,000 tons) were destroyed by aviation.[11]

During the spring of that year the Allies bolstered their strengths and reduced their vulnerabilities in other ways, too. Among the most important of

---

10 For coverage of U-boat operations off America's eastern seaboard in 1942 see Milner, *Atlantic*, pp. 79–81 and Barnett, *Engage the Enemy*, pp. 441–3.

11 Roskill, *The War at Sea 1939–1945, Volume One: The Defensive*, Appendix R, p. 616.

the initiatives taken was a reorganization of the command and control structure that oversaw anti-submarine warfare (ASW) in the pivotal theatre of the
North Atlantic. In February 1941 the Western Approaches Command (WAC)
was moved from Plymouth to Liverpool and, in April, issued refined convoy
instructions. The new headquarters – which came under the jurisdiction of the
Assistant Chief of the Naval Staff (Trade) at the Admiralty – took day-to-day
responsibility for the security of the Atlantic sea lanes, enacting plans that were
drawn up in consultation with several agencies. Among these were the Naval
Control Service, which grouped commercial traffic into convoys, and the Submarine Tracking Room of the Naval Intelligence Division. The WAC allotted
escort vessels to merchant shipping and, on shifting to Liverpool, acquired operational control over much of RAF Coastal Command. Soon after these developments new aerodromes in Iceland were activated and the US Navy extended its
security zone further into the Atlantic. By May, the first of the Royal Canadian
Navy's corvettes were making their presence felt, too.

The Allies' growing capacity to turn out and crew new merchantmen was a
key consideration as well. At the very least, supply had to keep abreast of losses.
The use of welding, as opposed to riveting, in ship construction had proliferated
during the interwar period, substantially reducing the time many yards needed
to assemble a given type of ship. The employment of modular designs in large,
commercial hulls especially also accelerated production and, thereby, cut costs.
As they honed their techniques, several American shipwrights proved capable
of completing a 'Liberty' freighter in a week or ten days. Some took as little
as four days. Journey times could also have a substantial impact on the overall
tonnage of disposable shipping. Convoys could only proceed as fast as their
slowest constituents and the desirability of avoiding exposure to observation and
attack could lengthen journeys appreciably. The passage of commercial shipping between the North Atlantic and the Indian Ocean offers the most obvious
illustration of this dilemma. Rather than take the most direct route through the
embattled Mediterranean, for so long as Italy remained hostile most Allied merchantmen went all the way round the Cape of Good Hope. This transformed
the Suez Canal into more of a terminus than a thoroughfare. Middle Eastern
oil, for instance, was seldom exported to metropolitan Britain, being consumed
instead by the garrison and population of British-held Egypt. The eradication,
early in 1941, of Mussolini's forces in Italian East Africa and the wider Horn
of Africa also had a pronounced effect on patterns of trade in this quarter, for
neutral powers, not least the USA, promptly ruled that the Red Sea and the Gulf
of Aden were no longer combat zones and thus safe for their shipping to transit.
The ensuing influx of neutral hulls eased the strain on Britain's own merchantmen here and elsewhere.

With the sinking of the *Bismarck* at the end of May 1941, another significant
threat to Britain's trade links was to be moderated. After mounting a fruitless
foray towards the all-important sea lanes of the North Atlantic, this formidable
surface raider, damaged by gunfire and aerial torpedoes, had been striving to put
into Brest for repairs and refuelling when her pursuers overtook her. As with

the case of the *Graf Spee* in 1939, this episode underscored the perils of operating large, complex warships far from logistical infrastructure. After the loss of *Bismarck*, Hitler became increasingly reluctant to commit Germany's remaining capital vessels, particularly the mighty *Tirpitz*, to operations beyond the reach of friendly, land-based aircraft. This inhibition, coupled with oil shortages and the scarcity of maintenance facilities capable of accommodating such an extraordinarily large ship, effectively confined the *Tirpitz* to Scandinavian waters, where she never actually fired on Allied merchantmen. In the meantime, the noose around her was tightened, not least through one of the war's boldest amphibious raids, namely that on St. Nazaire's '*Normandie*' complex in March 1942, whereby the British wrecked the largest dry dock on France's Atlantic seaboard.

The potential danger posed by battleships and large cruisers had long preoccupied the British. For much of the 1930s they anticipated that the greatest threat to their maritime communications would comprise an amalgam of surface raiders and aviation. As the Second World War loomed, it was widely if mistakenly expected that the danger posed by submarines would be marginalized through the use of Asdic. However, as fear of the U-boat had waned, concern about the possibility that the Germans might fuse big-gun warships with reconnaissance aircraft had begun to wax. Guided by a solitary seaplane, between 1916 and 1918 the German auxiliary cruiser *Wolf* had inflicted shocking, disproportionate damage on Allied trade in the Indian and Pacific Oceans especially. In January 1940 it seemed to Captain Charles Daniel, the Admiralty's Director of Plans, that the *Kriegsmarine* might send the *Graf Zeppelin* together with one or more capital ships into the Atlantic, where the carrier's numerous planes would not only search far and wide for merchantmen but would also warn of the approach of Allied hunting groups bent on intercepting the interlopers. 'The enemy's best course of action', he surmised, 'would probably be to retain the *Bismarck* at home to contain the maximum of our forces and to send a *Scharnhorst* [class vessel] with a carrier to the North Atlantic. To meet such a combination, and possibly a *Deutschland* in the South Atlantic, we ourselves should need every aircraft carrier that we could make available . . .'.[12]

If only because the *Graf Zeppelin* was never made operational, this particular nightmare was never actualized. Nevertheless, the fate of one of the Arctic convoys, PQ17, which sailed from Iceland barely a year after Hitler launched his invasion of the USSR, adequately illustrates just how intimidating and perplexing agglomerations of German sea and air power could prove for the Allies. Anticipating that U-boats and aircraft would make a major effort to interdict the convoy, the British Admiralty had allotted it a sizeable escort of destroyers, corvettes, minesweepers, auxiliary AA vessels and armed trawlers. It was also shadowed by a cluster of Royal and US Navy cruisers, just in case any surface

---

12 Document 29, 'Minute by Director of Plans: Threat posed by *Bismarck* and *Graf Zeppelin*', 29 January 1940, in *FAA in Second World War: Volume One*, p. 80.

raiders materialized. But a week into the voyage, 'Ultra' intelligence intercepts revealed that the *Tirpitz*, the *Admiral Hipper*, the *Admiral Scheer*, several destroyers and, possibly, the pocket battleship *Lützow* had been amassed in the Altafjord – between Narvik and the Nordkapp – from where they might easily head off PQ17 and the nearby cruiser group alike. Justifiably apprehensive that both flotillas would be cornered and ravaged in turn, Admiral Pound, the First Sea Lord, instructed the freighters to scatter and their escorts to turn back. Sure enough, on 5 July, having established through aerial reconnaissance that the British warships did not include anything more powerful than a cruiser and that PQ17 had progressed too far for the Home Fleet at Scapa Flow to come to its assistance, *Tirpitz* and her companions did set out with the intention of falling on the convoy the next day. By this point, however, PQ17 was already facing extermination. Whilst salvation could sometimes be found more in dispersal than in concentration, on this occasion the merchantmen were located and picked off one by one by submarines and bombers; only 11 out of the 36 that had originally set sail for the USSR ever reached their destination.[13]

Tirpitz duly aborted her sortie almost as soon as it had begun. This was not the first occasion on which she had sought to come to grips with the foe and failed. Nor would it prove to be the last. Ever since March 1942 – when, in trying to intercept an Arctic convoy, she had come under attack from torpedo planes launched by HMS *Victorious* – her active use had been subject to restrictions that were largely inspired by a dread of Allied aviation. Mindful of the fate that had befallen *Bismarck*, Hitler and Raeder were unwilling to risk sending Germany's most powerful battleship into waters that might conceal a carrier. *Tirpitz* was to be withheld unless – as in the case of PQ17 – the enemy's dispositions were known in detail and the *Luftwaffe* could provide significant support, notably in countering ship-borne reconnaissance and combat aircraft.[14] Such conditions could seldom be fulfilled, however, partly because of a shortage of viable targets, partly because dependable intelligence on the movements of Allied naval forces was far from comprehensive, and partly because of the competing and growing strains on the *Luftwaffe*. During the autumn of 1942, by way of an illustration, the Arctic convoys were temporarily suspended while the Allies focussed resources on Operation 'Torch', the amphibious invasion of the Vichyite enclaves in northern Africa. By the time the aid shipments to the USSR had resumed in December 1942, many of the planes that had formerly been stationed in Norway had been redeployed to the Mediterranean. This reduced the assistance the *Luftwaffe* could extend to the *Kriegsmarine* as well as the threat that it itself posed to Allied vessels plying the Arctic.

If only to promote the cause of his beloved *Luftwaffe* in a spiralling, inter-service competition for resources, *Reichsmarschall* Göring was openly critical of

13 See: Barnett, *Engage the Enemy*, pp. 710–22.

14 See: Keith Bird, *Erich Raeder: Admiral of the Third Reich* (Annapolis, 2006), p. 187; and Frederich Ruge, *Sea Warfare 1939–1945: A German Viewpoint* (London, 1957), p. 215.

the decision to recall *Tirpitz* and her companions from the incipient attack on PQ17, depicting the battleship's withdrawal as evidence of obsessive caution on the part of the *Kriegsmarine*.[15] But having frightened the convoy into scattering and its escort into retreat, it was already apparent that the German surface vessels had set the scene for a fight in which their big guns could only have turned out to be surplus to requirements. Raeder was, moreover, sticking to a policy stipulated by Hitler. The failure of the *Hipper* and the *Lützow* to overwhelm the cruisers and destroyers protecting another Arctic convoy, JW51B, at the turn of the year proved a much more controversial affair within the Nazi hierarchy. Indeed the rows and recriminations that followed the disappointing Battle of the Barents Sea culminated in Raeder's resignation as commander-in-chief of the *Kriegsmarine* and Hitler threatening to decommission all of its major vessels.[16]

The *Großadmiral* felt, with some justification, that the *Führer* was mishandling strategy and, with it, the allocation of resources. Raeder had after all expected to be granted the time and the wherewithal to fulfil Plan Z – the building of a world-class navy that would include the *Graf Zeppelin* and several other carriers. But Hitler's wider rearmament programme had become so burdensome in every sense that no sooner had Plan Z been approved in principle (in January 1939) than Germany had to either go to war with her immediate neighbours at least, or go bankrupt (again), or both. Although the gamble paid off initially, with the *Blitzkrieg* of 1939–40 securing the *Reich* better access to raw materials, not least Scandinavian iron ore, Romanian oil and Polish coal, the failure to subdue Britain through either diplomacy or conquest dictated that the issue of economic sustainability would soon take centre stage once more. Widening the conflict by turning on Stalin and embroiling the USA was, to Raeder, premature if not foolhardy. It was, he felt, imperative to neutralize the British Empire first, the prerequisites for which included the domination by the Axis of the Mediterranean – the principal conduit linking India and London's outposts in the Pacific and Indian Oceans with the motherland.

This called for investments in naval strength at a time when Germany was preparing to embark on hostilities against the USSR, a land power of daunting proportions. Trying to satisfy the demands imposed by the rapacious *Luftwaffe* and *Wehrmacht* left little material and human capital over for the *Kriegsmarine*. Construction of sizeable surface vessels all but ceased and, in place of Plan Z, short-term and stop-gap measures had to be adopted. Coming close on the heels of Operation 'Barbarossa', Hitler's formal declaration of war against the Americans – a major maritime nation – exacerbated Raeder's dilemma enormously. It is perhaps unsurprising that, within weeks of this momentous development, the arguments within the Nazi leadership regarding the allocation of resources in general and the shape, size and relative effectiveness of the

15 See: Bird, *Raeder*, p. 199.
16 See: ibid., pp. 202–7.

Kriegsmarine in particular came to a head. In January 1943 Raeder stepped down from the post he had held since 1928 and was superseded by Karl Dönitz, the head of the U-boat arm.

Raeder's *Kriegsmarine* had finished up as a rather lopsided force. The dearth of carrier-based aviation especially hindered the use of the fleet's capital ships in the *guerre de course* for which they had initially been envisaged. (At the war's conclusion, their tally of Allied merchantmen was to stand at 47.[17]) Unaccompanied by reconnaissance and combat planes that might guide and help protect them, Germany's surface raiders ran considerable risks whenever they ventured far from friendly shores. *Scharnhorst* and *Gneisenau*, for instance, executed several sorties in the first year of the war without destroying any commercial vessels, while *Bismarck* fared even worse, being cornered and sunk on her maiden voyage without having so much as glimpsed an enemy freighter. Although during 1941 *Scharnhorst*, together with *Gneisenau*, did cause appreciable harm and disruption to Britain's transatlantic trade, both ships were subsequently bottled up in Brest for the best part of eleven months. Here, they lay within easy reach of RAF bombers stationed in Britain until, along with the heavy cruiser *Prinz Eugen*, the two battlecruisers made a break for German waters by running the gauntlet of the English Channel. Both struck mines in the process. While undergoing repairs in Kiel on 26 February 1942, *Gneisenau* was again assailed from the skies, one bomb penetrating a magazine and inflicting almost irreparable damage.

Prinz Eugen was also in a bad way, having been hamstrung off Norway by the submarine HMS *Trident* just days earlier. After being patched up in Trondheim, by May she was sufficiently seaworthy to make for Kiel's repair yards, where a full restoration might be effected. Grasping a fleeting opportunity to finish her off with air power, the British risked dispatching two flights of *Beaufort* torpedo-bombers in pursuit. The strike went awry, however, largely because one of the attacking groups experienced difficulty in locating its quarry and became entangled in dogfights with *Luftwaffe* interceptors: several of the *Beauforts* were destroyed or damaged by the fighters; others had to jettison their weapons and take evasive action; and three more were brought down by AA fire as they approached *Prinz Eugen* and her screen of four destroyers. None of the torpedoes that were released at the target struck home. Nevertheless, so serious were the heavy cruiser's existing wounds that she was unable to return to service until the autumn.

Two months later, at the end of December 1943, *Scharnhorst*, after unsuccessfully trying to molest an Arctic convoy, JW55B, found herself encircled off Nordkapp by battleships of the Home Fleet; she was sunk with virtually all hands. *Tirpitz*, too, had made her last offensive foray. Very badly damaged at her moorings in a raid by X-craft miniature submarines during September, she was kept all but inoperable by a series of bold, aerial attacks, Churchill insisting that

---

17   Milner, *Atlantic*, p. 23.

'crippling this ship would alter the entire face of the naval war and . . . the loss of 100 machines and 500 airmen [in so doing] would be well compensated for'.[18]

Not least because of a scarcity of resources, the surface raider was quickly supplanted by the submarine – a platform that was less costly to produce, maintain and operate – as the commonest threat to Allied merchant shipping. In September 1939 the *Kriegsmarine* possessed just a couple of dozen ocean-going U-boats. Around eleven hundred more were to be constructed as the war unfolded.

The single most important factor in explaining the success of U-boats in the conflict's early years was that, when submerged at least, they were comparatively stealthy. Although hydrophones and Asdic could in principle detect craft concealed beneath the waves, in practice environmental variables – such as the contours of the seabed and irregularities in the depth, salinity and temperature of the water – could complicate matters significantly. (Later in the war especially, the Germans also strove to compound these difficulties by coating some of their U-boats in a thin layer of insulation. Known as 'Alberich', this sheathing muffled any sound emanating from the vessel and could also diminish its Asdic echo by absorbing rather than reflecting the pulsations.) In any case Dönitz – a submariner by experience as well as training – encouraged his crews to vitiate Asdic's usefulness through the simple expedient of mounting attacks on the surface and under the cover of darkness. He also favoured grouping submarines together into 'wolf packs', whereby they could pool their reconnaissance capabilities and, on pinpointing potential prey, their firepower.

Submarines during this era did, however, have grave, inherent weaknesses. First among their foes was their operational environment rather than any mortal enemy. Whilst hiding was their main form of defence, once under water they could neither transmit nor receive messages and could only picture their surroundings through instruments such as periscopes and audiological sensors, both of which had obvious limitations. Moreover, the longer they remained immersed, the more any occupants were exposed to the danger of asphyxiation. The electrical batteries that provided motive power, communications and light would likewise be drained over the course of a few hours or days, depending on the rate at which their energy reserves were consumed. Exploiting the third dimension incurred other risks, too: whereas torpedoes could not be launched below a certain depth, even slight damage to the vessel's ribbed pressure hull could trigger an implosion that would almost certainly have fatal consequences. A submarine's manoeuvrability was in any event dramatically reduced whenever it ventured below the waves. This not only compromised its ability to shake off pursuers but also hindered it when roaming the seas in search of targets. Submerged, it might plod along for several hours at three or four knots, after which time its batteries would be running commensurately low and, despite carbon dioxide absorption systems, the atmosphere within the compartments would be

18 Quoted in Roskill, *Churchill and the Admirals*, p. 131.

turning increasingly toxic. On the surface, by contrast, as well as being able to circulate fresh air through the hatches, recharge its storage batteries and utilize any deck-mounted armaments and devices, the craft's crew could engage its diesel engines. Capable though these were of thrusting it forward at 18 knots or more, such propulsion plants were also very efficient and could endow the boat with a range of several thousand miles. In fact submarines during this period were predominantly designed less with diving than rapid cruising on the surface in mind, with reach increasing in proportion to the vessel's size. The grandest ocean-going types had operating radii that were several times bigger than that of any destroyer and, in addition to acting as weapon platforms, were employed as transporters. Japan's I-400 class, for example, was so capacious that it could accommodate three floatplanes and still find room for a bulky cargo of fuel, food or munitions.

If, as 'Jacky' Fisher had insisted towards the end of the First World War, the advent of the long-range submarine necessitated revisions to the manual of maritime blockade, then so too did the maturation of aviation that coincided with the Second. It was not that there were significantly more Axis submarines in operation than London had anticipated. Rather, changes in the geopolitical environment had transformed the geostrategic situation. Thanks to Germany's conquest of France and Norway in 1940, Dönitz's U-boats and Göring's *Luftwaffe* had acquired numerous forward bases that gave them easier and far less circuitous access to the Atlantic sea lanes. During 1941 planes alone were to sink over a million tons of British shipping, which was nearly twice the total tonnage destroyed in the previous year;[19] just a single sortie by *Kondor* bombers against a convoy west of Ireland in February 1941 ended in seven ships foundering after being struck.[20] Together with the surface raider, the aircraft and the submarine posited a threat that was far more complex than that which any one of them could have posed in isolation, compelling their opponents to try to balance requirements that competed with one another voraciously. Neutralizing dangers that might lurk below the waves as well as on and high above them called for a wide spectrum of capabilities at a time when Britain's limited resources were subject to countless other demands.

The means available for countering either planes or U-boats were, furthermore, not just relatively meagre but also disconcertingly unreliable. Purely reactive air-defences had too often turned out to be deficient, while detecting, let alone destroying, underwater targets had frequently proved difficult. Much to the bewilderment of Churchill for one, Asdic – which had been regarded as the panacea for any submarine menace – was failing to live up to expectations.[21] There had been problems, too, with the small bombs earmarked for use against U-boats and with the training of some of the personnel who were

---

19 See: Von der Porten, *German Navy*, pp. 174–8.
20 Milner, *Atlantic*, p. 49.
21 See: Roskill, *Churchill and the Admirals*, p. 135.

called upon to employ these ill-conceived projectiles. Although in November 1939 RAF Coastal Command announced that it would henceforth allot as much importance to ASW as it did to its professed forte, reconnaissance, prior to this decision nobody had thought to give much instruction to the service's air crews in the minutiae of combat against submarines. The somewhat tragicomical outcome of the war's very first clash between a Coastal Command patrol and a U-boat highlighted this oversight: swooping low to maximize accuracy, the plane released a bomb that, after bouncing across the sea, detonated beneath the aircraft, bringing it down. The submarine was untouched. Two FAA *Skuas* later suffered the same ignominious fate in a single incident. More generally, British efforts to locate and destroy U-boats with either naval hunting groups or aviation or both were as fruitless as they were frustrating. Whereas Coastal Command's patrols clocked up roughly fifty-five million air miles between them and executed a total of 587 attacks during the first two years of the war, no more than five U-boats were definitely sunk as a result of these operations.[22]

Yet aviation was destined to spearhead the Allies' counteroffensive against the U-boat. Several overlapping developments help explain this turnaround, the first of which was the severe attrition suffered by the capital ships of the *Kriegsmarine* during the period up to June 1941. The loss of the *Graf Spee*; the damage sustained in the Norwegian campaign; the cornering of *Scharnhorst* and *Gneisenau* in Brest; the sinking of the *Bismarck* and the crippling of the *Lützow*: all of these events led to the threat posed by surface raiders abating, if only for a time, allowing the British to divert additional resources to ASW.

Whereas shipping might seek and find safety through dispersal, the rationale behind the concept of the convoy was the maximization of active rather than passive defences. If these were plainly inadequate in relation to perceived threats, agglomerating vessels was potentially counterproductive insofar that it risked presenting opponents with better targets. Although from the conflict's outset the British above all attached great importance to the principle of active defence, they were far from alone in struggling to perfect its practice in the face of changing technology and a style of warfare that rode roughshod over the international accords from which seafarers derived a degree of passive protection. On formally becoming a belligerent, the USA, for instance, was slow to react to the realities of the Battle of the Atlantic: sharply silhouetted against the twinkling lights of America's eastern seaboard, throughout the winter of 1941–42 coastal traffic was beset by U-boats that, with virtual impunity, slipped in and out of the sea lanes during the long hours of darkness.[23] Japan, by contrast, never really managed to put the safeguarding of her maritime trade on a firm footing. She did not formalize a convoy system until as late as November 1943 and was in any case to prove woefully unsuccessful in countering the blockades mounted by Allied

---

22 See: Milner, *Atlantic*, pp. 20, 65.
23 Overy, *Allies*, pp. 46–7.

submarines and aircraft.[24] In 1944 she lost around three quarters of a million tons of tankers alone, which left her with next to no oil reserves. This in turn paralysed much of her remaining industry, battle-fleet and aviation. Similarly, British torpedo-bombers and submarines especially took a heavy toll on Italy's waterborne commerce, slashing imports of coal, ores and oil as well as interdicting the supply of essentials to Axis forces in Africa and the Balkans.[25]

Vital though control over other conduits remained, the safeguarding of the North Atlantic was the fulcrum of the *guerre de course* as far as Britain was concerned. And it was here, perhaps more than in any other theatre and particularly between 1941 and 1943, that the pattern and nature of the contest came to be defined by aviation's fundamental attributes and the details of its capabilities. As Admiral Sir Max Horton – who was shortly to succeed Sir Percy Noble as the head of WAC – argued early in 1942:

> It is essential that we keep control of our sea communications, prevent the enemy from obtaining his raw material, and procure what we need for ourselves. . . . It is beyond the task of our much depleted and overstrained Fleet alone (with its present air auxiliaries) to provide adequate security to avoid defeat. Recent events have proved that Fleets cannot operate without the close co-operation of air power, and if we are to hold our own . . . and wear down the enemy before we are ourselves exhausted, it is essential that the whole of our naval and air strength should be concentrated and employed in the battle for sea-power. If we lose our sea-power, we lose the war.[26]

Horton's comments came at a juncture when arguments about resource allocations were peaking. Whereas safeguarding the sea lanes was predominantly a defensive undertaking, the RAF was trying to take the war to the enemy. However, as the Butt Report had recently confirmed, the attempt to cripple Germany's war industries from the air was not going well. Indeed in February 1942 RAF Bomber Command decided to change its tactics from precision strikes to area bombing in an effort to improve its performance. This policy called for even more planes, crews and ordnance. February 1942 also witnessed the 'Channel Dash', whereby *Prinz Eugen*, *Scharnhorst* and *Gneisenau* escaped from Brest and returned to Germany, much to the embarrassment of both the Admiralty and the RAF. Before the month was out, Captain Charles Lambe, the Royal Navy's Director of Plans, was insisting that:

> In order to overcome the threats that the enemy is now making, it is of vital importance that the co-operation of the R.A.F. should be of the most active and extensive nature. . . . It will be urged, no doubt, that if the R.A.F. devotes

24 See: H.P. Willmott, *The Second World War in the East* (London, 1999), pp. 159–68.
25 M.A. Bragadin, *The Italian Navy in World War Two* (Annapolis, MD, 1957), pp. 365–6.
26 Letter from Admiral (Submarines) to First Sea Lord, 26 February 1942, TNA, ADM 205/15.

to naval co-operation that proportion of its total effort which we are asking for, what has hitherto been considered its primary function . . . will be so weakened as to be rendered valueless. This is indeed the crux of the whole question. Is it possible for [strategic] bombing . . . ever to have a decisive effect on the result of the war? . . . It is submitted that all the evidence goes to show that such is not the case. . . .[27]

But simply transferring planes and personnel from one set of missions to another was unlikely to ameliorate the situation very much. Indeed Admiral Lyster, the Fifth Sea Lord, acknowledged that RAF Coastal Command's dilution at home was partly a result of 'the necessity to divert . . . aircraft to the Mediterranean and other theatres'. Obliged to confront not only German and Italian forces but also those of Japan as well, the Royal Navy, too, was feeling the pinch more than ever before. With insufficient escort vessels to cope should the *Kriegsmarine* intensify its *guerre de course*, the only potential solution to the dilemma that he could see was 'a greater air effort'.[28] To be effective against U-boats amidst the vast expanse of the Atlantic, however, airmen required special training and their planes needed to be endowed with not just reach and endurance but also extraordinary precision and lethality.

All of this demanded qualitative as well as quantitative improvements in the equipment and other assets available to RAF Coastal Command especially. These changes were already occurring, albeit at a slower pace than some in Britain's military and political hierarchy would have liked to see. By the autumn of 1940 the service was starting to receive significant deliveries of munitions that were far less inapt than the 100 lb anti-submarine bomb if only because, as depth-charges, they were detonated by hydrostatic pistols rather than contact fuses. Other new armaments followed, particularly over the next three years. Besides depth-charges that were more compact yet, packed with innovative explosives such as minol and torpex, were more potent than their bulkier forebears, solid-nosed rockets that could punch holes in submarine hulls were fitted to many maritime patrol aircraft. New navigational and targeting aids were also being introduced. The most important of these by far was ASV radar, but the Leigh Light – a powerful searchlight that both illuminated prey and dazzled any AA gunners protecting it – was another device that assisted in the execution of nocturnal attacks. By late 1942 small, disposable sonobuoys were also becoming available. Similarly, acoustic and electromagnetic sensors were linked to various projectiles to provide aircrew with novel means of engaging quarry, be it on or below the waves. Magnetic anomaly detectors – which register the distortions caused in the Earth's magnetic field by ferrous metal masses – could reveal a submerged U-boat's location to a circling bomber that might then lash out with either a

27 Paper B from Director of Plans to First Sea Lord, 27 February 1942, TNA, ADM 205/15.
28 Minute from the Fifth Sea Lord to the Chief of the Naval Staff, 9 February 1942, TNA, ADM 205/13.

rocket-propelled, 65 lb retro-bomb or 'Fido'. The former hurtled seawards as the aircraft pulled out of its dive; the latter was a guided torpedo that homed in on the sound radiating from the target's propulsion system.

In order to accommodate the expanding gamut of electronic gadgetry as well as large quantities of weapons and fuel, even the smallest of maritime patrol aircraft needed to be fairly big, powerful machines. But these characteristics could make them less stealthy, exacerbating the difficulties experienced in hunting elusive prey that, if threatened, would dive out of vision in a matter of moments. Just as 'dazzle' and other camouflage techniques were used to soften the silhouettes of merchantmen, planes were duly painted in bland greys and white so as to make them harder to espy in daylight. Likewise, adjustments were made to older ASV radar sets to lessen the chances of foes being alerted by their emissions. Britain's centimetric ASV was not only undetectable by the Germans' 'Metox' monitoring scanner but also, unlike the earlier variants, provided its operator with a 360° view of the surrounding sea.

As far as actual aircraft are concerned, initially the *Hudson* supplanted the Avro *Anson* as the backbone of the fleet, but a variety of machines came to swell the ranks, notably *Sunderlands, Catalinas, Wellingtons*, B.17s and even a few *Beaufighter* torpedo-bombers. Of the British-built planes, *Wellingtons* and *Sunderlands* could carry a 450 lb depth-charge. Also endowed with exceptional reach and endurance, the latter were particularly handy, as was the *Liberator*.[29] However, B.24s remained comparatively hard to come by if only because the USAAF wanted these machines for operations in the Pacific and for strategic bombing raids against European targets; 54 were lost in a single raid against the Ploesti oilfields in August 1943. By mid-1941, out of its total of approximately two hundred maritime patrol planes, RAF Coastal Command had just ten *Liberators* as opposed to 36 *Catalinas* and 80 *Hudsons*.

New, conveniently sited aerodromes were also needed. Several were erected, both within the UK – notably in the Hebrides and at Lough Erne in Ulster – and abroad. From May 1941 the web of maritime patrols over the Atlantic started expanding appreciably as better machines and more airfields entered service: planes probed westwards for roughly seven hundred miles from Northern Ireland, eastwards for approximately six hundred miles from Canada and southwards from Iceland for some four hundred miles. This left sizeable holes in the umbrella provided by land-based aircraft, notably off Greenland and around the Azores, an outpost of neutral Portugal. However, these gaps, too, began to shrink once the USA formally entered the war and more airstrips and very long range planes were activated. By the end of 1942 the CDC's efforts had banished U-boats from the Caribbean and, in the wake of an agreement reached at the Washington Convoy Conference the following March, several *Liberators* were assigned to the Royal Canadian Air Force with which it sealed up the gap south

---

29 See: Milner, *Atlantic*, pp. 43, 113.

of Greenland. Within six months that west of the Azores had disappeared as well after Lisbon controversially consented to the Allies setting up military installations on the islands. Similarly, the establishment of bases along Africa's western seaboard was to help eject the U-boat from the South Atlantic.

Choreographed from his headquarters in France by radio, the preferred tactic of Dönitz's wolf packs was to gather in front of or behind a convoy prior to assailing it under the cover of night. Attacks were frequently mounted on the surface, where a U-boat would not only be swifter and more manoeuvrable than when submerged but also able to employ any deck-mounted weaponry as well as its torpedoes. However, even here the craft's main means of protecting itself remained its passive defences, namely its relatively small size, low profile and dull colouring that, together, made it hard to distinguish from its natural surroundings. Since any punctures in the pressure hull could bar it from diving, collisions and enemy fire had to be avoided at all costs. This imperative placed a premium on remaining unseen by human and electronic eyes alike.

In the war's early stages this was easy enough to achieve, particularly in environments where poor visibility was a common phenomenon. The fog that tends to loiter over the Grand Banks off Newfoundland was, for instance, a haven for U-boats, enabling them to lurk unspotted within a couple of hundred miles of the Canadian coast. But the Allies' ongoing application of electronic monitoring devices to ASW created fearful problems for their adversaries insofar that these instruments threatened to erode the stealthiness from which submarines derived so much of their survivability. It was impossible for such craft to remain under water for long, yet electromagnetic sensors steadily made their usual habitat, the surface, increasingly dangerous, too. Whereas nearby vessels might track their movements beneath the waves with Asdic, ASV sets became sensitive enough to pick out submarines on the surface from many miles away. (From an altitude of 2,000 feet, the British Mark I and II suites could usually detect them up to twenty miles distant.) Such devices enabled aircraft to turn the tables on U-boats that had believed themselves to be safely out of sight; often oblivious to their predicament, they became unknowing targets for pre-emptive operations by either planes or warships if not both. Airborne radar thus came to form the foundation of the Allies' defensive and offensive systems, with aviation dictating where and when submarines might hope to function successfully.[30]

Spreading and then tightening the net ineluctably took the Allies some years, during which time their effectiveness at waging ASW gradually increased as more and better sensors, platforms and armaments were devoted to the task. 'Ultra' and other intelligence-gathering mechanisms, notably HF/DF ('huff-duff'), were also of fundamental importance, not least because the information they gathered helped to promote the most cost-effective distribution of reconnaissance and combat assets across an environment that was colossal. Whereas in

---

30 See: ibid., pp. 93–4.

1941 only two to three per cent of attacks on German submarines were likely to bear any fruit, by 1944 this figure had climbed to around forty per cent. In the war's final months U-boat skippers were painfully aware that, if detected by bespoke ASW forces, on paper their chances of survival were just fifty-fifty.

This was partly so because of refinements to the ASW capabilities of surface vessels. Besides the introduction of a broader range of depth-charges, these included large mortar systems, such as the 'Squid'. The simultaneous employment of assorted projectiles – which, because of their differing weights, sank at contrasting speeds – led to targets being enveloped by detonations, the overlapping pressure waves from which were that much more likely to rupture a submarine's hull. There were also innovative means of achieving the same end, notably Britain's 'Hedgehog', which lobbed suitably dense patterns of small bombs far over the bows of ships. This approach enabled an Asdic 'fix' to be maintained for more of the attacking vessel's advance on its quarry. Moreover, as the projectiles were armed with contact fuses, unlike depth-charges they only exploded on finding their mark.

Menacing though the combination of Asdic and such weaponry could prove, for submariners the greatest threat imaginable was that which arose from the integration of surface units with air power. In the Atlantic 'gaps' and other sea areas where cover from land-based aviation was either unavailable or too tenuous, carriers were frequently used to support other warships in protecting sea lanes from hostile planes and submarines alike. With regard to, for instance, three British undertakings in the Mediterranean during the summer of 1942 – Operations 'Harpoon', 'Vigorous' and 'Pedestal' – the availability (or otherwise) of aerial escorts proved to be the single most important factor in deciding the fate of the convoys involved. Whereas the merchantmen participating in 'Vigorous' had to retreat after two days of largely uncontested raids by Axis bombers, cocooned by a few fighters from HMS *Eagle* and *Argus*, those involved in 'Harpoon' were able to press on to their destination in spite of repeated aerial attacks.[31] Similarly, in the case of 'Pedestal' – the grandest effort ever made to succour beleaguered Malta – the convoy's 15 freighters pushed on through vigorous opposition with minimal loss until the core of their escort, the carriers *Victorious* and *Indomitable*, was obliged to withdraw. In the three days thereafter, nine of the remaining merchantmen were sunk.[32]

Whilst operating carriers amidst or close to large formations of commercial shipping could give rise to tactical and navigational complexities, the Allies, realizing that even a few aircraft could have a disproportionate impact in defending convoys especially, also made appreciable use of customized and makeshift escort types, such as the FCS, CAM and MAC, as well as their larger carriers. As early as July 1941 the Royal Navy finished transforming a (captured) merchantman

---

31 See: Letter from Commanding Officer, HMS *Eagle*, to Senior Officer, Force 'T', 18 June 1942, TNA, ADM 199/835; and Minute by Director of Operations Division (Foreign), 3 November 1942, TNA, ADM 199/1244.

32 Report by Commanding Officer, HMS *Victorious*, 15 August 1942, TNA, ADM 199/1242.

into its first escort carrier, HMS *Audacity*. Speedily and cheaply converted, this unpretentious vessel lacked a hangar but her flight-deck was spacious enough to accommodate several *Martlets*. Although she did not survive long, being torpedoed off Portugal six months after she was commissioned, in the interim *Audacity* proved her worth in escorting shipping to Gibraltar; her planes worried hostile submarines and downed five *Kondor* bombers and damaged three more.[33] The first MAC, the *Empire MacAlpine*, was to enter service in spring 1943, from which time onwards bespoke escort carriers also began to be attached to convoys plying routes, both in the Atlantic and the Arctic, that were particularly exposed to aerial attack. Whereas British escort carriers were primarily used to accompany convoys, those of the US Navy formed the centrepieces of the task forces that actively sought out U-boats, notably when they were en route to and from their bases in France or trying to rendezvous with replenishment craft, so-called '*Milchkühe*'. By the end of March 1943 there were five such hunter-killer flotillas in operation, one of which, led by the carrier USS *Guadalcanal*, was to actually capture a U-boat, the *U-505*, during June 1944 in an operation redolent of the Royal Navy's actions against the *U-110* and the *U-559*. Aboard the *U-505* were 'Enigma' consoles and concomitant documents and grid charts that provided the Allies with further insights into their adversaries' modus operandi. The submarine also contained one of Germany's latest weapons, the T5 '*Zaunkönig*'. Scrutinization of this acoustic homing torpedo promptly spawned the development of effective countermeasures.

Indeed both sides sought to make torpedoes more dependable through the incorporation of the 'influence' fuses that had long been a feature of mine warfare. Particularly in the Pacific, submarines had to take the lead in laying minefields in waters dominated by the enemy, but, once conveniently sited aerodromes became available, planes were able to take on this task, too. Besides using both land-based and carrier aircraft in conjunction with specialist surface vessels to first contain and then eliminate submarines seeking to operate far out at sea, in coastal waters the Allies made significant use of buoyancy mines to help keep underwater interlopers at bay. Often suspended sufficiently close to the seabed to let friendly surface vessels pass safely over them, these reactive weapons posed a grave threat to craft moving blindly through the depths. As in the First World War, in 1939 an immense barrage was laid around the confluence of the English Channel and the North Sea, for example. Areas of Britain's western approaches, such as the Bristol Channel's fringes, were also extensively mined as the war against the U-boat progressed. The *guerre de course* in the Pacific also involved the widespread use of such barriers, with the Americans sowing hundreds of mines to tighten the blockade of Japan and her possessions overseas while the Japanese simultaneously strove to cordon off key sea lanes around Borneo, New Guinea, Celebes and Java in particular. Although these obstacles did impede the

---

33 See for instance: Document 171, 'Report by Anti-Submarine Warfare Division: Operations by HMS *Audacity* in Defence of Convoy HG76, 14–21 December 1941', in *FAA in Second World War: Volume One*, pp. 544–52.

US Navy's submarines somewhat, overall Japan's efforts to protect her maritime commerce proved fatally ineffectual, her merchant fleet slumping from a total of over five million tons in 1942 to well under a million by 1945.[34]

This was a proportionately worse rate of attrition than that suffered by Allied shipping in the Atlantic, the battle for which was unique among the Second World War's major campaigns insofar that it dragged on from the wider conflict's beginning until its very end. Dönitz reasoned that the sinking of 700,000 tons of merchant shipping every month would be more than sufficient to leave much of Britain's population and industry on the brink of starvation as well as thwarting any Anglo-American plans to mount large amphibious landings in Europe or Africa.[35] But fulfilling this goal would require an average of around a hundred U-boats to be kept on station. Allowing for craft in transit or tied up in refitting or training, this suggested that a fleet of some three hundred submarines would have to be maintained overall, although refinements to intelligence-gathering and dissemination, logistics, and the design and construction of U-boats promised to yield some efficiencies and savings.

However, whilst the *Kriegsmarine* did enjoy some success in extending the operating radii and depths of its U-boats, other rudimentary limitations remained stubbornly static. To counter the growing ubiquity of hostile air power, it was imperative that submarines be able to manoeuvre under water both for the majority of the time and at far greater speeds than had been hitherto possible. *Schnorchel* tubes that permitted them to ingest air while at periscope depth were eventually to be incorporated into many German craft, but this only occurred from 1944 onwards. In any event the *Schnorchel* was a mixed blessing, as was 'Alberich', which had first been tested in 1940. The former could not be safely employed at any great speed lest it be damaged. Indeed the device's valves especially added to the U-boat's many weak spots. When running submerged on diesel engines, the vessel was also that much noisier. Above all, although efforts were made to camouflage the head of the *Schnorchel*, it left a wake that human if not electromagnetic eyes might readily spot. 'Alberich', too, could prove counterproductive if the cladding ripped or was peeled away by currents. The first adhesives used to bond the coating to the metalwork of submarines proved insufficient to the task, and it was not until 1944 that this fundamental problem was overcome. Even then, if applying the insulation to begin with was a lengthy, painstaking job, maintaining it was harder still, particularly in circumstances where time, labour and raw materials were in increasingly short supply. For similar reasons, although streamlined 'Electro' U-boats were perfected, few could be manufactured and those that were only started entering service after the Battle of the Atlantic was irretrievably lost.

Imaginative though they were, such technical innovations came too late and had too marginal an impact on the wider war, so much of which was being shaped by

34 See: Overy, *Allies*, p. 61; Willmott, *War in East*, pp. 167–8, 204.
35 Overy, *Allies*, p. 45.

aviation, a highly adaptable tool that might be employed almost universally, unlike the submarine. By 1943 the Allies' aerial capabilities in maritime environments alone were substantially better than they had been just a year or two before. Particularly when employed in concert with increasingly versatile ASW ships, air power that was endowed with not only reach, speed, endurance and flexibility but also precision and lethality promised to thwart even the most sophisticated submarines that mechatronic engineering could create. Ground down by the ceaseless and spiralling need to support the *Wehrmacht*, not least on the immense Russian front, and simultaneously shield Germany's industries, towns and military infrastructure from strategic bombing, the *Luftwaffe* could offer the *Kriegsmarine* negligible help. If only because ever fewer planes other than fighters were being turned out, only a handful of aircraft at all suitable for long-range maritime operations were made available. Although in autumn 1943 the Ju.290, which was intended to succeed the *Kondor*, started to appear, it was essentially just a more muscular version of its predecessor, not a brand new, bespoke design. In any event only a score of these planes materialized. Likewise, only one of the huge BV.238 flying boats devised by Blohm & Voss ever flew. As Dönitz himself observed: 'As we ourselves had no organized air reconnaissance at our disposal, my opposite number, Admiral Horton, was able to take a look at my cards without my being able to look at his. . . . Germany was waging war at sea without an air arm: that was one of the salient features of our naval operations, a feature that was as much out of line with contemporary conditions as it was decisive in its effect'.[36]

Matters began to come to a head in the first half of 1943, partly because the Casablanca Conference in January set new parameters for the Allies' incipient Combined Bomber Offensive, namely: 'The progressive destruction and dislocation of the German military, industrial and economic system, and the undermining of the morale of the German people to a point where their capacity for armed resistance is fatally weakened'.[37] This had immense ramifications for both sides with regard to the availability and allocation of resources. Dönitz's submarine bases in France especially, together with their overland supply corridors, which were rather more vulnerable than the reinforced-concrete pens themselves,[38] were repeatedly attacked during the winter of 1942–43, sucking personnel and materiel galore into the maintenance and defence of these and countless other potential targets. Similarly, as suitable aerodromes and aircraft became available in the Mediterranean theatre, round-the-clock bombing sent Italy's industrial output into free fall, paving the way for the Allies' invasion of Sicily in July.

Coincident with and influenced by these wider developments, the Washington Convoy Conference – a spin-off of the Casablanca summit – predictably had a dramatic impact, too. Anxious to move their own destroyer flotillas and

36 Doenitz, *Memoirs*, pp. 325–6.
37 See: Biddle, *Rhetoric and Reality*, p. 215. Also see: 'Note on Air Policy' by the Prime Minister, 22 October 1942, TNA, ADM 205/14.
38 See: S.W Roskill, *The Navy at War 1939–1945* (Ware, 1998), p. 270.

long-range aviation to other hot spots, the Americans insisted on handing primary responsibility for the Battle of the Atlantic to British and Canadian forces. Besides the establishment of a new North-West Atlantic Command under a Canadian admiral, extra assets were to be committed to the fight, notably escort carriers and some additional *Liberators*. Indeed the number of B.24s equipped with centimetric radar at the disposal of RAF Coastal Command more than doubled within a month. There were promising signs that the tussle for reliable intelligence at the strategic level was also starting to turn in the Allies' favour after a lengthy run of German successes in cryptanalysis. Not only had both Britain's Naval Cypher Two and its replacement been broken during 1941, in February 1942 the U-boats had adopted a new code, 'Triton', through the addition of a fourth rotor to their 'Enigma' consoles. This encryption mechanism was finally unravelled by Bletchley Park at the turn of the year. Together with the expanding HF/DF network and the ongoing efforts of the Submarine Tracking Room, this breakthrough provided invaluable insights into Dönitz's intentions and preparations.

Whereas in March 1943 Germany's U-boats sank 108 vessels in all, their tally for May was to be no more than 50. Dönitz's losses for the latter month alone were, on the other hand, as unprecedented as they were unsustainable: 41 submarines foundered, among them that of his son Peter, the *U-954*. Over fifty more were sunk during June and July as the remnants of the wolf packs struggled back to their bases, leaving most of the Allies' transatlantic convoys wholly unmolested; over the next few weeks many criss-crossed the ocean without incident, let alone loss.[39] Horton and his colleagues had deliberately sought to confront the U-boats rather than try to dodge them. Effectively using merchant shipping as bait, the Allies had lured Dönitz's forces into combat against units that blended air with sea power to very good effect.[40] Reporting to the Commons on the four months up to 18 September, Churchill stated:

> No merchant vessel was sunk by enemy action in the North Atlantic. [Losses in] the month of August [were] . . . the lowest . . . we have ever had since the United States entered the war, and [were] . . . less than half the average of . . . Allied sinkings in the 15 months preceding the American entry into the war. During the first fortnight [of this current month] no Allied ships were sunk by U-boat action in any part of the world. This is altogether unprecedented in the whole history of the . . . struggle. . . . Naturally I do not suggest . . . that this immunity or anything like it could possibly continue, [if only because] a new herd of U-boats have been coming out in the last week or so into the Atlantic . . . and they have no doubt been fitted with what is thought to be the best and latest apparatus.[41]

39 See: ibid., pp. 274–7; Milner, *Atlantic*, p. 179; Overy, *Allies*, p. 58.
40 Doenitz, *Memoirs*, p. 340.
41 *Commons Debates*, 21 September 1943, vol. 392, col. 75.

Sure enough, that autumn the Germans did gingerly return to the offensive, having regained some of their quantitative strength. But their adversaries' technology, tactics and sheer numbers again overwhelmed them: whereas only 47 Allied vessels were sunk across the entire Atlantic during the second half of 1943, no fewer than 141 submarines were destroyed. 'The new means which the enemy introduced to the fight and the proven ones which he used in considerably increased numbers – escort carriers, support groups, very long-range aircraft – all owed their success primarily to the help they received from a very short-range radar', Dönitz concluded. 'With it the enemy was able, by day or night, in any weather, in darkness, fog and bad visibility, to locate U-boats on the surface, to fly his aircraft directly on to their targets and engage them'.[42] In 1944 shipping losses in the Atlantic were to plummet to just three per cent of the levels experienced in 1942.[43]

After D-Day, 1944 also witnessed the slow but steady reduction of Dönitz's forward naval bases along France's western periphery. Long besieged by the Allied air forces, although the garrisons of these ports offered stiff resistance to the American troops fanning out from Normandy, once their overland supply routes were severed these enclaves struggled to sustain U-boat operations in the Atlantic. The Allies' strategic bombers duly refocused their efforts on *Kriegsmarine* installations further east, not least the historic outpost of Heligoland, which, in the autumn of 1939, had been the target of numerous, costly and essentially futile sorties by the RAF. A new wave of attacks commenced during the spring of 1944, peaking in a spectacular raid on 18 April 1945 in which almost a thousand planes participated. Much of Heligoland was reduced to ruins. Only three of the bombers failed to return.

Germany surrendered unconditionally on 7 May. On a couple of occasions Churchill admitted that he had been particularly fearful of the U-boat menace.[44] Hitler, too, was impressed by the submarine, suggesting that it would ultimately decide the entire war.[45] Yet, as we have seen throughout this study, the aircraft's ability to exploit the third dimension over both land and sea endowed it with a flexibility that no other platform could fully match. This helps explain the frequently bitter disputes within the military and political hierarchies of the Axis and the Allied powers alike over the acquisition, maintenance, refinement and employment of aviation. Although there can be no doubt that the submarine and the armoured vehicle played important parts in shaping both the World Wars, that of the aircraft was greater still.

42 Doenitz, *Memoirs*, p. 333.

43 Overy, *Allies*, p. 58.

44 See: Roskill, *Churchill and the Admirals*, p. 138; and Winston S. Churchill, *The Second World War: Volume Two: Their Finest Hour* (London, 1949), p. 529.

45 Overy, *Allies*, p. 45.

Bibliography

Manuscript sources

Bundesarchiv, Freiburg, Germany:

Imperial German Navy Admiralty papers (Planungsunterlagen zu einem Krieg mit den USA.): RM 5/879, RM 5/879k (Karten) and RM 5/885.

The National Archives (TNA), Kew, United Kingdom:

Admiralty (ADM) papers: 1/8433/270B, 1/8478/10, 1/8525/136, 1/9720, 1/11971, 116/3432, 116/3477, 116/4038, 116/4351, 116/4352, 116/4528, 116/5150, 116/6158, 137/1012, 137/2706, 156/195, 199/124, 199/473, 199/474, 199/480, 199/835, 199/844, 199/1187, 199/1242, 199/1244, 199/1302, 205/13, 205/14, 205/15, 205/43, 267/84, 267/106, 267/111.

Air Ministry (AIR) papers: 1/145/15/66, 1/149/15/104, 1/308/15/226/188, 1/344/15/226/285, 1/436/15/279/1, 1/625/17/12, 1/631/17/122/44, 1/678/21/13/2138, 1/678/21/13/2186, 2/1071, 2/1101, 2/1748, 2/1749, 2/1910, 2/2080, 2/2608, 2/2715, 2/3018, 2/8875, 14/415, 14/1218, 20/1309.

Cabinet Office (CAB) papers: 66/11, 69/4, 70/5, 102/536, 121/537.

Prime Ministerial (PREM) papers: 3/163/2, 3/173/4.

Published primary sources

Foreign Office, *British Documents on the Origins of the War, 1898–1914: Anglo-German Tension: Armaments and Negotiations 1907–12*, Gooch, G.P., and Temperley, H. (eds), (London: HMSO, 1928).

Historical Section, Admiralty, *The Defeat of an Enemy Attack on Shipping: A Study of Policy and Operations, Volume 1A* (London: HMSO, 1957).

Jones, Ben (ed.), *The Fleet Air Arm in the Second World War: Volume One, 1939–1941* (Farnham: Ashgate for the Navy Records Society, 2012).

Jones, H.A., *The War in the Air: Being the Story of the Part Played in the Great War by the Royal Air Force* (5 vols, Oxford: Clarendon, 1928–37).

Mellor, W. Franklin (ed.), *History of the Second World War: Medical Series* (London: HMSO, 1972).

Ministry of Defence (Navy), *War With Japan: Volume II, Defensive Phase* (London: HMSO, 1995).

Parliament of the United Kingdom, (*Hansard*) *Parliamentary Debates* (London: House of Commons and House of Lords, 1900–1945).

Raleigh, W., *The War in the Air: Being the Story of the Part Played in the Great War by the Royal Air Force: Volume One* (Oxford: Clarendon, 1922).

Ranft, Bryan (ed.), *The Beatty Papers:Vol. I, 1902–1908* (Aldershot: Scolar Press for the Navy Records Society, 1989).

Richards, Denis, *Royal Air Force 1939–1945:Volume 1:The Fight at Odds, 1939–1941* (London: HMSO, 1953).

Roskill, S.W. (ed.), *Documents Relating to the Naval Air Service:Volume I, 1908–1918* (London: Spottiswoode, Ballantyne & Co for the Navy Records Society, 1969).

———, *Naval Policy Between the Wars,Volume 2:The Period of Reluctant Rearmament, 1930–1939* (London: Collins, 1976).

———, *The War at Sea 1939–1945* (4 vols, London: HMSO, 1954–61).

Simpson, Michael (ed.), *The Cunningham Papers,Volume I: The Mediterranean Fleet, 1939–1942* (Aldershot: Ashgate for the Navy Records Society, 1999).

Secondary sources

Abbateillo, John J., *Anti-Submarine Warfare in World War I: British Naval Aviation and the Defeat of the U-Boats* (London: Routledge, 2006).

Acworth, Bernard, *The Navy and the Next War: A Vindication of Sea Power* (London: Eyre & Spottiswoode, 1934).

Axell, Albert, and Kase, Heideaki, *Kamikaze: Japan's Suicide Gods* (New York, NY: Longman, 2002).

Baldwin, Ralph B., *The Deadly Fuze:The Secret Weapon of World War II* (Princeton, NJ:Princeton University Press, 1947).

Barnett, Correlli, *The Audit of War* (London: Macmillan, 1986).

———, *Engage The Enemy More Closely: The Royal Navy in the Second World War* (London: Penguin, 2000).

Bergerud, Eric M., *Fire in the Sky: The Air War in the South Pacific* (Boulder, CO: Westview Press, 1999).

Biddle, Tami Davis, *Rhetoric and Reality in Air Warfare: The Evolution of British and American Ideas About Strategic Bombing, 1914–1945* (Princeton, NJ:Princeton University Press, 2004).

Bird, Keith, *Erich Raeder: Admiral of the Third Reich* (Annapolis, MD: Naval Institute Press, 2006).

Bond, Gordon C., *The Grand Expedition: The British Invasion of Holland in 1809* (Athens, GA: University of Georgia Press, 1979).

Bragadin, M.A., *The Italian Navy in World War Two* (Annapolis, MD: Naval Institute Press, 1957).

Brown, David, *Carrier Operations in World War II: Volume One: The Royal Navy* (London: Ian Allan, 1974).

Brown, David K., *The Grand Fleet: Warship Design and Development, 1906–1922* (London: Chatham Publishing, 1999).

———, *Nelson to Vanguard: Warship Design and Development, 1923–1945* (London: Chatham Publishing, 2000).

Buckley, John, *The RAF and Trade Defence, 1919–1945: Constant Endeavour* (Keele: Ryburn Publishing, 1995).

Burlingame, Burl, *Advance Force Pearl Harbor* (Annapolis, MD: Naval Institute Press, 2002).

Burt, R.A., *British Battleships 1889–1904* (Annapolis, MD: Naval Institute Press, 1988).

Callwell, Charles Edward, *The Effect of Maritime Command on Land Campaigns Since Waterloo* (London and Edinburgh: W. Blackwood and Sons, 1897).

———, *Military Operations and Maritime Preponderance: Their Relations and Interdependence* (London and Edinburgh: W. Blackwood and Sons, 1905).

Carlgren, Wilhelm Mauritz, *Swedish Foreign Policy during the Second World War* (London: Ernest Benn, 1977).

Chesneau, Roger, *Aircraft Carriers of the World, 1914 to the Present: An Illustrated Encyclopedia* (London: Arms & Armour, 1992).

Churchill, Winston S., *The Second World War* (6 vols, London: Houghton Mifflin, 1948–53).

———, *The World Crisis 1911–1918: Volume 2* (London: Odhams, 1939).

Clements, J.A., 'Royal Navy Ship-Based Air Defence, 1939–1984', *RUSI Journal*, 129, (1984) pp. 19–24.

Colomb, John Charles, *The Defence of Great and Greater Britain: Sketches of the Naval, Military and Political Aspects* (London: Edward Stanford, 1880).

Conroy, Robert, *The Battle of Manila Bay: The Spanish-American War in the Philippines* (New York, NY: Macmillan, 1968).

Corbett, Julian S., *Some Principles of Maritime Strategy* (London: Longmans, Green & Co., 1911).

Derry, Thomas Kingston, *The Campaign in Norway: History of the Second World War: United Kingdom Military Series: Official Campaign History* (London: HMSO, 1952).

Doenitz, Karl, *Memoirs: Ten Years and Twenty Days* (London: Cassell, 2000).

Douhet, Giulio, *Diario critico di Guerra* (2 vols, Rome: Paravia, 1921–22).

———, *Il Dominio dell'aria* (Rome: L'Amministrazione Della Guerra, 1921 and second edition, 1926).

———, *Giulio Douhet: Scritti inediti*, (ed.) A. Monti (Gennaio: Scuola di Guerra Aria, 1951).

———, *Probabili aspetti della Guerra future* (Palermo: Sandron, 1928).

Dugan, James, and Stewart, Carroll, *Ploesti: The Great Ground-Air Battle of 1 August 1943* (New York, NY: Brassey's, 1998).

Epkenhans, Michael, 'Imperial Germany and the Importance of Sea Power', in N.A.M. Rodger (ed.), *Naval Power in the Twentieth Century* (Annapolis, MD: Naval Institute Press, 1996), pp. 27–40.

Erickson, John, *The Road to Stalingrad* (London: Panther Books, 1985).

Evans, David C., and Peattie, Mark R., *Kaigun: Strategy, Tactics and Technology in the Imperial Japanese Navy, 1887–1941* (Annapolis, MD: Naval Institute Press, 1987).

Fisher, John Arbuthnot, *Memories of Admiral of the Fleet Lord Fisher* (London: Hodder and Stoughton, 1919).

———, *Records by Admiral of the Fleet Lord Fisher* (London: Hodder and Stoughton, 1919).

Fisk, Robert, *In Time of War: Ireland, Ulster and the Price of Neutrality, 1939–1945* (London: Gill and Macmillan, 1996).

Friedman, Norman, *British Carrier Aviation: The Evolution of the Ships and their Aircraft* (London: Conway Maritime Press, 1988).

———, *US Naval Weapons: Every Gun, Missile, Mine and Torpedo Used by the US Navy from 1883 to the Present Day* (Annapolis, MD: Naval Institute Press, 1982).

Garzke, William H., Denlay, Kevin V., and Dulin, Robert O., 'Death of a Battleship: The Loss of HMS *Prince of Wales*, December 10, 1941: A Marine Forensic Analysis of the Sinking', (National Harbor, MD: Society of Naval Architects and Marine Engineers, International Marine Forensics Symposium, 2012).

Gat, Azar, *Fascist and Liberal Visions of War: Fuller, Liddell Hart, Douhet and other Modernists* (Oxford: Clarendon Press, 1998).

Gates, David, *The Napoleonic Wars, 1803–1815* (London: Pimlico, 2003).

———, *Sky Wars: A History of Military Aerospace Power* (London and Chicago, IL: Reaktion, 2003).

———, *Warfare in the Nineteenth Century* (London and New York, NY: Palgrave, 2001).

Glines, Carroll V., *The Doolittle Raid* (New York, NY: Crown Publishing Group, 1988).

Gollin, Alfred, *No Longer an Island: Britain and the Wright Brothers, 1902–1909* (London: Heinemann, 1984).

Gordon, Andrew, *British Seapower and Procurement Between the Wars* (Basingstoke: Macmillan, 1988).

———, *The Rules of the Game: Jutland and British Naval Command* (London: John Murray, 1996).

Grahame-White, Claude, and Harper, Harry, *Air Power: Naval, Military, Commercial* (London: Chapman & Hall, 1917).

Greene, Jack, and Massignani, Alessandro, *The Naval War in the Mediterranean 1940–1943* (London: Chatham Publishing, 2002).

Gröner, Erich, *German Warships, 1815–1945* (Annapolis, MD: Naval Institute Press, 1990).

Grose, Peter, *An awkward Trial: The Bombing of Darwin, February 1942* (Crows Nest: Allen & Unwin, 2009).

Hadjipateras, Costas N., and Fafalois, Maria S., *Crete 1941: Eyewitnessed* (Athens: Efstathiadis, 1989).

Hall, Christopher D., *British Strategy in the Napoleonic War, 1803–15* (Manchester: Manchester University Press, 1992).

———, *Wellington's Navy: Sea Power and the Peninsular War, 1807–1814* (London: Chatham Publishing, 2004).

Hansell, Haywood S., *The Air Plan That Defeated Hitler* (Atlanta, GA: Air Corps Tactical School, 1972).

Hastings, Max, *All Hell Let Loose: The World at War, 1939–1945* (London: Harper Press, 2011).

Hirst, Francis Wrigley, *The Six Panics and Other Essays* (London: Methuen, 1913).

Hone, Thomas C., Friedman, Norman, and Mandeles, Mark D., *American and British Aircraft Carrier Development, 1919–1941* (Annapolis, MD: Naval Institute Press, 1999).

Hone, Thomas C., and Hone, Trent, *Battle Line: The United States Navy, 1919–1939* (Annapolis, MD: Naval Institute Press, 2006).

Horne, Alistair, *To Lose a Battle: France 1940* (London: Penguin Books, 1979).

Howard, Michael, *The Franco-Prussian War* (London: Methuen & Co., 1981).

Howse, Derek, *Radar at Sea: The Royal Navy in World War 2* (London: Macmillan, 1993).

Jenkins, Roy, *Churchill* (London: Pan Books, 2002).

Kennedy, Paul, *The Rise and Fall of British Naval Mastery* (London: Penguin, 1976).

King, Ernest J., and Whitehall, Walter M., *Fleet Admiral King: A Naval Record* (New York, NY: W.W. Norton, 1952).

Kinglake, Alexander William, *The Invasion of the Crimea* (8 vols, London and Edinburgh: Blackwood & Sons, 1863–69).

Lambert, Andrew, *The Challenge: Britain Against America in the Naval War of 1812* (London: Faber and Faber, 2012).

———, *The Crimean War: British Grand Strategy Against Russia, 1853–56* (London: Ashgate, 2011).

Layman, R.D., *Before the Aircraft Carrier: The Development of Aviation Vessels, 1849–1922* (Annapolis, MD: Naval Institute Press, 1989).

———, *Naval Aviation in the First World War: Its Impact and Influence* (Annapolis, MD: Naval Institute Press, 1996).

Lundstrom, John B., *The First Team: Pacific Naval Air Combat from Pearl Harbor to Midway* (Annapolis, MD: Naval Institute Press, 1984).

Mackinder, Halford John, *Britain and the British Seas* (Oxford, Clarendon Press, 1907).

———, *Democratic Ideals and Reality: A Study of the Politics of Reconstruction* (New York, NY: Henry Holt, 1919).

Mallett, Robert, *The Italian Navy and Fascist Expansionism, 1935–1940* (London: Frank Cass, 1998).

Maltin, Tim, *Titanic: A Very Deceiving Night* (Wilsford: Malt House Books, 2012).

March, Daniel M., *British Warplanes of World War II: Combat Aircraft of the Royal Air Force and the Fleet Air Arm, 1939–1945* (London: Aerospace Publishing, 1998).

Marder, Arthur, *Old Friends, New Enemies: The Royal Navy and the Imperial Japanese Navy* (2 vols, New York, NY: Oxford University Press, 1981–90).

Mason, Francis K., *The British Bomber Since 1914* (London: Putnam, 1994).

McFarland, Stephen, *America's Pursuit of Precision Bombing, 1910–1945* (Washington DC: Smithsonian Institute Press, 1995).

Melhorn, C., *Two-block Fox: the Rise of the Aircraft Carrier, 1911–1929* (Annapolis, MD: Naval Institute Press, 1974).

Middlebrook, Martin, and Mahoney, Patrick, *Battleship: The Loss of the Prince of Wales and the Repulse* (London: Allen Lane, 1977).

Milner, Marc, *The Battle of the Atlantic* (Stroud: Tempus, 2005).

Mitchell, William, *Our Air Force: The Keystone of National Defense* (New York, NY: E.P. Dutton, 1921).

———, *Winged Defense: The Development and Possibilities of Modern Air Power, Economic and Military* (New York, NY and London: G.P. Putnam's Sons, 1925).

Morgan, Howard Wayne, *America's Road to Empire: The War With Spain and Overseas Expansion* (New York, NY: John Wiley & Sons, 1965).

Morison, Samuel E., Commager, Henry Steele, and Leuchtenburg, William E., *A Concise History of the American Republic* (New York, NY: Oxford University Press, 1977).

Morris, A.J.A., *The Scaremongers: The Advocacy of War and Rearmament, 1896–1914* (London: Routledge and Kegan Paul, 1984).

Morrow, John H. Jr, *The Great War in the Air: Military Aviation from 1909 to 1922* (Washington DC: Smithsonian Institute Press, 1993).

Moulton, James Louis, *A Study of Warfare in Three Dimensions: The Norwegian Campaign of 1940* (Athens, OH: Ohio University Press, 1967).

Mowthorpe, Ces, *Battlebags: British Airships of the First World War* (Stroud: Sutton Publishing, 1998).

Munch-Petersen, Thomas, *The Strategy of Phoney War: Britain, Sweden and the Iron Ore Question, 1939–1940* (Stockholm: Militärhistorika Forlaget, 1981).

Murray, Gilbert, *The Foreign Policy of Sir Edward Grey, 1906–1915* (Oxford: Clarendon Press, 1915).

Murray, Williamson, *Strategy for Defeat: The Luftwaffe, 1933–1945* (Maxwell, AL: Air University Press, 1983).

Neon, *The Great Delusion: A Study of Aircraft in Peace and War* (London: Ernest Benn, 1927).

O'Brien, Phillips Payson, 'Politics, Arms Control and the US Naval Development in the Interwar Period', in Phillips Payson O'Brien (ed.), *Technology and Naval Combat in the Twentieth Century and Beyond* (London: Frank Cass, 2001), pp. 148–64.

O'Connell, Robert L., *Sacred Vessels: The Cult of the Battleship and the Rise of the U.S. Navy* (Oxford: Oxford University Press, 1991).

O'Keefe, David R., *Dieppe Decoded: The Remarkable True Story Behind the Greatest Raid of World War II* (Toronto: Knopf, 2014).

O'Toole, George J.A., *The Spanish War: An American Epic, 1898* (New York, NY: W.W. Norton, 1986).

Overy, Richard, *Why The Allies Won* (London: Jonathan Cape, 1995).

Parker, R.A.C., 'British Rearmament 1936–9: Treasury, Trade Unions and Skilled Labour', *English Historical Review*, 96 (1981), pp. 306–18.

Parshall, Jonathan, and Tully, Anthony, *Shattered Sword: The Untold Story of the Battle of Midway* (Dulles, VA: Potomac Books, 2007).

Peattie, Mark R., *Sunburst: The Rise of Japanese Naval Air Power, 1909–1941* (Annapolis, MD: Naval Institute Press, 2001).

Peden, G.C., *British Rearmament and the Treasury, 1932–1939* (Edinburgh: Scottish Academic Press, 1979).

———, 'Keynes, the Economics of Rearmament and Appeasement', in W. Mommsen and L. Kettenacker (eds), *The Fascist Challenge and the Policy of Appeasement* (London: Allen & Unwin, 1983), pp. 142–56.

Pemberton-Billing, Noel, *Air War: How to Wage It: With Some Suggestions for the Defence of the Great Cities* (London: Gale & Polden, 1916).

———, *Defence Against the Night Bomber* (London: Robert Hale, 1941).

Petrow, R., *The Invasion and Occupation of Denmark and Norway, April 1940 – May 1945* (London: William Morrow, 1974).

Philip's Great World Atlas (Third edition, London: George Philip Limited, 1993).

Pineau, Roger, 'Admiral Isoroku Yamamoto', in Sir Michael Carver (ed.), *The War Lords* (London: Weidenfeld and Nicolson, 1976), pp. 390–403.

P.I.X. [Hallam, Douglas], *The Spider Web: The Romance of a Flying-Boat War Flight* (London: William Blackwood, 1919).

Polmar, Norman, *Aircraft Carriers: A History of Carrier Aviation and its Influence on World Events, Volume One, 1909–1945* (Washington: Potomac Books, 2006).

Poolman, Kenneth, *Focke-Wulf Condor: Scourge of the Atlantic* (London: MacDonald & James, 1978).

Porten, Edward P. von der, *The German Navy in World War II* (London: Pan, 1972).

Powers, Barry D., *Strategy Without Slide-Rule: British Air Strategy, 1914–1939* (London: Croom Helm, 1976).

Prince, Stephen, 'Air Power and Evacuations: Crete 1941', in I. Speller (ed.), *The Royal Navy and Maritime Power in the Twentieth Century* (London and New York: Frank Cass, 2005), pp. 67–87.

Redford, Duncan, *A History of the Royal Navy: World War II* (London: I.B. Tauris, 2014).

Rohwer, Jürgen, *Chronology of the War at Sea, 1939–1945* (Annapolis, MD: Naval Institute Press, 2005).

Roskill, S.W., *Churchill and the Admirals* (Barnsley: Pen & Sword, 2004).

———, *The Navy at War 1939–1945* (Ware: Wordsworth, 1998).

Ruge, Frederich, *Sea Warfare 1939–1945: A German Viewpoint* (London: Cassell, 1957).

Sadkovich, James J., *The Italian Navy in World War II* (Westport, CT: Greenwood Press, 1994).

Salisbury, Harrison E., *The 900 Days: The Siege of Leningrad* (Cambridge, MA: Da Capo, 2003).

Schreiber, Gerhard, Stegemann, Bernd, and Vogel, Detlef, *Germany and the Second World War: Volume Three: The Mediterranean, South-East Europe and North Africa, 1939–1941* (Oxford: Clarendon Press, 1995).

Shay, R.P., *British Rearmament in the Thirties* (Princeton, NJ: Princeton University Press, 1977).

Sherry, Michael, *The Rise of American Air Power* (New Haven, CT: Yale University Press, 1987).

Smith, Malcolm, *British Air Strategy Between the Wars* (Oxford: Clarendon Press, 1984).

Spanner, E.F., *The Broken Trident* (London: William and Norgate, 1929).

Strachan, Hew, *The First World War: Volume One: Call to Arms* (Oxford: Oxford University Press, 2001).

Sumida, Jon T., 'The Best Laid Plans: The Development of British Battle-Fleet Tactics, 1919–1942', in *International History Review*, Vol. XIV, 1992.

————, 'British Naval Procurement and Technological Change, 1919–1939', in Phillips Payson O'Brien (ed.), *Technology and Naval Combat in the Twentieth Century and Beyond* (London: Frank Cass, 2001), pp. 128–47.

————, 'Sturdee: Falkland Islands, 1914', in Eric Grove (ed.), *Great Battles of the Royal Navy* (London: Arms & Armour Press, 1994), pp. 161–8.

Thetford, Owen, *British Naval Aircraft since 1912* (London: Putnam, 1982).

Till, Geoffrey, 'Airpower and the Battleship in the 1920s', in Bryan Ranft (ed.), *Technical Change and British Naval Policy, 1860–1939* (London: Hodder & Stoughton, 1977), pp. 108–22.

————, *Air Power and the Royal Navy, 1914–1945: A Historical Survey* (London: Jane's, 1979).

Trask, David F., *The War With Spain in 1898* (New York, NY: Macmillan, 1981).

Willmott, H.P., *The Battle of Leyte Gulf: The Last Fleet Action* (Bloomington, IN: Indiana University Press, 2005).

————, *Pearl Harbor* (London: Cassell & Co, 2001).

————, *The Second World War in the East* (London: Cassell, 1999).

Zimm, Alan D., *Attack on Pearl Harbor: Strategy, Combat, Myths, Deceptions* (Havertown, PA: Casemate, 2011).

Index